The Nature of Ore Deposits
Volume 1

by Dr. John Beck

with an introduction by Kerby Jackson

Introduction

It has often been said that "*gold is where you find it*", but even beginning prospectors understand that their chances for finding something of value in the earth or in the streams of the Golden West are dramatically increased by going back to those places where gold and other minerals were once mined by our forerunners. Despite this, much of the contemporary information on local mining history that is currently available is mostly a result of mere local folklore and persistent rumors of major strikes, the details and facts of which, have long been distorted. Long gone are the old timers and with them, the days of first hand knowledge of the mines of the area and how they operated. Also long gone are most of their notes, their assay reports, their mine maps and personal scrapbooks, along with most of the surveys and reports that were performed for them by private and government geologists. Even published books such as this one are often retired to the local landfill or backyard burn pile by the descendents of those old timers and disappear at an alarming rate. Despite the fact that we live in the so-called "Information Age" where information is supposedly only the push of a button on a keyboard away, true insight into mining properties remains illusive and hard to come by, even to those of us who seek out this sort of information as if our lives depend upon it. Without this type of information readily available to the average independent miner, there is little hope that our metal mining industry will ever recover.

This important volume and others like it, are being presented in their entirety again, in the hope that the average prospector will no longer stumble through the overgrown hills and the tailing strewn creeks without being well informed enough to have a chance to succeed at his ventures.

Kerby Jackson
Josephine County, Oregon
May 2018

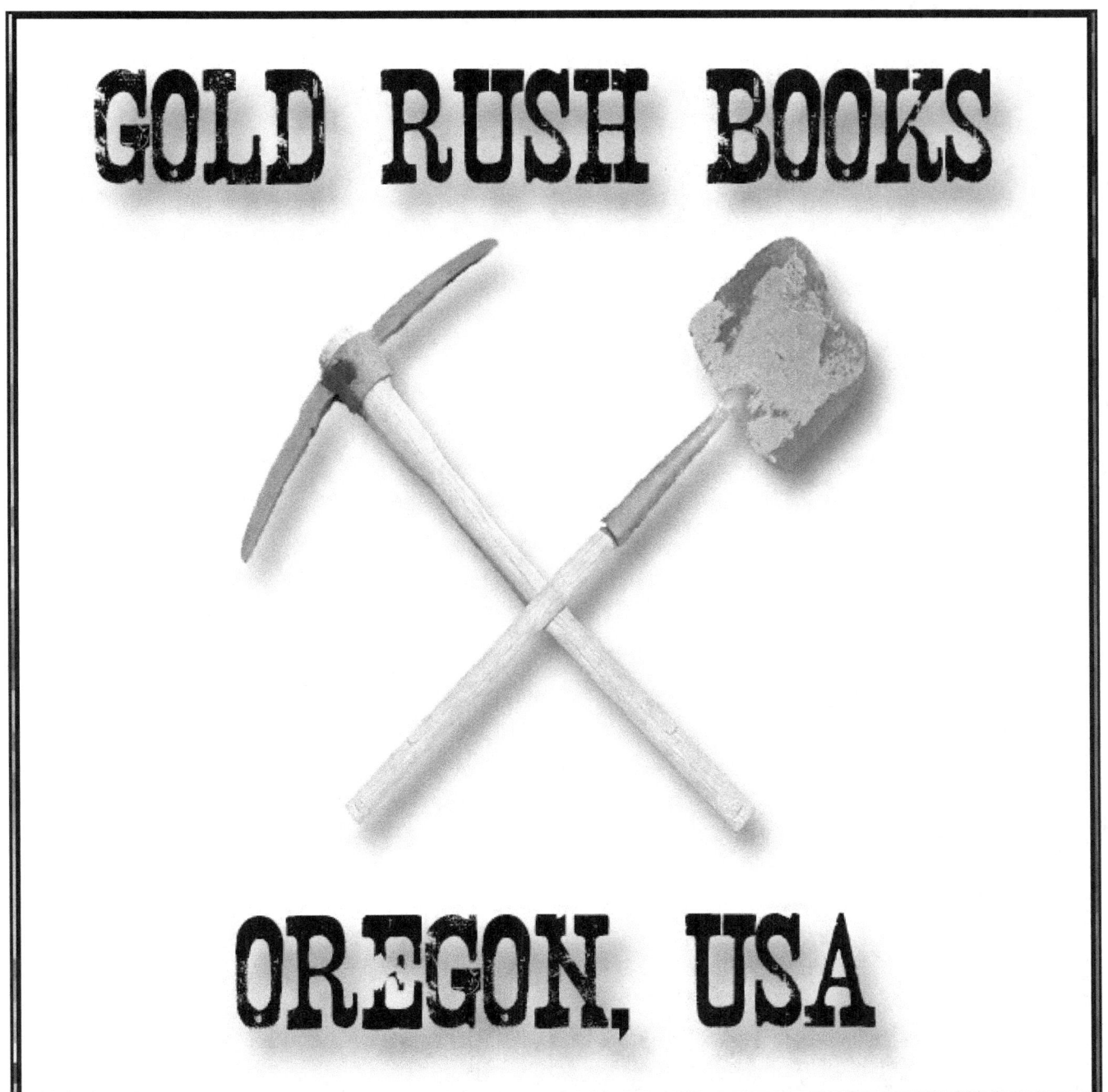

GOLD RUSH BOOKS

OREGON, USA

www.GoldMiningBooks.com

TABLE OF CONTENTS

CONTENTS.

CONTENTS.

SECTION III.

Epigenetic Deposits.

CONTENTS.

CONTENTS.

CONTENTS.

CONTENTS.

CONTENTS.

INTRODUCTION.

DEFINITION OF AN ORE AND OF AN ORE DEPOSIT.

In a mineralogical sense an ore is a metalliferous mineral or a mixture of such minerals. Practically, however, this definition of an ore must be qualified by the statement that only those minerals and mixtures of minerals are ores *from which metals or metallic compounds may be produced on a commercial scale and at a profit.*

Two examples of equal mineralogic or petrographic value may differ materially: a basalt carrying enough magnetite to influence the magnetic needle, but yet containing less than 10 per cent of iron, is far from being an iron ore. On the contrary, a vein with a silver content of only 0.5 per cent is an ore deposit, since with this content it is commercially valuable. In the case of a gold deposit the amount sufficient to distinguish a gold ore from barren rock may be even less, for in California and Dakota gold ores with only from 4 to 6 grains per ton[1] of gold are exploited. Hence it is the economic point of view that must always be borne in mind, the profit of working being subject to variation in the course of time. While nickel and cobalt were formerly nicknames for materials which were thrown upon mine dumps as useless, and were considered as a mere nuisance in silver mining, today the compounds of these metals and the associated minerals are in great demand as ores.

To a certain degree a mineralized material may be an ore in one locality and yet not be an ore in another place, the cost of reduction depending on the proximity to lines of traffic and cheap freights.

The science of ore deposits is, in other words, the study and consideration of the deposition, distribution and origin of rock bodies containing ores in such quantities that they may be extracted profitably by mining operations.

The science of ore deposits is a branch of geology which can only be understood and practiced by those who have some knowledge of that science, particularly of petrography, and of course, also, a knowledge of mineralogy and chemistry.

---

[1] The ton as used in this work is the metric ton of 2,205 ℔.

THE MOST IMPORTANT TREATISES ON ORE DEPOSITS, AND THE LIST OF JOURNALS IN WHICH ORE DEPOSITS ARE DESCRIBED OR DISCUSSED.

Bernhard von Cotta. 'Die Lehre von den Erzlagerstätten.' Two parts: First part, second edition; second part exhausted. Freiberg, 1859–1861. (Translated into English by F. Prime, New York, 1870.) Antiquated in many respects.

Albrecht von Groddeck. 'Die Lehre von den Lagerstätten der Erze.' Leipzig, 1879. (Translated into French by M. Küss.) Out of print and in many respects antiquated.

E. Fuchs and L. De Launay. 'Traité des Gîtes Minéraux et Métalliféres.' Volumes I and II. Paris, 1893. Large encyclopedic handbook.

L. De Launay. 'Formation des Gîtes Métalliféres.' Paris, 1893. Encyclopædia Scientifique des Aide Mémoires. Small manual.

J. A. Phillips and H. Louis. 'A Treatise on Ore Deposits.' Second edition. London, 1896.

J. F. Kemp. 'The Ore Deposits of the United States and Canada.' Third edition. New York and London, 1900.

Branner, J. C. 'Syllabus of Economic Geology.' Stanford University, California.

B. Lotti. 'I Depositi de i Minerali Metalli.' Milan, 1903.

Posepny, Van Hise, Emmons, Weed, Le Conte, Lindgren. 'Genesis of Ore Deposits.' Am. Inst. Min. Eng. New York, 1900.

Rickard, Weed and others. 'Ore Deposits—a Discussion,' New York, 1903.

The Engineering and Mining Journal, New York City.

Zeitschrift für Praktische Geologie. Edited by M. Krahman. Berlin.

Zeitschrift für das Berg-, Hütten- und Salinen-Wesen im Preussichen Staate. Published by the Minister for Commerce and Industry, Berlin.

Oesterreichische Zeitschrift für Berg- und Hüttenwesen. Vienna.

Jahrbuch der k. k. österreichischen Bergakademien. Vienna.

Jahrbuch für das Berg- und Hüttenwesen im Königreiche Sachsen, Freiberg.

Berg- und Hüttenmännische Zeitung. Published by G. Köhler (Clausthal) and F. Kolbeck (Freiberg). Leipzig.

Mining Journal. London.

Transactions of the Institution of Mining Engineers. Newcastle, England.

Transactions of the American Institute of Mining Engineers. New York.

Annales des Mines. Paris.

Geologiska Föreningens i Stockholm Förhandlingar. Stockholm.

Bergjournal. St. Petersburg. (In Russian.)

Revista minera periodico cientifico é industrial. Madrid.

Among general textbooks of geology which devote special attention to ore deposits we may mention: H. Credner, 'Elemente der Geologie,' ninth edition, Leipzig, 1903 ; M. Neumayr, 'Erdgeschichte,' second edition, Leipzig, 1900.

Many essays on ore deposits are also scattered through periodicals dealing with general geology, especially *Zeitschrift* der Deutschen geologischen Gesellschaft, Berlin; *Journal of Geology*, Chicago University, Chicago, Illinois; *American Journal of Science*, New Haven, Connecticut; *American Geologist*, Minneapolis, Minnesota; *Neues Jahrbuch* für Mineralogie, Geologie and Paläontologie, Stuttgart; and Tschermak's Mineralog u. Petrographische, Mittheilungun, Vienna.

The statistical data on the different mining districts of the world are best summarized in the following works, which appear every year: 'The Mineral Industry,' published by *The Engineering and Mining Journal*, New York and London; 'Mineral Resources,' published by the U. S. Geological Survey, Washington, D. C.

CLASSIFICATION OF ORE DEPOSITS.

A historical sketch of the different attempts at a classification of ore deposits has been given by J. F. Kemp[1], to which little need be added.

[1] J. F. Kemp. 'The Classification of Ore Deposits.' A review and a proposed scheme based on origin. Contributions from the geological department of Columbia College, No. 5. New York.

The practical miner even nowadays names and classifies his deposits according to their form and appearance, and this was the method used by the first scientific writers. Waldauf von Waldtenstein[1] (1824), and B. von Cotta[2] (1859), subsequently Lottner-Serlo[3] (1869).

G. Köhler[4] (1884) and lately H. Hofer[5] (1897) based their groupings solely on form. J. D. Whitney[6] (1854), J. Grimm[7] (1869), J. S. Newberry (1880), and J. A. Phillips[8] (1884), laid stress on the mode of origin. A. von Groddeck[9] (1879) on the other hand, introduced a mainly genetic system, and advocated this principle again in his essay of 1885[10].

Genetic principles were followed by H. S. Munroe, J. F. Kemp[11], F. Posepny[12], G. Gürich[13] and W. H. Weed[14].

A. W. Stelzner's system, used in his lectures, briefly stated in the program of the Freiberg Mining Academy and employed by him in arranging the collection of ore deposits in that institution, is also purely genetic. Finally, J. H. L. Vogt, among the more recent writers, furnished many important data for a scientific classification in this field, especially in his essay of 1894[15]. The classification of W. H. Weed follows a logical sequence from magmatic segregations to placer gravels, and utilizes the results of modern petrologic research and chemistry to a greater extent than preceding schemes. It entirely abandons form as a factor of importance and is thus purely genetic. Following Stelzner's system in many ways we have in the present work used the following classification:

[1] J. Waldauf von Waldenstein: 'Die besonderen Lagerstätten der nutzbaren Mineralien.' Volumes I and II. Vienna, 1824.

[2] B. von Cotta: 'Lehre von den Erzlagerstätten.' Volumes I and II. Freiberg, 1859.

[3] E. H. Lottner and A. Serlo: 'Leitfaden zur Bergbaukunde.' Volumes II and III. Berlin, 1869-1872.

[4] G. Köhler: 'Lehrbuch der Bergbaukunde.' Leipzig, 1884.

[5] H. Höfer: *Zeit.* f. Prak. Geol., 1897.

[6] J. D. Whitney: 'The Metallic Wealth of the United States.' Philadelphia, 1854.

[7] J. Grimm: 'Die Lagerstätten der nutzbaren Mineralien.' Prague, 1869.

[8] J. A. Phillips: 'A Treatise on Ore Deposits.' First edition. London, 1884.

[9] A. von Groddeck: 'Die Lehre von den Lagerstätten der Erze.' Leipzig, 1869.

[10] 'Bemerkungen zur Classification der Erzlagerstätten.' *B. u. H. Z.*, 1885, p.217.

[11] J. F. Kemp: 'The Ore Deposits of the United States.' New York, 1893.

[12] F. Posepny: 'Die Genesis der Erzlagerstätten.' *Jahrb.* d. k.k. osterr. Bergakademien. Vienna, 1895.

[13] G. Gürich: 'Ueber die Eintheilung der Erzlagerstätten.' Schlessche Ges. f. vaterl. Kultur. Breslau, 1899.

[14] 'Classification of Ore Deposits—a Proposal and a Discussion.' *The Engineering and Mining Journal*, Feb., 1903. Also 'Ore Deposits—a Discussion.'

[15] J. H. L. Vogt: 'Beiträge zur genetischen Classification.' 1894, p. 381.

CLASSIFICATION OF ORE DEPOSITS.

(COMPARE WITH TABLE OF CONTENTS OF THIS WORK.)

I. PRIMARY ORE DEPOSITS.

A. SYNGENETIC; formed simultaneously with the country rock.

1. *Magmatic segregations;* for example, magnetic iron ores in orthóclase porphyries.

2. *Sedimentary ores;* in part approximately in the same condition as at the time of their deposition, for example, bog iron ores; in part altered by metamorphic processes, as, for example, the magnetic iron ores of the crystalline schists.

B. EPIGENETIC DEPOSITS; formed later than the country rock.

1. *Veins;* fillings of fissures, together with kindred formations, for example, tin veins, in which filling of the fissures was accompanied by replacement of the wall rock.

2. *Epigenetic ore deposits other than veins.*

(a) Epigenetic deposits, formed essentially by an impregnation of non-calcareous rocks, the deposits being generally in distinct beds.

(b) Epigenetic stocks; formed essentially by a metasomatic replacement of calcareous rock, mostly in the form of stocks[1], pockets, or stringers; for example, the cadmium deposits in the Muschelkalk.

(c) Contact metamorphic ore deposits; ore-beds and stocks formed through contact metamorphism caused by Plutonic intrusive masses; for example, contact metamorphic magnetic iron ores.

(d) Ore-bearing cavity fillings; deposits formed essentially by a simple filling of pre-existing cavities mostly in the form of stocks or stringers; for example, pisolitic iron ores.

II. SECONDARY DEPOSITS.

Those formed by the destruction and transposition of primary deposits:

1. *Residual deposits* formed essentially by chemical alteration of primary deposits.

2. *Placer deposits* formed essentially by mechanical degradation of primary deposits or placers.

[1] *i.e.,* irregular masses, with or without definite boundaries.

TABLE OF THE MOST IMPORTANT ORE MINERALS.

Ores.	Cryst. System.	Chemical Composition.	Percentage of	
			Gold.	Other Metals.
Gold				
Gold	isom.	Au, mostly (Au, Ag)	....	0.16—39 Ag
Krennerite	orth.	(Ag, Au) Te$_2$	35	....
Nagyagite	"	Au$_2$Sb$_2$Pb$_{10}$Te$_6$S$_{15}$	5.9—7.6	57.2—60.5 Pb
Petzite.	"	(Ag, Au)$_2$Te	3.3—25.6	40.8—59.6 Ag
Sylvanite.	mono.	Au Ag Te$_4$	26.5—40.6	2.24—11.3 Ag
Calaverite	tri.	(Au, Ag) Te$_2$	39.5—	3.1 Ag

The following minerals are sometimes auriferous: pyrite, arsenopyrite, chalcopyrite, stibnite, zinc-blende.

Silver.			Silver.	
Argentite.	isom.	Ag$_2$S	87.1	....
Bromyrite	"	Ag Br	57.4	....
Chlorargyrite	"	Ag Cl	75.2	....
Embolite.	"	Ag (Cl, Br)	61—69.8	....
Dyskrasite...........	orth.	Ag$_2$Sb (?)	63.9—94.1	....
Iodyrite	hex.	Ag I	45.9	....
Polybasite.	mono.	Sb S$_6$ (Ag, Cu)$_9$	64—72	3—10 Cu
Proustite.	hex.	As S$_3$ Ag$_3$	65.5	....
Pyrargyrite (dark ruby silver)	hex.	3 Ag$_2$S, Sb$_2$S$_3$	59.8	....
Silver, native	isom.	Ag	90—100	Often holding Sb, As, Hg, Co, Fe, Cu and Au
Stephanite............	orth.	5 Ag$_2$S, Sb$_2$S$_3$	68.4	....
Stromeyerite.	rh.	Cu$_2$S + Ag$_2$S	53.1	31.2 Cu

The minerals given in this table include only those of economic importance in mining and smelting or those which are particularly characteristic of certain deposits. The chemical formulas are chiefly taken from 'Tabellarischen Nebersicht der Mineralien' of P. Groth, 4th edition, 1898; revised in this edition to accord with Dana's Mineralogy.' The metallic percentages are from F. Kolbeck's new revision of Plattner's 'Blowpipe Analysis,' Leipzig, 1897.

The following minerals often hold silver: tetrahedrite, galena, zinc blende, pyrite, chalcopyrite, chalcocite:

Ores.	Cryst. System.	Chemical Composition.	Contents in Percentages of	
			Silver.	Other Metals.
Platinum.				
Platinum, native	isom.	Pt	70—90	Nearly always holds Fe, Cu, Rh, Ir, Pd and Os

Iridium, osmiridium, etc., also contain platinum.

			Mercury.	
Mercury.				
Cinnabar	hex.	Hg S	86.2	. . . .
Mercury, native	. . . . . .	Hg	. . . .	. . . .

Mercurial tetrahedrite also contains 1.56—1.73% quicksilver.

			Lead.	
Lead.				
Anglesite.	orth.	Pb SO$_4$	68.3	
Boulangerite.	"	$2 Pb S + Sb_2 S_3$	53.9—59.5	
Bournonite	orth.	$2 PbS + Cu_2 S + Sb_2 S_3$	42.3	
Cerussite	"	Pb CO$_3$	77.6	
Galena.	isom.	Pb S	86.6	
Jamesonite	orth.	$2 PbS + Sb_2 S_3$	50.6	
Mimetite	hex.	$3 Pb_2 As_2 O_8 + Pb Cl_2$	69.5	
Phosgenite	tetr.	$Pb Cl_2 + Pb CO_3$	73.8	
Pyromorphite.	hex.	$3 Pb_2 P_2 O_8 + Pb Cl_2$	76.2	
Wulfenite.	tetr.	Pb Mo O$_4$	55.8	
Croicite	mono.	Pb Cr O$_4$	64.6	
Copper.			Copper.	
Atacamite.	orth.	$Cu Cl_2, 3CuO, 3 H_2 O$	52.7—59.4	
Azurite	mono.	$3 Cu O, 2 CO_2 + aq.$	55.2	
Bournonite.	orth.	$2 Pb S + Cu_2 S + Sb_2 S_3$	13.0	42.3 Pb
Bornite	isom.	$x Cu_2 S + y Cu S + 2 FeS$	43—63.4	
Brochantite.	orth.	$4 Cu O, SO_3 + 3 aq.$	56.1	
Chalcopyrite (copper pyrite)	tetr.	$Cu_2 S + Fe_2 S_3$	34.5	

Ores.	Cryst. System.	Chemical Composition.	Contents in Percentages of	
			Copper.	Other Metals.
Copper.				
Chalcocite (cop'r glance)	orth.	Cu_2S	79.8	. . . .
Chrysocolla	. . . .	$Si O_3 Cu. 2 H_2O$	33.9	. . . .
Covellite	hex.	$Cu S$	66.4	. . . .
Cuprite	isom.	Cu_2O	88.7	. . . .
Enargite	orth.	$As S_4 Cu$	48.2	. . . .
Famatinite	"	$Sb S_4 Cu$	43.2	. . . .
Tetrahedrite	isom.	$\begin{cases} As_2S_7(Cu_7, \; Fe_2, \; Zn)_4 \\ Sb_2S_7(Cu_2, \;\; Ag_2, \;\; Fe, \\ \qquad\qquad\qquad\qquad Zn)_4 \end{cases}$ 13—43		0—31 Ag 0.5—17.2 Hg
Copper, native	isom.	. . . .	. . . .	. . . .
Malachite	mono.	$CO_3 [Cu.OH]_2$	57.4	. . . .
Tennantite	isom.	$As_2S_7(Cu_2) (Fe, Zn)_2$	47.7—51.6	. . . .

Pyrite and pyrrhotite are also cupriferous.

Nickel.			Nickel.	
Annabergite	mono.	$Ni_3 As_2 O_8 + 8$ aq.	29.2	. . . .
Chloanthite.	isom.	$Ni As_2$	28.2	Contains Co.
Garnierite.	. . . .	$Si O_4 (Ni, Mg) H_2 + $ aq.	3—33	. . . .
Gersdorffite	isom.	$(Ni Fe) As S$	30—35.1	. . . .
Linnæite.	"	$[(Ni, Co, Fe)S_2]_2(Ni, Co)$	14.6—42.6	11—40.7 Co.
Niccolite.	hex.	$Ni As$	43.6	. . . .
Genthite	amor.	$Si_3 O_{10} Ni_2 H_4$	39.1	. . . .
Ullmannite.	. . . .	$Ni Sb S$	27.6	. . . .

Pyrite and pyrrhotite are also nickeliferous.

Cobalt.			Cobalt.	
Asbolite	amor.	$Mn O_2, Co O, Cu O, H_2O$	19.4	. . . .
Erythrite.	. . . .	$Co_2 As_2O_8 + 8$ aq.	29.5	. . . .
Glaukodot.	orth.	$(Fe, Co) (As, S)_2$	4.5—24.8	. . . .
Cobaltite.	isom.	$(Co, Fe) As S$	35,5	. . . .
Linnæite.	"	$[(Ni, Co, Fe)S_2] (Ni, Co)$	11—40.7	14.6—42.6 Ni
Skutterudite.	"	$Co As_3$	20.8	. . . .
Smaltite	"	$(Co, Fe, Ni) (As S)_2$	8.2—23	. . . .

The following minerals also contain cobalt: pyrrhotite, white nickel pyrite, gray nickel pyrite, arsenopyrite, danaite, cobalt-arsenic-pyrite.

	Cryst. System.	Chemical Composition.	Percentage of	
			Iron.	Other Metals.
Iron Ores.				
Chamosite.................	....	Fe O, Mg O, Al$_2$ O$_5$, Si O$_2$, H$_2$O	33—47	
Hematite (red, specular iron)................	hex.	Fe$_2$O$_3$	70	
Ilmenite (menacconite).	"	Fe Ti O$_2$ + x Fe$_2$ O$_3$ x = 0—5	20.6—68	
Limonite.............	ortho.	2 Fe$_2$ O$_3$ + 3 H$_2$O	59.9	
Magnetite	isom.	Fe O + Fe$_2$O$_3$	72.4	
Titanite.............	"	[(Fe Ti) O$_2$] $_2$Fe	....	
Siderite	hex.	Fe CO$_3$	48.2	

Mixed with other substances forms:
    Argillaceous sphärosiderite = iron carbonate + clay.
    Blackband = iron carbonate + coal.
    (Pisolitic ore bohnerz, minette) = oolitic limonite with clay and quartz
    Seerze = limonite, earthy; often oolitic.

	Cryst. System.	Chemical Composition.	Iron.	Other Metals.
Iron Sulphides.				
Arsenopyrite.........	orth.	Fe S$_2$ + Fe As$_2$	34.3	
Chalcopyrite.........	tetr.	Cu$_2$S + Fe$_2$S$_3$	30.5	
Pyrrhotite...........	hex.	Fe$_n$ S$_n$ $_+$ $_1$	60—61.6	
Marcasite............	orth.	Fe S$_2$	46.6	
Pyrite	isom.	Fe S$_2$	46.6	
Chromium Ores.			Chromium	
Chromite	isom.	(Fe, Mg, Cr) O + (Cr$_2$, Al$_2$, Fe$_2$) O$_3$	40—60 usually 40—53	15—25 Fe O
Manganese Ores.			Manganese	
Braunite	tetr.	Mn$_2$O$_3$	69.6	
Asbolite	....	Mn O$_2$, Co O, Cu O, H$_2$O	c. 19	
Hausmannite	tetr.	Mn$_3$O$_4$	72.1	
Manganite............	orth.	Mn$_2$O$_3$ + aq	62.5	
Polianite.............	orth.	Mn O$_2$	63.2	
Psilomelane..........	amor.	Combined with Mn O$_2$, Mn O, Ba O, K$_2$O and H$_2$O.	49.2—62.9	
Pyrolusite...........	orth.	Mn O$_2$	63.2	
Rhodochrosite	hex.	Mn C O$_3$	47.8	
Rhodonite...........	tric.	[Si O$_3$]$_2$Mn$_2$ commonly also [Si O$_3$]$_2$ (Mn Fe Ca Mg)$_2$	42	
Wad................	not cryst.	Predominant: Mn O$_2$ MnO and H$_2$O	....	

	Cryst. System.	Chemical Composition.	Percentage of	
			Zinc.	Other Metals.
Zinc.				
Calamine..............	orth.	Si O$_3$ [Zn OH]$_2$	53.7	....
Hydrozincite..........	amor.	Co$_3$ [Zn OH]$_2$	57.1	....
Smithsonite...........	hex.	Zn CO$_6$	52.0	....
Willemite	"	Si O$_4$Zn$_2$	58.1	....
Wurtzite..............	orth.	Zn S	to 67	....
Sphalerite............	isom.	Zn S	67	The colored species up to 8% Fe S
Zincite...............	hex.	Zn O	80.2	....
Bismuth.			Bismuth	
Bismutite	....	CO$_3$ [Bi O] Bi (OH)$_2$	87.1	....
Eulytite.	isom.	[Si O$_4$] $_3$ Bi$_4$	83.8	....
Bismuth, native.......	hex.	Bi	95—99.9	....
Bismuthinite · (bismuth glance).	isom.	Bi$_2$S$_3$	81.2	....
Bismite (Bismuth ochre)	isom.	Bi$_2$O$_3$	89.6	....
Tin.			Tin.	
Cassiterite (tinstone)...	tetr.	Sn O$_2$	78.6	....
Stannite (tin pyrite) ...	isom.	Sn S$_4$ Cu$_2$ Fe	24.1—31.6	23.6—29.8 Cu
Tungsten.			Tungsten.	
Scheelite	tetr.	Ca WO$_4$	63.9	....
Wolframite	mono.	WO$_4$(Mn, Fe)	57.9—60.3	....
Hubnerite............	"	Mn WO$_4$	60.7	
Molybdenum.			Molybde'm	
Molybdenite..........	hex.	Mo S$_2$	59.0	....
Antimony.			Antimony	
Stibnite (antimonite) ...	orth.	Sb$_2$S$_3$	71.7	....
Cervantite	cryst.	Sb$_2$O$_4$Sb	79.2	....
Stibiconite	....	H$_2$Sb$_2$O$_5$	74.6	
Valentinite	orth.	Sb$_2$O$_3$	83.5	
Arsenic.			Arsenic	
Arsenic, native........	hex.	As (Sb, Ni, Fe, Mn, S)	90—100	....
Arsenopyrite..........	orth.	Fe As S	46	....
Orpiment.............	"	As$_2$S$_3$	61.0	....
Löllingite............	"	Fe As$_2$	72.8	....
Realgar..............	mono.	As$_2$S$_2$	70.1	....

TABLE OF UNITS IN WHICH METALLIC CONTENTS OF ORES ARE MOST COMMONLY EXPRESSED.

In this work the metric ton of 2,205 pounds is used, unless otherwise stated.
100 kg. = 1 centner = 1 quintal.

GERMANY.

The contents are mostly given in percentages. In Saxony for silver ores "pound parts" are still in use. 1 pound part = 0.01 per cent or 100 grams per 1,000 kilograms.

The contents of gold ores are given in grams (g) per metric ton (t) of 1,000 kilograms. For statements of production the metric ton is used: 1 ton = 1,000 kilograms.

ENGLAND.

1 long ton = 2,240 lb. = 1,016 kg.
1 short ton = 2,000 lb. = 907.2 kg.
1 pound avoirdupois (lb.) = 453.6 g.
1 lb. = 16 ounces (oz.)
1 ounce (oz.) = 28.3 g.

The contents of gold and silver ores are entered by means of

1 ounce troy (oz.) = 31.1 g

1 pennyweight (dwt.) = 1-20 oz. = 1.5 gram.
1 grain (grn.) = 1-24 dwt. = 0.06 gram.
1 grain troy = 1 grain avoirdupois.

UNITED STATES.

Like England, except that the short ton is almost exclusively used, 1 ton 2,000 pounds = 907.2 kilograms.

1 kilogram (kg.) = 2.20462 pounds avoirdupois.
1 flask mercury = 76.5 pounds avoir. = 34.7 kg.

Gold contents in dollars:

1 dollar ($) = 100 cents (cts.).
1 oz. troy = $20.6718 ⎫   = $1.2929 ⎫
1 dwt.   = $1.0386 ⎬ for gold.   ⎬ for silver.
1 grain   = $0.04306 ⎪   ⎪
1 gram   = $0.6646 ⎭   = 0.04157 ⎭

IN MANY STATES OF SOUTH AMERICA.

Contents in marcos per cajon:

1 marco = 230 grams = 7.398 troy ounces.
1 cajon = 64 quintals.
1 quintal = 46 kg.

RUSSIA.

Gold contents are given in zolotniks per 100 poods or 10 berkowitz:

1 berkowitz = 10 poods = 163.8 kg.
1 pood   = 16.379 kg.
1 zolotnik   = 4.265 g.
1 dole   = 0.044 g.

SECTION I.

MAGMATIC SEGREGATIONS.

It is known that all eruptive rocks, especially those with but little silica, in rising from the unknown depths of the earth's interior to higher regions of the crust or to the surface, bring with them metals and metallic compounds, mostly, it is true, in very small amount. Magnetite and titanic iron ore, also pyrite, magnetic pyrite and chromite, are found in the form of small grains and crystals, scattered rather uniformly through many such eruptive rocks, in which the nature, microscopic intergrowth and relation to other constituents show conclusively that these small ore particles are primary constituents of the rock. It is true, as a rule, that such examples do not form true ore deposits, since the amount of these primary ores contained in such rocks is too small to pay for mining. However, weathering and erosion occasionally lead to a concentration on the earth's surface of these uniformly disseminated metallic minerals, as will be explained in the section on placers.

In some cases a concentration of the ores either into stock-like masses or into bands (schlieren) has taken place in the rock either before or during its solidification from the molten condition; these magmatic segregations or secretions, being of primary origin, will be considered first in this work.

Although concentrated in compact masses, the ore of these magmatic deposits is exactly the same as that which occurs in sparsely scattered particles through the enclosing rock, in which the ore minerals are accessory constituents. This, as will be shown in specific cases, is proven by the conditions of microscopic structure and intergrowth. This particular fact is the most important argument for the truly primary nature of such deposits, and enables one to discriminate between magmatic segregations and those accumulations of ore that have been formed through secondary processes in an eruptive mass.

The exact physico-chemical processes involved in the segregation of definite components of the magma into distinct portions of different composition, the so-called process of magmatic differentiation, is not as yet understood, although this problem has, in recent decades, occupied the attention of many investigators.[1]

[1] A very comprehensive presentation of these relations was given by J. P. Iddings: 'The Origin of Igneous Rocks.' Phil. Soc. of Washington, *Bull.* XII, pp. 89-214,

As regards this particular theory of the origin of ore deposits, the work of J. H. L. Vogt is most important. The process had long been recognized in examples of lean ore; notably the remarkable "mixed lodes" described by H. Bücking and others, where, in an eruptive dike, the more basic material of the magma has become concentrated in the two borders in sharply defined zones, making three parallel lodes or dikes side by side. Attention has also been devoted for a good while to the more basic, and often more highly iron-bearing marginal facies of eruptive stocks and large massives. In this segregation, processes of diffusion seem to have played a leading part. Teall, Lagorio and Brögger long ago noted the experiments of Soret (1879-1881), who, upon an unequal heating of different parts of saline solutions, obtained a different distribution of the salts, the solutions becoming more concentrated in the colder part than in the warmer. This is a consequence of the principle subsequently developed by van t'Hoff that the osmotic pressure increases in proportion to the absolute temperature. No doubt, however, other causes not yet known co-operate in producing this result.

Following J. H. L. Vogt, we will use the following grouping of ore deposits formed by magmatic differentiation according to the character of the ores predominant in each case:

    A. Segregations of native metals.

    B. Segregations of oxidic ores.

    C. Segregations of sulphide and arsenical ores.

(A) Segregations of Native Metals in Eruptive Rocks.

1. *Segregation of Native Iron in Basalts.*

The famous occurrence of native iron in the porphyritic feldspar basalt of Ovifak on Disko Island on the west coast of Greenland[1] is of great scientific, but of no economic importance.

Since the time of Captain Ross's visit it had been known that the Esquimaux worked up natural native iron into tools, but the place whence they obtained the raw material was not known. Finally A. E. Nordenskiold, in August, 1870, discovered the most important place to be Blaafjeld, on the

---

1892. Also in other publications. Compare also J. H. L. Vogt: 'Formation of Ore Deposits Through Processes of Differentiation in Basic Eruptive Magma.' *Zeit. f. Prak. Geol.*, 1893, p. 4, etc. The researches of J. Morozewitz are also of the greatest importance in this connection. 'Exper. Untersuch. über die Bild. d. Min. im Magma.' Tschermak's *Min. Mitth.*, 1899, Vol. XVIII, pp. 1-240.

[1] G. Nauckhoff: 'On the Occurrence of Native Iron in the Basalt Lode at Ovifak.' Tschermak: *Min. Mitth.*, 1874, pp. 109-136. J. K. V. Steenstrup: 'On the Iron of Greenland.' *Zeit. d. Deutch. Geol. Gesell.* 1876, pp. 225-233, and in the *Meddelelser fra Gronland* IV. Copenhagen, 1882. A. E. Törnebohm: 'On the Iron Bearing Rocks of Ovifak and Assuk.' Appendix to *K. Svenska Vet. Ak. Handl.*, Vol. V, No. 10, Stockholm, 1878, also contains further papers.

south side of Disko Island. In 1871 a special expedition, with G. Nauck-hoff as geologist, undertook a detailed investigation of the locality, and carried to Europe not only several very large blocks of iron found lying loose on the shore, but also abundant material still enclosed in the rock. It is now generally believed that the basalt of Ovifak is not a dike, but forms a sheetlike flow such as is common in the Tertiary coal-bearing formation of that region. This basalt porphyry encloses, according to A. E. Törnebohm, masses of a doleritic rock belonging to an earlier period of formation, hold-ing inclusions of a highly graphitic anorthite rock. This dolerite consists of labradorite, augite, olivine, titanic iron ore, magnetite and a glassy groundmass. Furthermore, it contains native iron (Schreibersite), usually near the boundary with the anorthite inclusions, also troilite, magnetic py-rite, graphite and a silicate, for the most part greatly altered and resembling hisingerite. The metallic iron appears in flakes, grains, globular masses and large lumps. On etching, it shows the Widmanstatten figures and con-tains nickel and cobalt, according to a new analysis by A. Iwanoff[1].

$$
\begin{array}{ll}
\text{Fe} & 92.91 \\
\text{Ni} & 2.66 \\
\text{Co} & 0.69 \\
\text{Cu} & 0.19 \\
\text{C} & 3.29 \\
\text{S} & 0.26 \\
\hline
& 100.00
\end{array}
$$

The rust-like decomposition crust of the ore consists essentially of basic hydroxides, basic oxychlorides and basic sulphates of iron. Many observers have noted, as shown in the illustration of thin sections of the rock given in Törnebohm's report, that in the smaller nests of iron the metal apparently must have been segregated after the other constituents. Nauckhoff and Törnebohm demonstrated that the iron probably occurred in the form of phosphor-nickel iron (Schreibersite) in breccia-like fissure fillings in the midst of the basalt, and Törnebohm inferred from this, and from the above mentioned microstructure of the iron-bearing dolerite, that the iron of Disko must have been formed secondarily from solutions. This inference, however, may be questioned, especially in view of the above-mentioned large blocks. The breccia in fact may be due to a basalt zone, broken up by dis-locations after the iron had been segregated. From the chemical point of view Törnebohm's theory has recently been strongly supported by C. Winkler[2],

[1] C. Winkler: 'Zur Zusammensetzung des Eisens von Ovifak,' etc. *Kongl. Vetensk Ak. Förh*, 1901, No. 7, Stockholm, p. 495.

[2] C. Winkler: 'On the Possibility of the Immigration of Metals into Eruptive Rocks through the Agency of Carbon Dioxide.' *Ber. d. Math.-Phys., Kl. d. Kgl. S. Ak. d. W.* Leipzig, 1900, p. 9.

who has called attention to the freely fluid compounds which, according to recent investigations, carbon dioxide forms with nickel-iron on even moderate heating, and which decompose again, with separation of the native metals. Such heating might take place for a considerable time at the contact of such gases with an eruptive body still undergoing congelation. The author himself, it is true, raises the difficult question as to the origin of such iron carbon compounds which must have been formed in a cooler zone. The geologic conditions of these and other closely related occurrences indicate that the original seat of the native metals was in the eruptive hearths themselves.

R. J. V. Steenstrup subsequently found in basalts, at several other points of Disko, inclusions of native iron, graphite and nickeliferous magnetic pyrite, which renders it very likely that the iron and nickel content was an original constituent of those basaltic magmas, and the masses are not of cosmic origin, as Nordenskiöld had originally conjectured.

2. *Segregations of Nickel Iron in Olivine Rock and Serpentine of Awarua in New Zealand.*

A very interesting occurrence of nickel iron, awaruite, discovered on the west coast of the south island of New Zealand, by W. Skey, in 1885, has been described by G. H. F. Ulrich[1]. This region consists of gneisses, mica schists and chloritic schists, which are broken through by vast stocks of an olivine rock of the composition of saxonite (olivine $+$ enstatite), in part transformed to serpentine. In the river valleys, descending from the mountains formed of serpentine and saxonite, the nickel iron was found in loose grains, and it was intended to wash it from the gravel. Subsequently it was also found in small particles intergrown with the eruptive rocks just named. The composition of awaruite, according to Skey, is as follows:

$$
\begin{aligned}
&\text{Ni} \dots\dots\dots 67.63 \\
&\text{Co} \dots\dots\dots 0.70 \\
&\text{Fe} \dots\dots\dots 31.02 \\
&\text{S} \dots\dots\dots 0.22 \\
&\text{Si O}^2 \dots\dots 0.43 \\
&\quad\quad\quad\quad\overline{100.00}
\end{aligned}
$$

3. *Segregations of Platinum in Olivine Rocks.*

It had long been suspected that the platinum of the Ural placer gravels (see later) must have its original source in the serpentines and olivine rocks found in the region drained by the upper courses of the platinum-carrying rivers. This inference became a certainty when an olivine gabbro

[1] G. H. F. Ulrich: 'On the Discovery, Mode of Occurrence, and Distribution of the Nickel Iron Alloy, Awaruite, on the West Coast of the South Island of New Zealand.' *Quart. Journ.* Geol. Soc. London, 1890, Vol. XLVI, pp. 619-633.

with grains of intergrown platinum was discovered on the Krestovozdri-schensky property, in the western Ural, and later, in the eastern part of the mountains in the Goroblagodatsk district, nests of chromite with platinum were found in an olivine rock[1].

Extensive experiments with crushed samples of olivine rock and olivine gabbro from the vicinity of Mount Soloviov, near Nizhni Tagilsk, have shown a slight platinum content in these rocks.[2]

The platinum of the Ural contains 5 to 13% of iron, besides some iridium, rhodium, palladium, osmium and copper. (See platinum gravels.)

St. Meunier[3] calls attention to the very irregular and often branched forms of the grains of iron-platinum in the olivine rocks of the Ural, and compares it with the peculiar structure of the small nests of native iron in the dolerite of Ovifak and the iron grains in meteoric magnesian silicate rocks. By experiment he obtained similar structures by using gases to reduce the respective metals from a heap of granules of such silicates. A jet of platinum chloride, hydrogen and some ferrous chloride passing through the heap at red heat produced iron platinum in quite analogous development. These observations are important. The conclusion he draws from them, however, that the rocks in question are products of primitive congelation, after the manner of meteorites, is not borne out by the conditions under which the platinum-bearing eruptive rocks actually occur.

The platinum contents of the olivine rocks of the Ural have, with few exceptions, been found much too low to give to these primary platinum deposits an economic significance; only the residual and alluvial placers pay for the working.

4. *Gold as a Primary Constituent of Eruptive Rocks.*

Though it may be difficult in an individual case to decide whether the particles of native gold found in eruptive rock are primary and really due to segregations or have been introduced by secondary processes, yet a series of occurrences seems to indicate the probability of the former alternative. Most of the observations relate to granites and other acid eruptive rocks.

G. P. Merrill[4] describes primary free gold occurring as an inclusion of

[1] A. Inostranzeff: 'Primary Site of Platinum in the Ural.' *Mitth.* an d. Ges. d. Naturf, in St. Petersburg. Nov. 7, 1892. (Reviewed by R. Helmhacker. *Z. f. pr. G.*, 1893, p. 87.)

[2] Oral communication of Mining Engineer Hamilton of Nizhni Tagilsk to R. Beck when on a visit to Mount Soloviov in 1897.

[3] St. Meunier: 'Study of the Matrix of the Platinum of the Ural,' etc. *Compte Rendu du VII. Congr. Geol. Intern.*, 1898, p. 157.

[4] G. P. Merrill: 'Occurrence of Free Gold in Granite.' *Am. Jour. Sci.*, 1896, I, p. 309. H. Schultze: 'Gold Mining,' in H. Kunz's 'Chile.' 1890, p. 78. W. Möricke: 'The Gold, Silver, and Copper Ore Deposits of Chile.' Freiburg, 1898, Vol. I, p. 16.

feldspar and quartz in a granite from Sonora, Mexico, as "a product of cooling and crystallization from the original magma." H. Schultze demonstrated the primary nature of gold in many granites of the coast Cordillera of Chile; W. Möricke proved the same for glassy and crystalline quartz trachytes of that country; W. P. Blake added examples from Arizona; Forbes, examples from Bolivia. It also seems that the primary nature of the gold in the granites in the Ekaterinburg region in the Ural, as, for example, on Lake Shartash, has been satisfactorily demonstrated. These occurrences are, however, only of economic importance because payable placer gravels may develop from them.

On the other hand, the primary content of free gold in diorites or amphibolitized diabases, so often referred to in different publications, seems of very doubtful authenticity. The occurrences of this kind examined by us, for example, from Mashonaland, rather suggest the assumption of a secondary immigration of gold. The reader will find further statements on this subject on a subsequent page.

In connection with this it may be mentioned that according to R. Daintree[1] gold-bearing pyrites are found widely distributed in the diorites which occur in the Upper Silurian or Devonian of New South Wales, Victoria and Queensland. The primary nature of these pyrites is not, however, evident from the accompanying plates of thin sections.

A remarkable case is that mentioned by K. Schmeisser[2], of gold in a basalt from Richmond river in New Zealand, which is said to assay as high as 18 grams per ton. This may be a local occurrence and due to highly auriferous sands gathered from gravels which were broken through by the basalt.

(B) SEGREGATIONS OF METALLIC OXIDES IN ERUPTIVE ROCKS.

Table of the Important Iron Ores.

Ore.	Fe.	H_2O.	CO_2.	S.
Magnetite FeO, Fe_2O_3	72.4	.....	.....	.....
Hematite (Specular ore) Fe_2O_3	70.0	.....	.....	.....
Siderite (Spathic iron ore) $FeCO_3$	48.27	.....	37.92	.....
Limonite, $2Fe_2O_3.3H_2O$	59.89	14.4	.....	.....
Pyrite, FeS_2	46.7	.....	.....	53.3

[1] R. Daintree: 'Occurrence of Gold in Australia.' *Quart. Journ.* Geol. Soc., 1878, Vol. XXXIV, p. 431.

[2] K. Schmeisser: 'Australasien,' p. 92.

1. *Segregations of Magnetic Iron Ore in Quartz-free Orthoclase Porphyries and Syenites.*

This group includes the best studied and authenticated magmatic deposits known, especially the famous deposits of the Ural[1] and of Lapland, which we will describe in detail.

(a) The Vysokaya Gora.

The famous iron ore deposit of Vysokaya Gora (High Mountain) lies directly west of Nizhni Tagilsk, the most important mining locality of the central Ural. It is merely one link in a chain of similar magnetite deposits of these mountains, which are all associated with a series of syenite and accompanying porphyritic rocks covering an area about 70 kilometers long and 15 kilometers broad, and striking north-south. The main mass of the Vysokaya intrusion consists of augite-syenite, which in many instances alternates with bands of a quartz-free orthoclase porphyry. The latter rocks consist of orthoclase feldspar and a little plagioclase; it is only here and there that augite (or uralite) and biotite take part in their composition. Besides the accessory ingredients (titanite, zircon and apatite), magnetite is also found in small granules or rounded crystals, which may constitute as much as 20% of the entire rock mass, until finally whole lumps of this mineral make their appearance, thus gradually effecting the transition to true orebodies and forming an irregular stock. At times, as, for example, in the fragments of a breccia due to later disturbance, the result of igneous energy, the orthoclase porphyry is amygdaloidal and shows primary magnetite concentrated around vesicular cavities now filled with calcite (Högbom). According to Tschernyschew, there is a connection between the dislocations and the epidote-zoisite-garnet rocks of the Vysokaya, which, besides these minerals, contain also zeolites, chlorite, quartz and calcspar. According to Högbom their structure shows them to be metamorphosed porphyries and syenites, since they still show feldspar lathes arranged in a fluidal structure. These eruptive masses are intruded in Devonian limestones, the rocks being well exposed in the open-cuts made in mining. Greatly decomposed interbedded tuffs occur with these limestones, especially in the eastern part of the mountain. The sedimentary strata are faulted, bringing the eruptive rocks and orebodies in contact with the tuffs and epidote-garnet rock masses along the fault plane.

---

[1] H. Müller: 'Ueber den Magnetberg Gora Blagodat.' *Berg u. Hutten Zeit.*, 1866. p. 54. P. Jéréméew: 'Les minerais de fer dans les districts miniers de la chaîne de l'Oural.' *Journ. d. Mines*, 1859, II., p. 313. Th. Tschernyschew: 'Guidebook of the Excursions of the VII. International Geological Congress, 1897.' Vol. IX. A. H. Högbom : 'Om de vid svenit bergarter bundna jernmalmerna i Ostra Ural.' *Geol. Foren.* i Stockholm Förh. Vol. XX, part 4, 1898.

The magnetic iron ore of Vysokaya is remarkable for its great purity. The ore has averaged 65 per cent during the past ten years. A large part of the ore that is mined has the composition of martite.[1] The ore is low in phosphoric acid compared with the analogous deposits at Lébiáshaya situated farther northeast, near Nizhni Tagilsk. The Vysokaya iron ores at times contain scattered grains of copper pyrite. To the southeast of this mountain of magnetic iron ore is the famous copper mine of Médnorudiansk. The characteristic analysis of the ore of the Vysokaya given below is from the official report of the company operating the mines:

$$
\begin{aligned}
&Fe_2O_3 \ldots \ldots \ldots \ldots \ldots \quad 74.40 \; \big\} \; Fe \; 66 \; per \; cent \\
&FeO \ldots \ldots \ldots \ldots \ldots \quad 16.71 \\
&Mn_3O_4 \ldots \ldots \ldots \ldots \quad 1.30 \\
&Cu \ldots \ldots \ldots \ldots \ldots \quad 0.06 \\
&S \ldots \ldots \ldots \ldots \ldots \ldots \quad Trace \\
&P \ldots \ldots \ldots \ldots \ldots \ldots \quad 0.03 \\
&SiO_2 \ldots \ldots \ldots \ldots \ldots \quad 2.85 \\
&Al_2O_3 \ldots \ldots \ldots \ldots \quad 1.80 \\
&CaO \ldots \ldots \ldots \ldots \ldots \quad 0.99 \\
&MgO \ldots \ldots \ldots \ldots \ldots \quad 0.98
\end{aligned}
$$

100.12 per cent.

The iron industry of Nizhni Tagilsk was founded about 1725 during the reign of Peter the Great. The mines supply half a dozen furnaces in the Ural district with ore, the average annual production during the past decade being a little over 100,000 tons.

(b) The Goroblagodat Iron Deposit.[2]

This famous mountain of magnetic iron ore, whose name means "blessed mountain," is an isolated peak rising to a height of 156 meters above the Kushva plain. The deposit supplies extensive reduction works, and is worked by vast open-cuts which run around the south side of the summit, and in terraces far down the west slope. A high pillar of rock has been left at the summit which is crowned by a chapel. (See Fig. 1.) The view from this point includes the range of the Ural, which here is not very high, to the west, while eastward the Tura and Tagil rivers are seen traversing the gently undulating west Siberian lowlands.

At Blagodat the same conditions seen at Vysokaya are exactly repeated, except that the arrangement of the orebodies proper is somewhat different. Here they have the form of streaks arranged in benches, one above the other, as shown in Fig. 1. This figure also exhibits the two main faults of the deposit. The transition between ore and normal eruptive rock is gradual and not sharp as represented in the figure.

[1] G. Lebedew: 'Lehrbuch der Mineralgi' (Russian), St. Petersburg, 1900, p. 126.
[2] Literature the same as under Vysokaya Gora.

In contrast with the Vysokaya ores, the magnetite contains much pyroxene and its chloritic decomposition products, as well as spinel, so that the percentage of iron is considerably lower, being on an average only 55%. On the other hand the quality is excellent because of the very small percentage of phosphorus and pyrite.

Mining began at Blagodat the early part of the last century, and the output reached 16,000 tons as early as 1770. From 1813 to 1898, 2,720,000

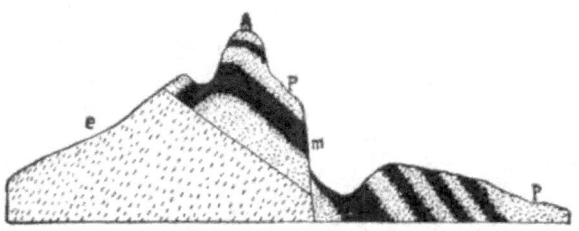

Fig. 1.—Section of Goroblagodat after Th. Tschernyschew.
p, orthoclase porphyry; m, magnetic iron ore; e, epidote-garnet rock.

tons of ore were taken from this mountain. In 1898 the production was 66,000 tons.

A characteristic analysis of the ore according to D. J. Mendelejeff[1] is given below:

SiO_2	9.40 per cent.
Al_2O_3	7.18
Mn_2O_3	0.20
FeO	21.65 ⎫ Fe 53.89 per cent.
Fe_2O_3	52.93 ⎭
CaO	6.00
MgO	1.62
S	0.05
SO_3	0.12
P_2O_5	0.33
Cu	0.01
Moisture	0.20

The amount of ore exposed and ready for extraction in 1890 was estimated to be 15,000,000 tons. The ore is treated in the furnaces of Kushva, Barancha and Verhneturye, the pig iron being shipped to Nizhni Turinsk and Perm. The chapel on the summit is dedicated to the memory of Vogul Stephen Chumpin, who revealed to the Russians the wealth of the mountain, and for doing so was burned alive by his tribe on the summit.

A third iron deposit in the southern Ural, 60 to 70 kilometers south of Verhne Uralsk, called Gora Magnitmaya, or Atach, seems to be connected with porphyritic and syenitic masses, but detailed geologic investigations have not yet been made. The ores contain 67 to 68% of iron and are almost free from phosphorus and sulphur.

[1] 'Iron Furnace Industry of the Ural in 1899.' II, p. 174. (In Russian.)

The iroń deposits of Kiirunavaara and Luossavaara in Lapland are of the Ural type just described[1]. They have been recently investigated in detail and actively developed.

(c) Kiirunavaara and Luossavaara.

These two great deposits are situated in the Swedish province of Norrbottom, in about 67° 50' north latitude. The ores form great stocks, reminding one of strata by the manner in which they outcrop within a vast mass of quartz-free orthoclase porphyry. This rock is bounded on the east by graywacke slate, clay slate and conglomerates, overlain by quartzitic sandstones, while on the west it abuts against strongly metamorphosed conglomerates.

The ore stock of Kiirunavaara forms a barren mountain ridge, the outcrop extending for 1½ miles, while magnetic observations prove its continuation northward to the shore of Lake Luossajarvi, and out as far as an island, so that the entire length must be nearly three miles. Throughout its entire extent the deposit dips uniformly eastward at from 45° to 80°, as determined by drilling. The thickness of the ore sheet is between 34 and 150 m. (111 and 500 feet) and averages 70 m. (230 feet). The total amount of ore available for extraction is estimated by Hjalmar Lundbohm to be at least 215,000,000 tons. This ore is not everywhere of the same quality. The author just named distinguishes five types with many intervening transitions: (1) Magnetic iron ore, poor in phosphorus; (2) the same, mixed with hematite; (3) phosphorus-bearing magnetic iron ore, with a little apatite; (4) highly phosphoritic magnetic iron ore, with apatite, in numerous pockets, stringers and stratiform bodies; (5) highly phosphoritic magnetic iron ore with minutely divided apatite. In more than 60% of the samples from the very numerous prospects the iron exceeded 67%, while the phosphorus varied between 0.05 to 6%.

The ore mountain of Luossavaara lies north of Lake Luossajarvi, and forms a broad stratiform stock about ¾ of a mile long by about 180 feet broad, dipping steeply east within the same porphyry mass. This deposit is estimated to contain at least 4,700,000 tons of ore carrying 67 to 70.5% iron and a very low phosphorus content. These occurrences will become of great importance upon the completion of the railway which will extend from the Lapland iron ore region across to the Ofoten Fjord, on the west coast of Norway.

[1] K. A. Fredholm: 'Rocks and Ores in Luossavaara and Kiirunavaara.' *Geol. Fören. Förh.*, 1891, Vol. XIII, p. 266. Hj. Lundbohm: 'Kiirunavaara and Luossavaara Iron Ore Field.' Geol. Surv. of Sweden, series C, No. 175, 1898, with 3 plates and 1 map.

2. *Segregations of Titaniferous Magnetite[1] in Gabbro.*

The most important examples of this class are the deposits of Taberg and Routivare in Sweden, Valimaki in Finland and the Adirondacks in New York. Kemp has published a review of all known deposits of this class[2].

The genesis of these segregation deposits has been discussed by Vogt[3]. He has attacked the problem by a study of numerous analyses of rocks and ores which have been platted in tables giving a graphic representation of the relations of the different elements. One of these tables, reproduced in Fig. 2, shows how, in the transition stages from normal rock to segregated ore, the amount of Fe_2O_3 and TiO_2 increases as the silica, lime and alumina decrease. The magnesia increases, attains a maximum in the middle transition forms, and afterwards decreases. Vogt infers from these relations that the real solvent of the ores must be an alumina-lime-sodium-silicate, and that this silicate plays the same part as the mother liquor of saline solutions. A part of the silica is, however, supposed to have taken part in the differentiation as a silicate of magnesia and iron. Petrographic investigations confirm these conclusions. Kemp has recently reviewed the evidence and presented his conclusions[4].

(a) The Taberg Near Jonkoping.

The iron deposit of Taberg[5] in Smaland, southwest of Lake Wetter, near Jonkoping, in Sweden, is a belt or streak of ore-rich rock forming a part of a mass of olivine gabbro (olivine hyperite or Törnebohm). This basic rock has by reason of its greater resistance been left during the degradation of the region as a residual mountain. The iron ore forming this mountain consists of titaniferous magnetite and olivine, with subordinate admixtures of biotite and plagioclase. Transitions may, however, be observed, showing all gradations between the magnetite-olivenite occupying the center of the igneous stock, to the ordinary olivine-gabbro, poor in magnetite, forming the border of the intrusive mass. The stock is intruded

---

[1] J. F. Kemp: 'Titaniferous Iron Ores of the Adirondacks.' 19th *Ann. Rept.* U. S. Geol. Surv., III, 1899, p. 387.

[2] J. F. Kemp: 'A Brief Review of the Titaniferous Magnetites,' *School of Mines Quarterly*, 1899, July, pp. 323-356, Nov., pp. 56-65.

[3] J. H. L. Vogt: 'Further Researches upon the Segregation of Titaniferous Iron Ores in Gabbro.' *Zeit. f. Prak. Geol.*, 1900-1901.

[4] *The Engineering and Mining Journal*, November 28, 1903.

[5] A. Sjögren: 'On the Occurrence of the Iron Ore of Taberg in Småland.' *Geol. Fören. Förh.*, 1876 and 1877, p. 42, with older literature. *Ibid.*, 1882-1883, p. 264. A. E. Törnebohm: 'On Taberg in Småland,' etc. *Ibid.*, 1880-1881, p. 610, plates 25 and 26. J. H. L. Vogt: 'Formation of Ore Deposits by Differentiation,' etc. *Zeit. f. Prak. Geol.*, 1893, p. 8.

in a complex of gneisses and gneiss granites, and is accompanied by hornblende schists, probably patches of dynamo-metamorphosed gabbro.

The Taberg ore is characterized by the presence of 0.12 to 0.40% of vanadic acid; in fact, the element vanadium was first discovered by Seffstrom

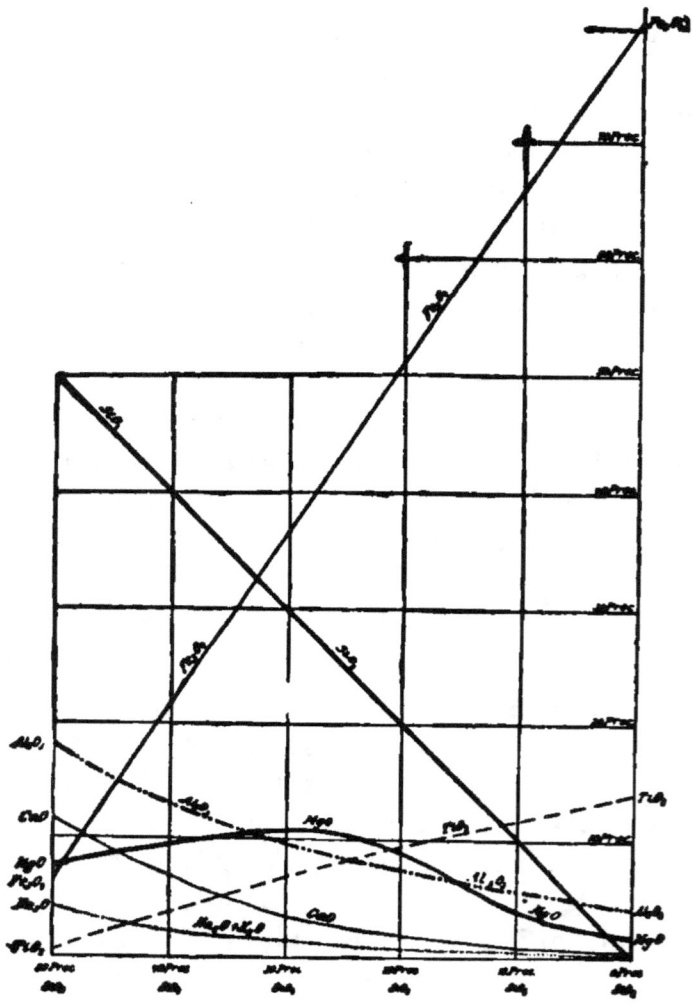

Fig. 2.—Diagram showing Normal Differentiation. (Vogt.)

in 1830 in the crude iron of Taberg. The high percentage of titanium greatly lowers the value of the Taberg ore; moreover, the iron content, even at the richest spots, is low, not exceeding 32 per cent.

(b) The Titaniferous Magnetites of the Adirondacks.

Among the North American representatives of this type of deposit the Adirondack iron deposits are the best known, and have been carefully studied and described by Professor Kemp.[1] The mines of the Adirondack region are in Essex county, New York, southwest of Lake Champlain.

The older gneisses and crystalline limestone of this region contain many intrusive masses of gabbro and closely related rocks which have been altered by regional metamorphism to schistose rocks.

These rocks include both anorthosites (coarse-grained rocks of almost pure labradorite) and the dark-colored gabbros and norites in which the ferromagnesian silicates predominate. These extreme types are both connected with masses of titaniferous magnetite and menaccanite, the former at Newcomb, Wilmington and North Hudson, the latter at Crown Point, Elizabethtown, etc. The minerals of the normal gabbro occur disseminated through these ores, and all transitions occur from gabbro to ore.

The petrographic features as described by Kemp correspond closely to those of Välimäki, Finland. The following analysis by Hillebrand, of the typical ore from Oak hill, represents the normal composition of this class of ores:

Fe	38.98	Al_2O_3	7.03
Fe_2O_3	30.34	CaO	3.59
FeO	22.81	MgO	6.92
TiO_2	5.21	P_2O_5	0.14
SiO_2	21.42	S	0.04

The iron deposits of the Mesabi range, Minnesota[2], Port Henry and Lake Champlain[3], New York; Iron Hill, Cumberland, Rhode Island[4], and those of various gabbro areas of Quebec[5] and Ontario[6], all belong to this class.

The magnetite and chromite deposits of the Zwartkoppies range and elsewhere in the Transvaal, Africa, occur in norites forming the contact facies of a red granite. They are regarded as magmatic segregations by Molengraaf[7].

Similar occurrences, though of less extent, have been found at several points in Sweden; for example, in the gabbro of Langhult in Smaland and of Ransberg in Vestergotland.

[1] J. F. Kemp: 'The Titaniferous Iron Ores of the Adirondacks.' 19th *Ann. Rept.* U. S. Geol. Survey, Part III, 1899, p. 383.

[2] N. H. and H. V. Winchell: 'The Iron Ores of Minnesota.' Geol. Surv. of Minn., 1891.

[3] Kemp in *Trans.* Am. Inst. Min. Eng., Vol. XXVII, pp. 146-203, 1898.

[4] M. E. Wadsworth: 'A Microscopic Study of the Iron Ore Peridotite of Ironmine Hill.' *Bull.* Mus. Comp. Zool. Harvard College, Vol. VII, 1881, p. 183.

[5] F. D. Adams: 'On the Igneous Origin of Certain Ores,' Montreal, 1894. *Proc.* Gen. Min. Assoc. of the Province of Quebec, Jan. 12, 1894.

[6] E. J. Chapman: 'On Some Iron Ores of Central Ontario.' *Trans.* Royal Soc. Canada, 1885, p. 9.

[7] 'Geol. Aufnahme der Süd-Afrikanischen Republik.' *Jahresber.* for the year 1898. Pretoria, 1900, p. 14. 'Géol. de la Rép. Sud-Africaine,' 1901, pp. 48 and 52.

(c) Deposit of Routivare in Norrbotten, Sweden[1].

This great deposit occurs in a gabbro (gabbro diorite) which has been subjected to strong regional metamorphism. The deposit forms a band 1,600 meters long by as much as 300 meters broad, consisting of titaniferous magnetite, menacconite, spinel, olivine, a pyroxene and some magnetic pyrite. The ore differs from that of other similar deposits, as it contains a large amount of spinel. It contains between 48 and 52% of iron and between 11 and 13% of titanic acid. Other magmatic deposits of similar ores rich in spinel occur in Norway and have been described by Vogt[2]. From his study of these ores in thin section under the microscope he has reached the important conclusion that the iron magnesian silicate has been the first to separate out of the magma, followed by the titaniferous magnetite. If the ores also contain pyrite and spinel the order of crystallization is: Pyrite, spinel, titaniferous magnetite, an exception to the usual order observable in eruption rocks, in which magnetite is the first mineral to separate out.

Between Sordavala and Pitkaranta, on the northeast shore of Lake Ladoga, near the village of Valimaki, a rock-complex consisting of phyllite, quartzitic schist, nodular and hornblende schist, together with mica schist carrying andalusite and staurolite, is penetrated by a coarse-grained diorite which is apparently an altered gabbro.

According to Blankett[3] this gabbro diorite forms a stock, having a north-south direction, cutting the schists and enclosing metamorphosed portions of them. The rock consists of plagioclase, hornblende (at times with augite core) with some epidote, apatite, magnetite, muscovite and hematite. Within this intrusive mass at some distance from its contact with the schists there are basic streaks of a rock very rich in hornblende, diallage, bronzite, olivine and titaniferous magnetic iron ore, with subordinate amounts of green spinel, pyrite, magnetic pyrite, copper pyrite, mica and apatite. These dark streaks as a rule contain 15 to 30% of iron, while richer portions run as high as 40%. One specimen contained 63.40% of ferric oxide and 2.90% of titanic acid. All possible transitions from poor to rich ore may be found within these very extensive segregation streaks, whose distribution is usually quite regular. The iron ore is crushed and the magnetite extracted by magnetic separators, the concentrate being briquetted and smelted at

---

[1] W. Peterson: 'On the Iron Ore Field of Routivare in Norrbotten Lane.' *Geol. Fören. Förh.*, Vol. XV, 1893, p. 45, with map. H. Sjögren: 'Additional Notes on Routivare Iron Ore.' *Ibid.*, p. 140.

[2] Vogt: *Zeit. f. Prak. Geol.*, 1900, pp. 233-242. Kolderup: 'Petrog Beschreibung Lofatens,' etc. *Bergens Mus. Aarbog*, VII, 1898, pp. 1-54.

[3] Blankett: 'Ore Field of Välimäki, etc.' *Geol. Fören. i Stockholm Förh.*, 1896, p. 201, with map.

Widlits on the opposite side of the lake. The mines produced some 7,764 tons of ore fit for smelting in 1896.

In other cases the magnetite is not only accompanied by spinel, but also by corundum and sillimanite, as in the norites of Westchester county described by G. H. Williams.[1]

3. *Segregations of Titaniferous Magnetite Iron Ore in Nepheline-Syenites.*

The iron deposits of Alnö, near the east coast of Sweden, in the Gulf of Bothnia, are described by A. G. Högbom[2] as basic segregations of nepheline-syenite. The nepheline-syenite of that locality shows an extraordinary number of varieties, and is closely associated with remarkable masses of crystalline limestone which, according to Högbom, are possible primary segregations from the extremely basic magma. The surrounding rocks are gneisses which have been altered by contact metamorphism. The basic streaks of the nepheline-syenite are distinguished by large amounts of titano-magnetite, abundant apatite, many silicates of iron and magnesium, as well as the absence of feldspar and, ordinarily, also of titanite, nepheline and cancrinite. Some portions are so rich in titano-magnetite as to be workable. Such an ore from the Tryg mine had, according to A. Tamm, the following composition:

SiO_2	3.10
TiO_2	12.14
Al_2O_3	Trace.
FeO	8.95
Fe_2O_3	64.38
MnO	1.15
CaO	2.30
MgO	8.00
S	0.07
P_2O_5	Trace.
Total	100.09

An occurrence in Brazil, which possibly may be much younger in age, is most naturally mentioned here. According to O. A. Derby[3] and E. Hussak,[4] the magnetic deposits of the Jacupiranga and Ipanema mines in the province of Sao Paulo are segregations in granular pyroxenic rocks. The deposits of magnetite-pyroxenite (jacupirangite) with 60 to 70% of magnetite are, moreover, closely connected by transition forms with nepheline

---

[1] G. H. Williams: 'The Iron Ore and Emery in the Cortlandt Norites.' *Am. Jour. Sci.*, Ser. III, Vol. XXXIII, 1887, p. 194.

[2] A. G. Högbom: 'On the Nepheline-Syenite Area of the Island of Alnö.' *Geol. Fören. Förh.*, Vol. XVII, 1895, pp. 100-214.

[3] O. A. Derby: 'On Nepheline Rocks in Brazil.' *Quart. Journ.* Geol. Soc., 1891, Vol. XLVII, p. 251.

[4] E. Hussak: 'On Brazilite,' etc. *Neues Jahrbuch f. Min.*, 1892, Vol. II.

rocks. The iron-bearing streaks also contain, besides magnetite and pyroxene, nepheline, perofskite and apatite, as well as brazilite, a tantalo-niobate.

4. *Segregations of Titanic Iron Ore in Gabbro Rocks.*

Segregations of titanic iron ore in gabbro are less numerous than the magnetites just described. Such occurrences have been described in great detail by J. H. L. Vogt[1] and Kolderup, from whose description the following abstract is given:

The Titanic Iron Ore Deposits of Ekersundsoggendal.

In southern Norway, in the Ekersund-Soggendal district, the crystalline schists are broken through by a norite intrusion forming a mass 1,200 square kilometers in extent. The eruptive rock shows a great variety of development with all possible transitions between pure labradorite and norites rich in hypersthene, apatite and enstatite-granite or bronzite-granite.

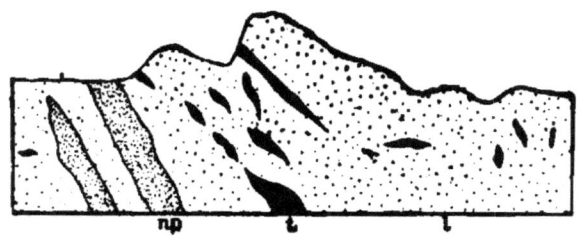

Fig. 3.—Section through the Blaafjeld. (Vogt.)
l, labradorite; np, dikes of norite-pegmatite; t, titanic iron ore.

The labradorite contains veins and nests of pure titanic iron ore up to masses of 11 meters in thickness and 50 meters in length, as at Blaafjeld, illustrated in Fig. 3 after Vogt. The boundary between the ore and the labradorite is sometimes sharp, sometimes the reverse. Furthermore, the labradorite is traversed by dikes of a norite very rich in titanic iron ore. The rock of these menacconite-norite lodes consists of a granular crystalline aggregate of titanic iron ore, hypersthene and some labradorite. It contains as accessories, chrome-spinel, iron pyrite and very scanty apatite. The large lode (Storgangen) about 3 kilometers long and 30 to 70 kilometers wide, which is represented in Fig. 4, contains on an average 40% of titanic iron ore, and in some of its dark bands parallel to the selvage even 70 to 80 per cent.

[1] J. H. L. Vogt: 'Norske erts Forekomster.' *Arch. f. math. og naturv.* Christiania, 1887, Vols. X and XII, p. 101. Reviewed in *Neues Jahrbuch f. Min.*, 1899, Vol. II, p. 97. Also, 'Om dannelse af jernmalm forekomster.' Christiania, 1892. Also Bildung von Erzlagerstätten.' *Zeit. f. Prak. Geol.*, 1896, p. 6. Kolderup: 'The Labrador rocks of West Norway.' Bergen's *Mus. Aårbog*, 1896. No. 5, pp. 159-181.

5. *Segregations of Chrome Iron Ore in Olivine Rock and Serpentines.*

It has been known for a long time that all large deposits of chrome iron ore[1] are associated with serpentine. It might be readily imagined that all these ores were segregated during the serpentinization of olivine rocks. Indeed, it is probable that a part of the chromite masses enclosed in serpentines was formed during this process of rock alteration, since it is known that certain original ingredients common in olivine rocks, especially chrome diopside and chrome spinel, or picotite, contain chromine in considerable amount. Just as magnetite is segregated during the alteration of iron-

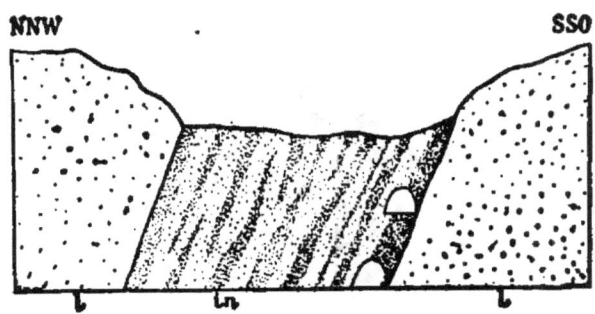

Fig. 4.—Profile through the Storgang. (Vogt.)
l, laboradorite; in, ilmenite-norite with dark streaks very rich in ore.

bearing silicates, as, for example, of olivine into serpentine, so in the decomposition of chrome diopside, and other chromine containing silicates, chrome iron ore may be formed. However, this origin, as already noted, holds true for only a part of these ores, not for all. On the contrary, the discovery of segregations of chromium ores in entirely fresh olivine rocks proves that in the main the chrome iron ore represents a segregation in the original magma. This has been shown very convincingly by Vogt[2].

The most important chrome iron mines, arranged according to their approximate figures of production, are scattered over the earth at the following places: In Asia Minor[3] (Harmandjyk, Brussa, vicinity of Smyrna); in the Ural (district of Ekaterinburg and Nizhni Tagilsk); in California; in North Carolina (Ashe and Clay counties); in Pennsylvania and Maryland; in New Caledonia (Tiebaghi Mountains at Mount Dore); in New Zealand; in Greece (Burdaly and Thessaly on Skyros); in Styria (Krau-

[1] According to Vogt the total annual production of chrome iron ore for the entire world is only 20,000 tons. It is used for chrome steel, dyeing, etc. (The U. S. imports for 1902 amounted to 20,000 tons; the world's production to 30,000 tons.—W. H. W.)

[2] J. H. L. Vogt: 'Zur Classification der Erzvorkommen,' *Zeit. f. Prak. Geol.*, 1894, p. 384.

[3] K. E. Weiss: 'Lagerstätten im West. Anatolien,' *Zeit. f. Prak. Geol.*, 1901, p. 250.

bath); in Bosnia (Dubostica); and in Norway (island of Hestmando, vicinity of Röros). Within the German Empire the inconsiderable occurrences of Silberg and Grochau in Silesia are the only ones known.

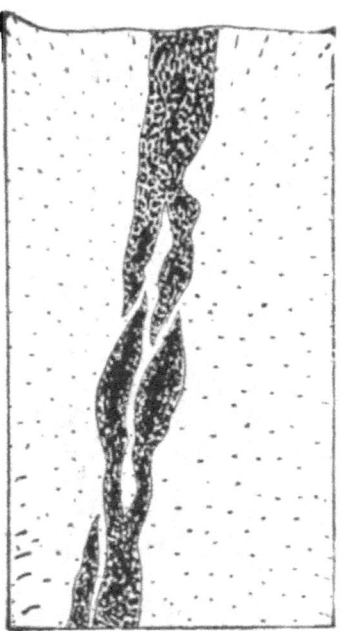

Fig. 5.—Section of a segregation band of chromitiferous olivine rock in the saxonite of Varnasfjeld. (Vogt.)
Length of lode 40 to 50 meters; thickness 0.1–0.8 meters .

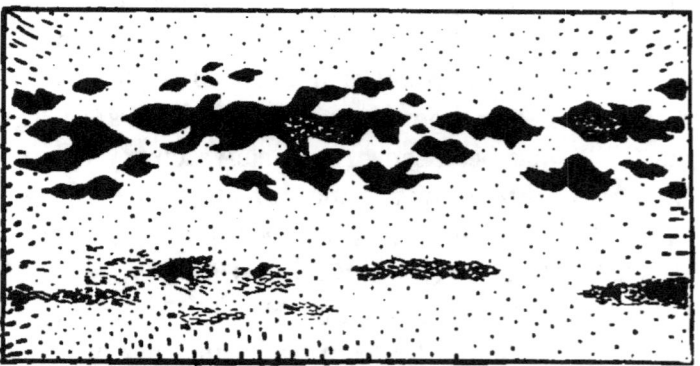

Fig. 6.—Profile of a segregation band (streak) of pure chromite and chromitiferous olivine rock in saxonite; streaks rich in enstatite with nests of chromite seen below. Ramberg on Hestmandö. (Vogt.)
Streak lode, 5 to 7 meters long; 0.5 to 0.8 meters thick.

Under this heading we will merely mention a few instances that have been accurately investigated.

From a genetic point of view decidedly the most interesting are the Norwegian occurrences studied by Vogt[1].

(a) The Chrome Deposits of Hestmando and Other Localities in Norway.

The island of Hestmando lies in northern Norway under the Polar Circle. Here the gneiss rock is intruded by at least a dozen small stocks of olivine rock, at most, 0.3 mile in diameter. They consist of olivine, enstatite, chrome-spinel and chromite. Sometimes they also contain a colorless or greenish actinolite, various mica-like minerals and as secondary

Fig. 7.—chromitiferous olivine rock of Ramberg. (Enlarged 50 times.)
o, olivine; ch, chromite.

formations magnesite and talc. According to the usual petrographic classification the rock would have to be called a saxonite. It varies from enstatite-poor to enstatite-rich rocks, so that the extreme types might be designated as enstatite-bearing dunite and enstatite rock. Within these olivine rocks the chrome iron ore forms streak-like, that is to say, ill-defined bands and lumps, or lode-like zones consisting of many such lumps as shown in Fig. 5 and 6, after Vogt. On the boundaries of such segregations, where the chromite masses are shown to be connected by transitions with the normal olivine rock, the microscope reveals the chromite as shown in Fig. 7, with well developed crystalline forms defined against the olivine. The segregation of ore must undoubtedly have taken place before the crystallization of the olivine.

From observations upon about 40 chromite deposits in olivine rock or

[1] J. H. L. Vogt: *Zeit. f. Prak. Geol.*, 1894. p. 389.

serpentine in Norway, Vogt infers that the size of the ore occurrences is about proportional to the magnitude of the respective eruptive stocks; thus the largest orebodies are found at Feragen and Rödhammer, near Röros, where they have been traced to a depth of about 50 meters, being at the same time within the largest serpentine fields.[1]

(b) *The Chromite Ore Deposit of Kraubath in Upper Styria.*

These conditions also obtain at the well known Kraubath[2] deposits whose genesis has been recently studied and explained.

The chromite ores of Kraubath, according to F. Ryba, are associated with an olivine rock, forming the remnant of an intrusive mass that rests unconformably on hornblende gneiss. The rock consists mainly of olivine and chromite, and hence must be called dunite. At some points, however, owing to the occurrence of bronzite, it passes over into chromite-harzburgite, which also contains some chrome diopside. As secondary alteration products serpentine masses occur with magnetite and a little brown hematite.

On the right bank of the Mur in the Sommergraben, where the chrome mining industry was carried on, the dunite or the chrome-harzburgite is so little serpentinized that it shows perfectly the character of the olivine rock even to the naked eye. The chromite is scattered through it irregularly, for the most part in octahedrons with more or less rounded edges; more rarely these crystals are crowded closer together, or the chrome iron ore occurs in irregular compact segregations or nests, which have been the objects of much prospect work and long-continued but unsuccessful mining.

The occurrence of the chromite in perfectly fresh olivine rock is also good proof of the magmatic segregation of the ore, the thin sections showing distinctly that the olivine must have been crystallized out later than the larger chromite grains, which look as if they were broken.

6. *Tin-Stone as a Primary Segregation in Granitic Rocks.*

Besides the occurrences of tin-stone as a constant ingredient of greisen (granite without feldspar) which is a rock of secondary origin, this ore is also a primary component of many granites, as, for example, the tourmaline-granite of Eibenstock of the Greifenstein and of Altenberg in the Erzge-

---

[1] Compare A. Helland: 'Om Kromjernsten i Serpentine,' *Vidensk. Selk. Förh.*, 1873.

[2] A. Miller von Hauenfels: 'Bericht über die geogn. Erforschung der Umgebung von St. Michael u. Kraubath, V. *Jahresber.* die geogn. mont. V. f. Obersteiermark, pp. 53-76. Also ' Nutzbaren Min.,' etc. *B. u. H. Jahrb.* der k. k. Bergakademie, 1864, Vol. XIII, pp. 214-217. H. Höfer: 'Anal. mehrerer Magnesia ges. der Ober'k,' *Jahrb.* d. k. k. geol. R. A., 1866, pp. 443-446. A. Kahl: 'Der Chrombergbau von Kraubath,' *Berg u. Hütten Jahrb*, 1869, Vol. XVII. pp. 266-281. H. Wieser: 'Olivin fels von Kraubath,' Tschermak's *Min. Mitth.*, 1872, p. 79. F. Ryba: 'Beitrag zur Genesis der Chromeisenerzlagerstätten bei Kraubath,' *Zeit. f. Prak. Geol.*, 1900, pp. 337-341.

birge. However, these examples have as yet no economic importance, except in so far as the small original granules of these granites, upon the decay of the mother rock, concentrate in and enrich the residual gravels whose chief tin content is, however, derived mainly from the greisen zones. There are no occurrences of primary tin-stone in a granitic rock which is rich enough to mine. A supposed example of this kind which proved an economic failure is found in South Dakota.

The Etta Knob Tin Deposit of the Black Hills, Dakota.

Etta Knob, situated in about the middle of the Black Hills, consists, according to Blake[1], of garnetiferous mica schist, with intervening quartzitic beds, broken through by a granite stock whose nearly circular outcrop measures 30 to 60 meters in diameter. By prospecting it was ascertained that this stock is an upright pillar-shaped mass with a pronounced concentric structure. Next to the country rock there is a sharply defined peripheral zone of dark mica rock with large plates of muscovite, which are workable.

Next inward follows a band of compact quartz, with irregular bunches of compact albite and orthoclase, as well as enormous crystals (up to 12 meters in length!) of spodumene, which lie in the midst of a fine-grained, scaly mass consisting essentially of mica and albite. In this complex mass the cassiterite is interspersed in granules or imperfectly developed crystals, the ore forming about 2.5% of the aggregate. The association of a lithion-pyroxene with tin-stone is also reported from several localities in Maine (E. Richards). Besides the minerals named, the tin-stone of Etta Knob is also accompanied by the following; Apatite, triphyline, tantalite, columbite, arsenical pyrite and copper glance. Besides tin ore spodumene has also been mined.

The Etta Knob rock is a pegmatite which cannot be assumed to have originated as a segregation of the normal granite, but shows, on the contrary, evidence that water had a decided part in its formation.

(C) Segregations of Sulphidic and Arsenical Ores.

The sporadic occurrence of small particles of pyrite in volcanic and plutonic rocks is well known. Iron pyrite in particular is one of the most widely distributed accessory constituents of many eruptive rocks, especially in diabase, diorite and syenite. It appears in entirely fresh undecomposed rocks of this kind under such conditions of microscopic association with

---

[1] W. P. Blake: 'Tin Ore Deposits of the Black Hills,' *Trans.* Am. Inst. Min. Eng., 1885, Vol. XIII, p. 691. W. O. Crosby: 'Geology of the Black Hills,' etc., *Proc.* Boston Society Nat. Hist., Vol. XXIII.

the other constituents that there can be no doubt as to its primary nature. At times the iron pyrite is replaced by copper pyrite. Thus, the tourmaline-granite of Predazzo, the diabases of the eastern Harz and even the younger lavas of the Alban Mountains at Capo di Bove carry granules of copper pyrite. Variegated copper ore is known from the tourmaline-granite of Kittlisvand in the Norwegian district of Nummedalen. Various sulphidic copper ores also occur disseminated through the syenite of the Plauen area near Dresden, more especially in the dark streaks rich in augite and hornblende, in which native copper, copper pyrite and copper glance often occur, besides other secondary products. Pyrrhotite is often found interspersed in gabbros; it is also observed sometimes in basalts, as at the Landberg, near Tharandt, and on the west coast of Greenland, on Disko, etc. Since, moreover, small crystals of copper pyrite, and, according to J. Scheerer, also another mineral of the same chemical composition, but rhombic form, become segregated before our eyes in smelting operations upon a slow cooling of a siliceous slag, it appears that the magmatic segregation of large bunches, stocks, primary stringers or streaks in the midst of eruptive rocks is *a priori* both possible and probable. According to L. De Launay[1] certain solvents existing under pressure in the magma, such as liquid carbonic acid, alkaline chlorides, fluorides or sulphates, perhaps also water vapor, have taken part in the concentration of such ores from the magma. It is true, however, that the evidence that the deposits of this class are direct segregations from a molten magma, is not as clear and conclusive as it is in the examples of the groups already noted. For many occurrences which are regarded by various authors as direct segregations from eruptive rocks, it will probably be necessary to allow at least a secondary concentration of the ores in the aqueous way, with a partial local transfer of material and an impregnation of the country rock, etc. The great mobility of the sulphide ores, as known from our observation upon them in veins, renders it exceedingly difficult to obtain a clear insight into this matter, especially as a special and detailed microscopic study of the problem has not been made. For these reasons there is grave doubt about the true genesis of such deposits. This is true particularly of the first of the deposits described in this chapter, as magmatic deposits. Recent microscopic study has indeed proved the Sudbury Canada deposits to be metasomatic replacements. There is, therefore, not a single example of magmatic copper deposits known in North America.

[1] *Ann. d. Mines,* 1897, Vol. XII, p. 119 *et seq.*

1. *Deposits of Nickel and Copper Ores in Connection with Gabbro Rocks or Diabases and Their Metamorphic Derivatives.*

The deposits of this class, consisting of nickel-bearing pyrrhotite, chalcopyrite and iron pyrite found in Norway, Canada and some other countries, have in recent years attracted general attention, particularly as a result of the work of J. H. L. Vogt[1]. A brief description of the Norwegian and Swedish occurrences studied by this author is given first, although from an economic point of view they are far inferior to those of Canada.

(a) The Norwegian Nickel Ore Deposits.

Norway contains, according to Vogt, some 40 intrusive masses of gabbro which are distributed uniformly over the country. These rocks are characterized by the presence of nickel ore, whose genetic connection with the gabbro was first described by Th. Kyerulf and T. and J. Dahll. These rock masses, nearly all of which have been strongly altered by regional metamorphism, were, in most cases, formerly regarded as crystalline schists of the same age as the gneisses about them, and the theories concerning the ore deposits were based on this supposition. It is necessary, therefore, to note the decisive observations which convinced the Norwegian investigators of the eruptive nature of this gabbro. The facts are as follows:

1. The boundary between these gabbros and the gneisses or the Cambro-Silurian regional-metamorphic schists is in places very sharply defined.

2. The other rock strata are often truncated and stop at the boundaries of the gabbro.

3. Blocks of schist occur enclosed in the gabbro.

4. Marginal facies, such as orbicular norites, have sometimes been observed.

5. Finally, younger dikes of normal and of pegmatitic gabbro occurring in the gabbro itself bear witness as "subsequent intrusions" and prove the eruptive nature of all these rocks. Their present petrographic character is, it is true, very unlike the prototype, being commonly that of an amphibolite or garnet amphibolite.

Where the original character of these gabbros has not been entirely obliterated by regional metamorphism it is seen that they were at first either olivine gabbros with ophitic structure or true granular-crystalline norites. The ores show a preference for the latter rocks which may or may not carry olivine and show transitions to olivine gabbros.

[1] J. H. L. Vogt: 'Nickel forekomster og nickel producktion,' *Geol. Fören. Förh*, 1892. pp. 315-338. Also 'Bildung von Erzlagerstätten durch Differentiationsprocesse in basischen Eruptivmagmata,' II, Nickelsulfiderze. *Zeit. f. Prak. Geol.*, 1893, pp. 125-143, 257-284.

The ores connected with the norites consist of:

(1) Magnetic pyrite with 2 to 5%, rarely as high as 10%, of nickel and some cobalt; (2) iron pyrite with some nickel and cobalt; (3) copper pyrite with up to 0.5% nickel and cobalt; (4) titanic iron ore.

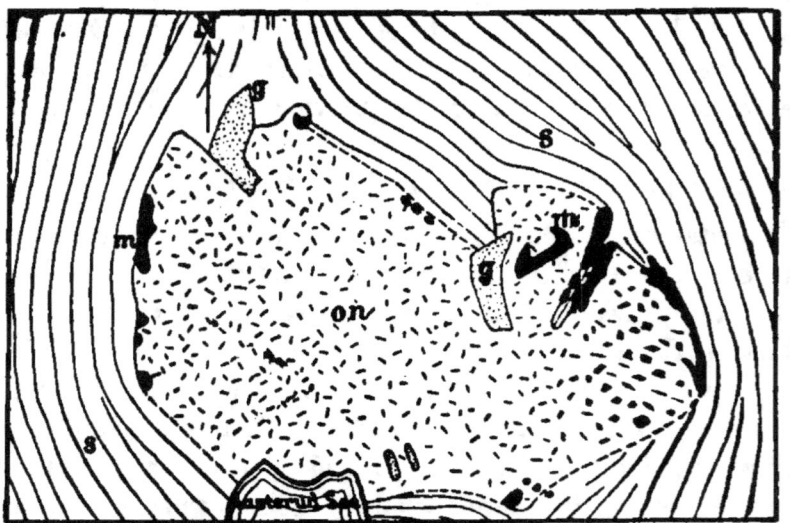

Fig. 8.—Sketch-map of the Erteli mineral region. (Vogt.)
s, crystalline schists; on, olivine norite and other varieties of norite; g, granite; m, ore masses.

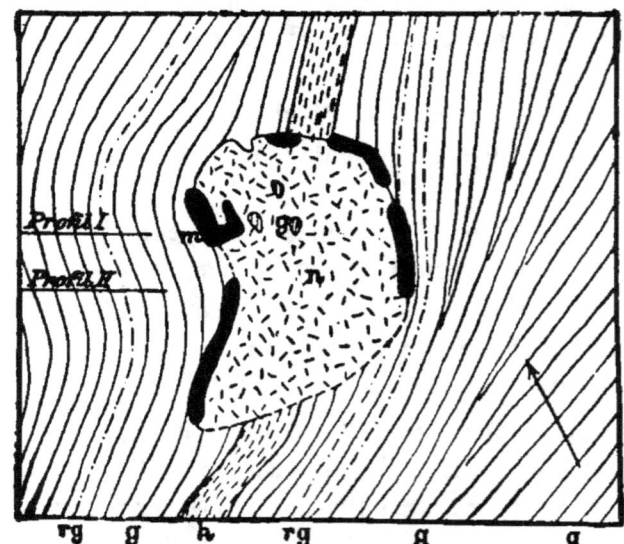

Fig. 9.—Sketch-map of the Meinkjar mining field. (Vogt.)
g, gray gneiss; rg, red gneiss; h, hornblende schist; n, norite; m, orebodies.

The magnetic pyrite always greatly predominates; the iron pyrite sometimes appears to have been segregated earlier than the magnetic pyrite, since it is embodied in the latter in the form of cubic crystals. Only in very rare cases do we also find iron-nickel pyrite and some other nickel sulphides proper.

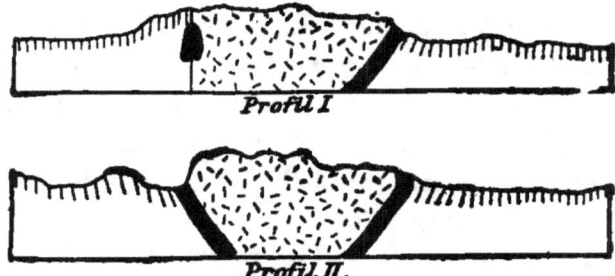

Fig. 10 and 11.—Sections through the norite massive of Meinkjär. (Compare Fig. 9.)

All the ores mentioned are also present in very small amount in fine particles disseminated through the normal norite, while in the deposits proper they form compact ore masses up to 120 or even 200 meters (393-656 ft.) in diameter. These masses have a very irregular shape and are evidently associated with the contact between the gabbros and the country

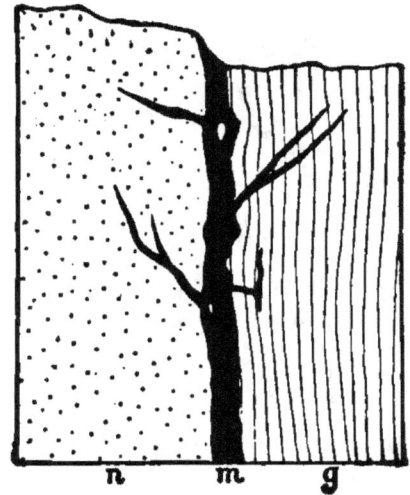

Fig. 12.—Section of a pyrite vein of Erteli mine No. 1. (Vogt.) n, norite; g, gneiss; m, pyrite mass. Height of section, 5 meters.

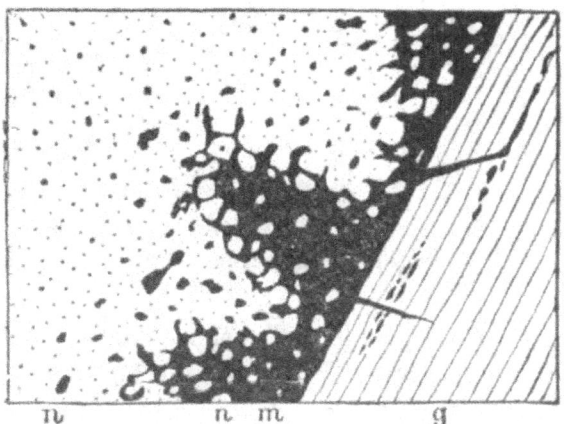

Fig. 13.—Section of a part of a norite contact in the Meinkjär field. (Vogt.) Explanation the same as in Fig. 12. Height of section, 1 meter.

rock. At times they are also found about the borders of enclosed blocks of country rock.

Both these facts are clearly illustrated in Figure 8 (after Vogt), which shows the distribution of the ore masses in the Erteli mining field in the district of Ringerike, northwest of Drammen. The norite massive of that locality has a longitudinal diameter of about 600 meters. Not less instructive are the figures 9-11 of the Meinkjär mine field (after Vogt).

In some places the pyrite mass passes very gradually into norites poor in pyrite, and finally into almost normal non-pyritiferous norites. In such transition rocks the pyrite often appears in the form of a very irregular venation in the midst of the mass of silicates. Furthermore, the pyrite, as pointed out by Vogt, forms real stringers and lodes which branch away from the orebodies proper, either into the norite or into the country rock, and may also enclose fragments of both. The detailed profile (from Vogt) of the Erteli mine No. 1, in Figure 12, shows these relations very distinctly.

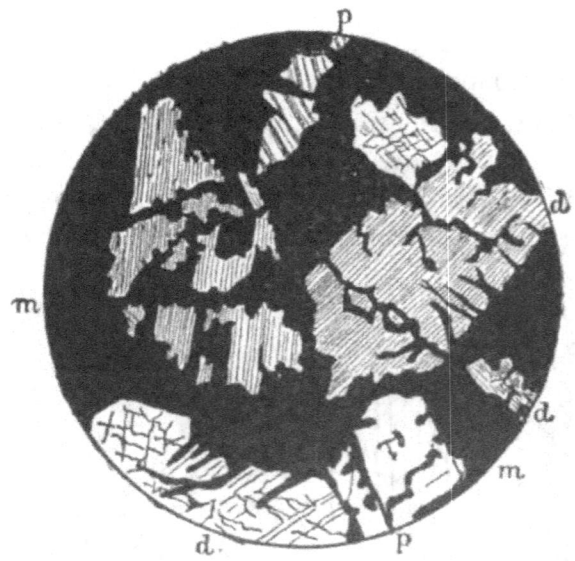

Fig. 14.—Nickel ore of Varallo under the microscope, with fifty-fold enlargement. d, diallage; p, plagioclase (with crossed nicols); m, magnetic pyrite.

A second profile of the same kind from Vogt, from the Meinkjär mine in Bamle, northwest of the island of Kragero, Figure 13, shows how the pyrite has penetrated the neighboring gneiss parallel to the stratification so that we here find perfectly isolated patches of ore and fahlband-like impregnations of larger rock bodies.

According to Vogt the pyrite is a direct differentiation product of the noritic magma, a fractional part of the magma whose melting point is somewhat lower than that of the silicates which compose the normal rock. The pyrrhotite of the original magma, with the other metallic constituents, is supposed to have been concentrated in about the same way as in metallurgical furnaces, the copper and lead separate from the slag. The silicates are supposed to have formed after this metallic segregation, and the pyrite, which was still fluid, filled the spaces between the crystals of silicate minerals and was itself congealed upon further lowering of temperature.

There are serious objections, however, to the theory that these ore masses in their present state are simple magmatic segregations. First of all the penetration of molten sulphides so far into the cooler country rock, as appears from Figures 11 and 12, is very difficult to explain in accordance with the laws of physics. Still more serious doubts are suggested by the condition of the ore-bearing rock bodies, as revealed by microscopic examination.

Cases occur, indeed, in which evidently corroded remnants of the normal silicates of the gabbros lie embedded in the mass of the ore as if they were

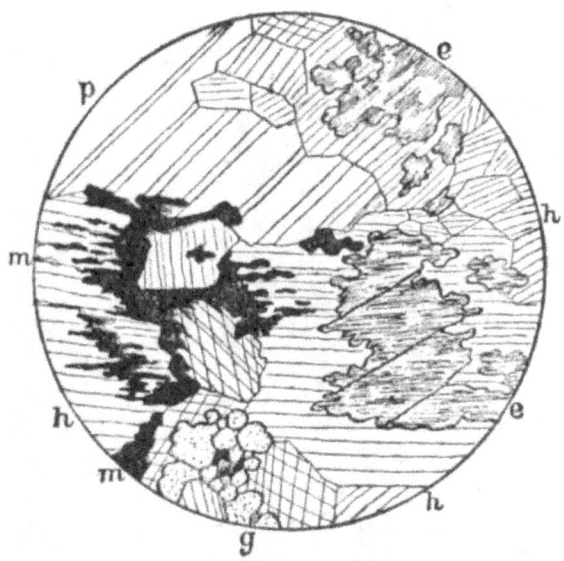

Fig. 15.—Ore-bearing norite from Solum Schurf, northwest of Erteli. (Enlarged 50 times.) e, enstatite; p, plagioclase; h, hornblende; g, garnet; m, magnetic pyrite.

earlier segregations which had been once more half melted down. In Figure 14 we give such an occurrence from Varallo, to be mentioned later on, and we admit that similar ones may be found in Norway, as in fact appears from Vogt's description. The corrosion of the plagioclase and diallage which we here observe, however, probably is due to solution by water, and this hypothesis is supported by the observation that in most cases the ore deposits are found in those parts of the gabbros that have been subjected to *very strong regional metamorphism,* and that in such cases the segregation of ore is shown to have taken place simultaneously with, or subsequent to, the regional metamorphism. The gabbro and norite patches against which the ores rest directly, or with which they are interspersed, have almost always been more or less completely transformed into amphibolites and garnetiferous amphibolites. Figures 15 and 16 show this process not quite terminated.

Here we still have remnants of enstatite or diallage. In Figure 15, however, the ores prove to be younger than the two products of this later transformation, the green hornblende and the garnet. In Figure 16 the ore appears to have been segregated simultaneously with the garnets, and penetrates the older diallage only along fissures. *We believe, therefore, that the real and economically important concentration of the sulphides in the gabbro rocks took place only during the period of regional metamorphism and that it was effected by aqueous solution.* What was the original

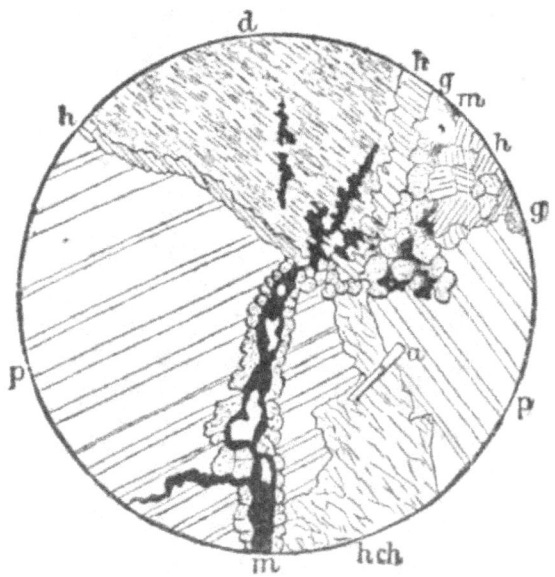

Fig. 16.—Ore-bearing gabbro from Erteli-Prästehaug.   (Enlarged 50 times.)
d, diallage; p, plagioclase; a, apatite; h, hornblende; hch, chloritized hornblende; g, garnet; m, magnetic pyrite.

form and distribution of the metallic compounds in question in these eruptive rocks; whether they were very finely and uniformly disseminated pyrite particles, or chemical components of the bisilicates, cannot at present be decided. This will require a much more thorough petrographic investigation than any yet made.

The production of Norway attained its maximum in 1876, with 42,500 tons of ore containing 360 tons of nickel, the favorable conditions at that time permitting the smelting of ores with only 0.9 to 1.5% nickel. According to Vogt, 330,000 tons of nickel ore were produced between 1850 and 1893.

Analogous deposits occur in the neighboring kingdom of Sweden. The best known is Klefva in Smaland, whose nickel ores, by the way, show a very small content in platinum, iridium, and rhodium.[1] To this class belong also Kusa in Dalarne, and Ruda in Ostergotland.

(b) The Nickel Ore Deposits of Varallo, Italy.

A second European occurrence of such deposits exists at Varallo[2] in Piedmont in the valley of Sesia. The gneisses and mica schists of the Monte Rosa area are traversed by an intrusive mass of norite about 20 kilometers long and 4 kilometers broad. The rock is an ordinary norite showing transitions to gabbro and to an olivine norite, but it has been largely altered by regional metamorphism into a hornblende rock like those of Norway. Large masses of magnetite found at the gneiss-gabbro contact in the Cevia and Sella Bassa mines were found to contain 4 to 5% of nickel and cobalt, and to hold pyrite and copper pyrite. The amount of nickel (including cobalt) in the ores formerly smelted in the Sesia and Scopello furnaces was 1.2 to 1.5%. The cobalt content is somewhat higher than in Norway, the ratio being 100 Ni: 50 Co: 40-50 Cu. The mines were worked from the end of the sixties to the end of the seventies. The Sesia mine belonged to the Saxon Blue Color Works.

From a genetic point of view what has been said of Norway is also true of Varallo. Compare also Figure 14.

(c) The Nickel Ore Deposits of Sudbury in Canada.

The Canadian nickel deposits are the most important deposits known of this kind. They have been described by numerous authors, but most of the papers deal with general and not the structural or genetic conditions[3].

The town of Sudbury is in northern Ontario on the line of the Canadian Pacific Railway, in the midst of an area formed mainly of Huronian schists,

[1] B. Santesson: 'Nikkdmalm fyndigheten vid Klefva.' *Geol. Fören Förh.*, Vol. IX, 107, 1887, pp. 66-73.

[2] A. W. Stelzner: 'Referat eines Vortrages,' *Berg u. Hütten Zeit.*, 1877, p. 87. M. Badoreaux: 'Mémoire sur la metallurgie du Nickel.' *Ann. d. Mines*, 1877, Mémoires, Vol. XII, p. 237. J. H. L. Vogt: 'Varallo in Piedmont,' *Zeit. f. Prak. Geol.*, 1893, p. 257.

[3] T. G. Bonney: 'Notes on a part of Huronian series of Sudbury,' etc., *Quart. Journ. Geol. Soc.*, 1888, pp. 32-44. W. H. Merritt: 'The Minerals of Ontario.' *Trans. Am. Inst. Min. Eng.*, 1889, Vol. XVII, p. 295. E. D. Peters: 'The Sudbury Ore-deposits,' *Trans. Am. Inst. Min. Eng.*, 1890, Vol. XVIII. A. E. Barlow: 'On the Nickel and Copper Deposits of Sudbury.' *Ottawa Naturalist*, March 6, 1891. R. Bell: 'The Nickel and Copper Deposits of the Sudbury district.' *Bull. Geol. Soc. Am.*, 1891, Vol. II, pp. 125-137. Also 'Report on the Sudbury mining district.' Geol. Survey of Canada, 1891. H. B. v. Foullon: 'Ueber einige Nickelerzvorkommen.' *Jahrb. d. k. k. geol. Reichsanst.*, Vienna, 1892. J. H. L. Vogt: 'Bildung von Erzlagerstätten,' etc. *Zeit. f. Prak. Geol.*, 1893, pp. 128, 257. G. R. Mickle: 'The relation between Pyrrhotite, Gangue, and the accompanying Rocks of the Sudbury District.' *Sudbury Journal*, 1894. F. A. Adams: 'On the igneous origin of certain ore deposits.' Min. Assoc. Prov. of Quebec, Jan., 1894. F. L. Walker: *Quart. Journ. Geol. Soc.*, Vol. LIII, No. 209, 1897, pp. 40-65. A. P. Coleman: *Report*, Bureau of Mines, Ontario, 1903, pp. 235-303. Also 1904, pp 192-224.

extending from the north shore of Lake Huron northeast to Lake Mistassini. According to Coleman, the nickel deposits are confined to the outer, basic edges and offshoots of a mass of igneous rock. A nearly continuous band of this rock, about 2 miles wide, forms a rude ellipse 36 miles long and 8 miles across, enclosing an area of tuffs, sandstones and slates of Cambrian or Upper Huronian age. Outside of this ellipse the rocks are of Laurentian or Lower Huronian age, a part of the so-called basal complex. The igneous rock presents a well defined gradation from norite (now largely altered to diorite) on its outer edge, to the quartz syenite and granite of the inner border. The norite (a hypersthene gabbro) contains scattered grains of bluish quartz and black mica. The ore deposits are of two types, (1) those "crowded into bay-like indentations in the adjacent rocks," (2) those "strung out along the narrow offshoots of the norite like sausages on a string, but with a long piece of string between them." The deposits are prevailingly lenticular, pinching out in both directions, and conforming to the general strike of the schists. The orebodies are much larger than those of the Scandinavian peninsula.

The Sudbury ores consist mainly of a mixture of pyrrhotite and chalcopyrite intimately associated with more or less country rock. The nickel occurs in the pyrrhotite, varying in amount from 2 to 5.5%, ordinarily 4 to 4.5%. The nickel occurs in the pyrrhotite as pentlandite $(Fe+Ni)S$. The copper pyrite is free from nickel. Occasionally polydymite (with 43% Ni) and millerite (64% Ni) occur in the ore, and at times patches of titaniferous magnetite have been observed. From the residue of weathered ores T. L. Sperry[1] isolated the interesting mineral sperrylite, a platinum arsenide, with traces of rhodium, palladium, and antimony. Furthermore. the Canadian matte, like that of Klefva and Ringerike[2], shows a small amount of platinum (0.25 to 0.50 oz. per ton) and palladium, as well as traces of iridium and osmium. Associated with the sperrylite some tinstone is found. Some ores contain small quantities of gold. The sperrylite occurs with the chalcopyrite in the unaltered sulphide ore.

According to A. McCharles[3] it has been found that the richest ores occur in the smaller offshoots of gabbro-diorite, and that the larger massives, with few exceptions, yield no workable ores at all. The gold. platinum, and the other accessory metals are contained not in the magnetic pyrite, but in the copper pyrite.

In their structure and in their relation to the country rock the Canadian deposits present a great similarity to those of Norway and Piedmont. The

[1] Am. Jour. Sci., Vol. XXXVII, 1899.
[2] Vogt: 'Platingehalt in Norwegischen Nickelerz.' Zeit. f. Prak. Geol., pp. 258-260.
[3] 'The Mineral Industry,' Vol. VII, 1899, p. 524.

Sudbury ores, moreover, contain peculiar breccias in which sharp-edged or rounded rock fragments lie imbedded in a paste consisting essentially of magnetic pyrite and some copper pyrite. The fragments consist of gabbro in highly varying stages of regional metamorphosis. Besides those which consist of a micro-breccia of minute diallage and plagioclase splinters, there are others which have been completely amphibolitized, or also show a distinct schistose structure. All are, moreover, more or less

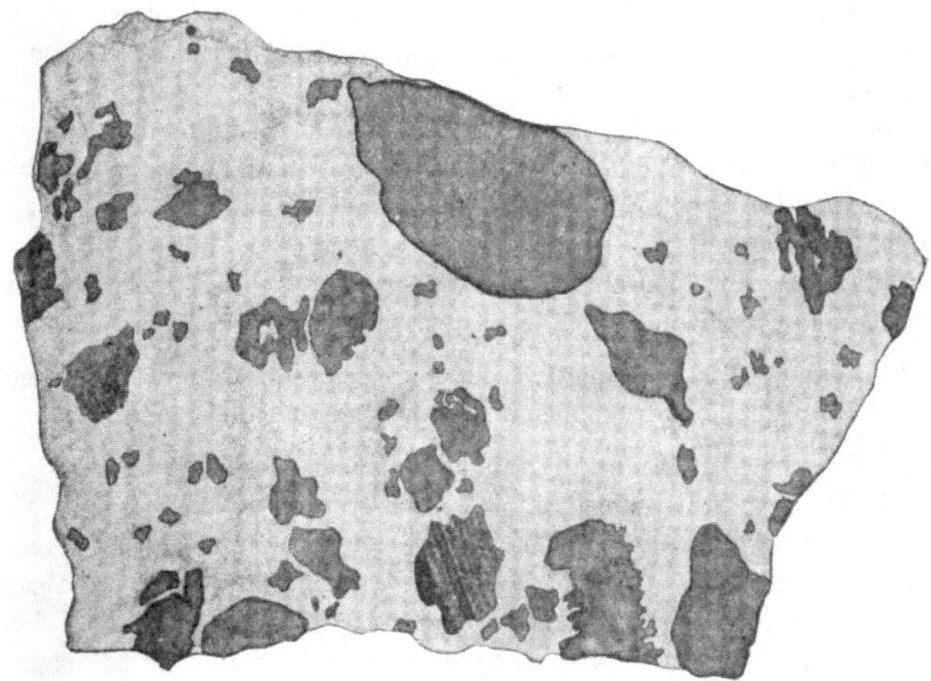

Fig. 17.—Polished ore specimen from Murray mine near Sudbury.
Two-thirds natural size. Gabbro fragments more or less transformed into hornblende schist in the midst of the ore mixture, which appears uniformly light colored.

impregnated with ore. Figure 17 shows an instance of such an occurrence, genetically interesting, from the Murray mine. Such specimens prove that the segregation of the magnetic pyrite and copper pyrite in their present condition *took place during or after the dynamo-metamorphosis of the gabbro rocks of that locality, and not by means of a magmatic differentiation.* This has been confirmed by the very careful detailed work of C. W. Dickson, who says that the evidence points to the secondary formation of the Sudbury orebodies as replacements along crushed and faulted zones[1].

---

[1] 'The Ore Deposits of Sudbury, Canada.' *Trans.* Am. Inst. Min. Eng., February, 1903.

The ore deposits of the Sudbury district are occasionally traversed by younger dikes of an olivine diabase, in which analysis has demonstrated the presence of NiO (0.027%) and CoO (0.005%) which was probably taken up secondarily (Walker).

The first deposit worked at Sudbury, the Copper Cliff mine, at first produced only copper, as indicated by the name. Later, toward the end of the eighties, the presence of a workable amount of nickel was found in the associated magnetic pyrite. This discovery in what had been waste ore at once caused the establishment of a very active nickel-ore mining industry. In the Sudbury district in 1903, a total of 220,937 tons of ore carrying 3.16% nickel were smelted, of which the greater portion came from the Creighton mine.

In the United States, the Lancaster Gap mine in Pennsylvania is, according to Kemp, the most important nickel deposit in the country. It seems to belong to the group just treated, being also situated at the boundary between a hornblende rock and crystalline schists.

Similar nickel ore deposits occur in other basic eruptive rocks altered by dynamo-metamorphism, especially in diabases.

(d) Deposits of Nickeliferous Pyrrhotite at Schweidrich near Schluckenau, in northeastern Bohemia, and at Sohland, in Saxony.

At Schweidrich the Lusatian granitite is traversed by a great west-northwest dike of a coarse-grained diabase, fine-grained at the borders, whose normal rock carries some primary hornblende and biotite besides the ordinary constituents. Near the northern wall of the dike this diabase is altered to a rock consisting mainly of secondary uralitic green hornblende, but at the same time more or less impregnated with nickeliferous magnetic pyrite and some iron pyrite. Under the microscope the segregation of the ores in many cases is distinctly seen to be younger than the formation of this secondary hornblende. The adjoining rock, a somewhat decomposed granite, is also impregnated with the same ores to a distance of more than one meter. Patches of compact ore occurring more rarely in the diabase contained, according to von Foullon[1], 7.8% nickel, 2.90% copper, 49.90% iron. No mining is going on there at present. At Aeusserst-Mittel-Sohland, north of Schluckenau, on the boundary between Saxony and Bohemia, the digging of a well in 1900 led to the discovery of an entirely analogous but much richer deposit. A dike 10 m. (33 ft.) thick of proterobase[2] rich in biotite traverses the Lusatian granite in a west-northwest

[1] H. B. v. Foullon: 'Ueber einige Nickelerzvorkommen.' *Jahrb.* k. k. geol. Reichanst, Vienna, 1892, p. 302. Compare O. Herrmann in 'Geol. Maps of Saxony,' 1897. Hinterhermdorf-Daubitz sheet, p. 19.

[2] A diabase rich in hornblende.

direction. Along the northern border, this rock has been found to be ore-bearing to a distance of about 700 m. (2,296 ft.) The ores consist of nickeliferous pyrrhotite, some copper pyrite and a little iron pyrite, forming either a mere impregnation, extending outward as much as half a meter into the granite, or as in the Herbergs Fund shaft, a compact ore up to 2.5 meters in thickness, dipping north at 75°, like the dike walls, and traceable for 20 meters along the strike, and yielding specimen ores with an average content of 5% of nickel and about 2% of copper. In the proterobase the sulphides are the last materials to crystallize, as in many cases fragments of pyroxene and of primary hornblende, which evidently have undergone corrosion, are surrounded by the pyrites. On the other hand, the pyrites are often seen in the midst of serpentinized pyroxenes, with a band of ore running parallel to their contour at a little distance from the edge. Very often, also, the pyrite is found in the form of thin lamellae wedged in between foliations of biotite.

In the hanging wall of the ore of the Fund shaft there are found in the proterobase very remarkable portions, quite rich in spinel and corundum. The rock also carries inclusions of sillimanite.

Immediately below the gossan of the deposit, which sometimes extends down as far as 10 meters, there has been found, as a secondary fissure filling, a rich blackish copper ore consisting essentially of copper glance. Recent investigations by R. Beck show that the ore was formed at the same time as the uraltic hornblende and the ore is regarded as a deposit from thermal waters rising as an aftermath of the diabase eruption[1].

The nickel deposits of Horbach in the Black Forest, concerning which but little scientific information is available, possibly should be included here, as also the nickel ore of the Hilfe Gottes mine, near Nanzenbach, in the Weyerheck, 4 miles (7 km.) northeast of Dillenburg in Nassau[2], where irregular stocklike masses of iron pyrite, copper pyrite, and nickel arsenide are associated with a greatly altered rock of the diabase group.

(e) Arsenious Nickel Ores in the Serpentine of Malaga, Spain.

While the widely distributed secondary nickel ores found in serpentines are treated under another heading, the Malaga deposits of arsenical nickel are most naturally included here.

In the Archean strata to the west of that town there are serpentine rocks which have resulted from the alteration of an olivine pyroxene rock, and from olivine norite, as well as from olivine rock (dunites) proper.

[1] *Zeit. f. Prak. Geol.,* Feb., 1904. *The Engineering and Mining Journal,* 1904, Vol. LXXVII, p. 363.

[2] H. Laspeyres: 'Das Vorkommen.' etc. *Verhandl.* d. Naturh. Ver zu Bonn, 1893, Vol. 50, Plate IV, and Pt. II, p. 451.

According to F. Gilman[1] the upper portions of these serpentines were found to contain secondary nickel ores of the pimelite and garnierite type, with 1 to 20% of nickel. At greater depth nickel arsenide (niccolite) made its appearance in the following manner:

1. Fine chromite grains are cemented by niccolite (see Figures 18 and 19); the bronze mosaic ore exists in nests and veins in fresh serpentine, and contains 5 to 20% of nickel.

2. Crystals of a dark greenish brown augite up to 1 centimeter long are cemented by niccolite and chrome iron.

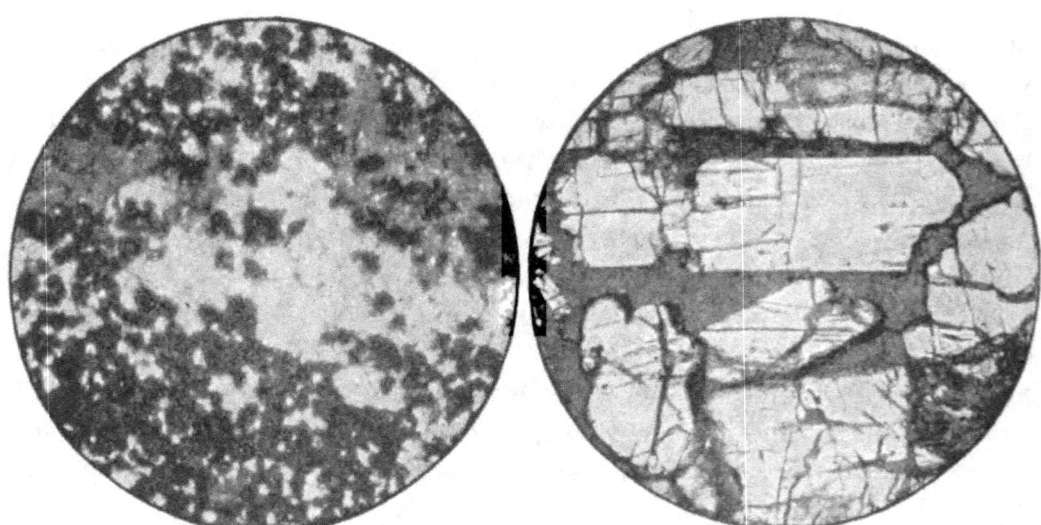

Fig. 18.—Chromite (dark) with niccolite (bright) from Primera mine; with light from above. (Enlarged 27 times.)

Fig. 19.—Augite crystals (bright) cemented by niccolite (dark); with incident and also transmitted light. (Enlarged 10 times.)

3. Masses of unaltered norite sometimes as large as an ostrich egg occur in the serpentine, which besides its main ingredients (plagioclase and rhombic pyroxene) also contains interspersed granules of niccolite and chromite and chrome iron ore. Sometimes the niccolite forms a regular groundmass about the silicate crystals; more rarely it has been secreted, together with chromite, in streaks and bands.

At times iron pyrite and copper pyrite are found in the ores. The above named author considers the niccolite (Ni As) as a product of magmatic differentiation. The example suggests that fresher rocks with arsenide nickel ore also may be expected, as other garnierite deposits are worked to a greater depth.

[1] F. Gilman: 'Notes on the Ore Deposits of the Malaga Serpentines.' Inst. Min. and Metal., London, January, 1896.

2. *The Copper Ore Deposits in the Serpentines of Italy.*

According to B. Lotti[1] the serpentines of Tuscany and Liguria almost all contain some copper sulphide, some of the serpentine masses containing deposits known for many centuries. According to Lotti three rock types are recognized in the strongly serpentinized rocks of this region: altered lherzoliths (olivine rock proper), decomposed olivine gabbros, and transformed olivine diabases (called by others melaphyres). These rocks form lenticular or stocklike intrusions in Eocene strata, occurring usually in an orderly sequence in which the former lherzolith is below, and the diabase on top. It is probable that there were two intrusions, first one of lherzolith, next a second whose magma congealed in the form of gabbro at the bottom, and as diabase at the top. In the middle member of this series, the gabbro, the following ores are found: pyrite, copper pyrite, bornite, copper glance, rarely also blende and galena, all in finely disseminated particles. At some points, however, these ores also occur as globular masses or large lumps, about which the decomposition of the country rock is apt to be particularly pronounced. Authors are at present fairly well agreed that the finely distributed ores are original segregations of the magma. On the other hand, in regard to the spherical masses and lumps the views are as yet wide apart. As described later the character and occurrence of such ore masses indicate their origin by secondary concentration of the sulphidic ores into such masses. Exact proof seems as yet impracticable. The best authority on these formations, B. Lotti, now regards them as original segregations. As a special example for this type of deposits we will select the famous locality of Monte Catini.

The Monte Catini Copper Deposit.

Between the Maremna Marshes and the Apennines in Tuscany there extends a hilly country whose numerous wooded summits of eruptive rock overtop the gently sloping areas of Tertiary marls. The mine of Caporciano, near Monte Catini, situated on one of these hills, is said to have been founded by the ancient Etruscans, and in fact the ancient Etruscan town of Felathri, now the walled town of Volterra, lies near the mine. The igneous rocks are intruded in the Tertiary strata in the form of a great lens or laccolith. Above the Eocene marls and limestones, which form the floor of the intrusion, is a strongly serpentinized olivine gabbro forming the base of the eruptive series. This occurs as a stock, which in cross-section is three-armed, and enveloped by the olivine diabases (melaphyres) of Monte Massi, and in part appears intercalated between the diabase and the sedimentary

[1] B. Lotti: 'La genèse des gisements cuprifères des depôts ophiolitiques tertiaires d l'Italié,' *Mem.* Geol. Soc. Belgium, Vol. III, 1899.

strata. A longitudinal layer of diabase divides the gabbro into two approximately parallel parts known as the *filone rosso* and the *filone blanco,* both wedging out downward at the point where the underlying sedimentary rocks begin (Figure 20). The serpentine stock is bounded on both sides by slickensides and its central portion is, moreover, exceedingly fissured and crushed to a fine breccia, probably as a result of the increase of volume on serpentinization. Certain parts of the stock resemble a breccia whose fragments are in part serpentinized olivine gabbro, in part decomposed olivine diabase (Fig. 21 and 22)[1]. In this breccia are also found concretionary secretions of quartz and chalcedony.

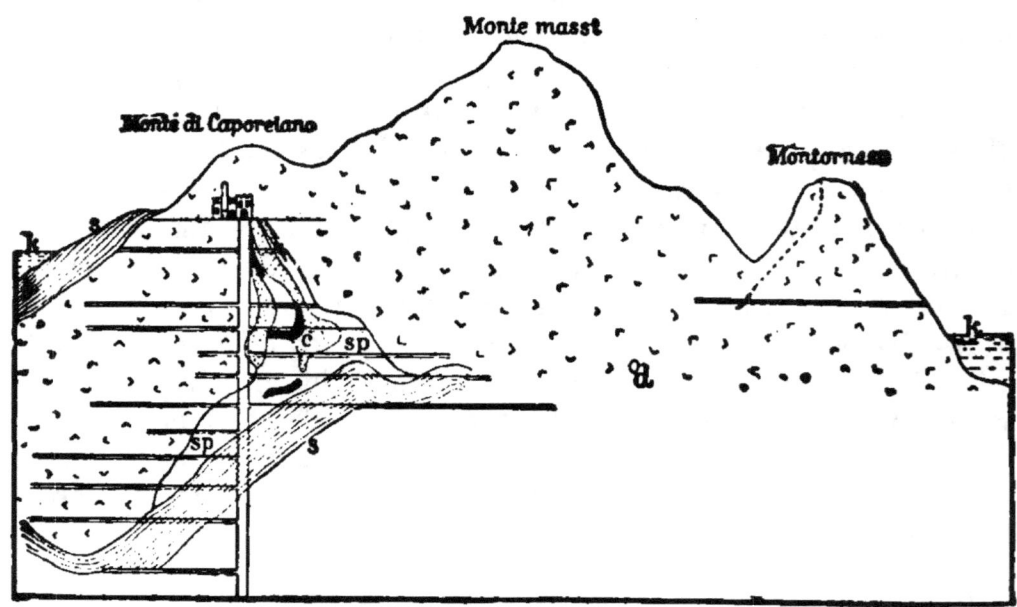

Fig. 20.—General profile through the deposit of Monte Catini from south-southwest to north-northeast. (Fuchs and De Launay.)

s, clay slates and calcareous marls of the Eocene; d, olivine diabase; sp. serpentine, c, conglomerate, the black parts representing ore; k, younger marls and limestones.

The ores form bodies of irregular shape and varying size, which are quite irregularly distributed in the serpentine, or occur at its contact with the above mentioned breccia; some of them are quite isolated (Figure 21); others are connected by ore stringers (Figure 22). The isolated ones attain a diameter of several meters, while occasional compact masses of copper

[1] G. vom Rath: 'Ein Besuch der Kupfergrube Monte Catini,' etc. *Zeit. d. D Geol.,* 1865, pp. 277-310. E. Reyer: 'Aus Toscana.' Vienna, 1884. B. Lotti: 'La miniera cuprifera di Montecatini.' *Boll.* Com. Geol., XV, 1884. A. Schneider: 'La miniera di Montecatini.' Append. *Riv. Mineraria del* 1889. B. Lotti : 'Considerazioni sintetiche sulla orografia e sulla geologia della Catena Metallifera in Toscana,' *Boll.* Com. Geol., Vol. XXIII, 1892.

pyrite have been met, having a volume of 6 to 10 cubic meters. The majority are rounded bodies coated with a brilliantly polished serpentine-like crust. This crust often shows cracks and streaks which record a movement in the soft country rock. Such streaks lead often to other ore-bodies.

In the main, these ore boulders or lumps consist of copper pyrite. Fre-

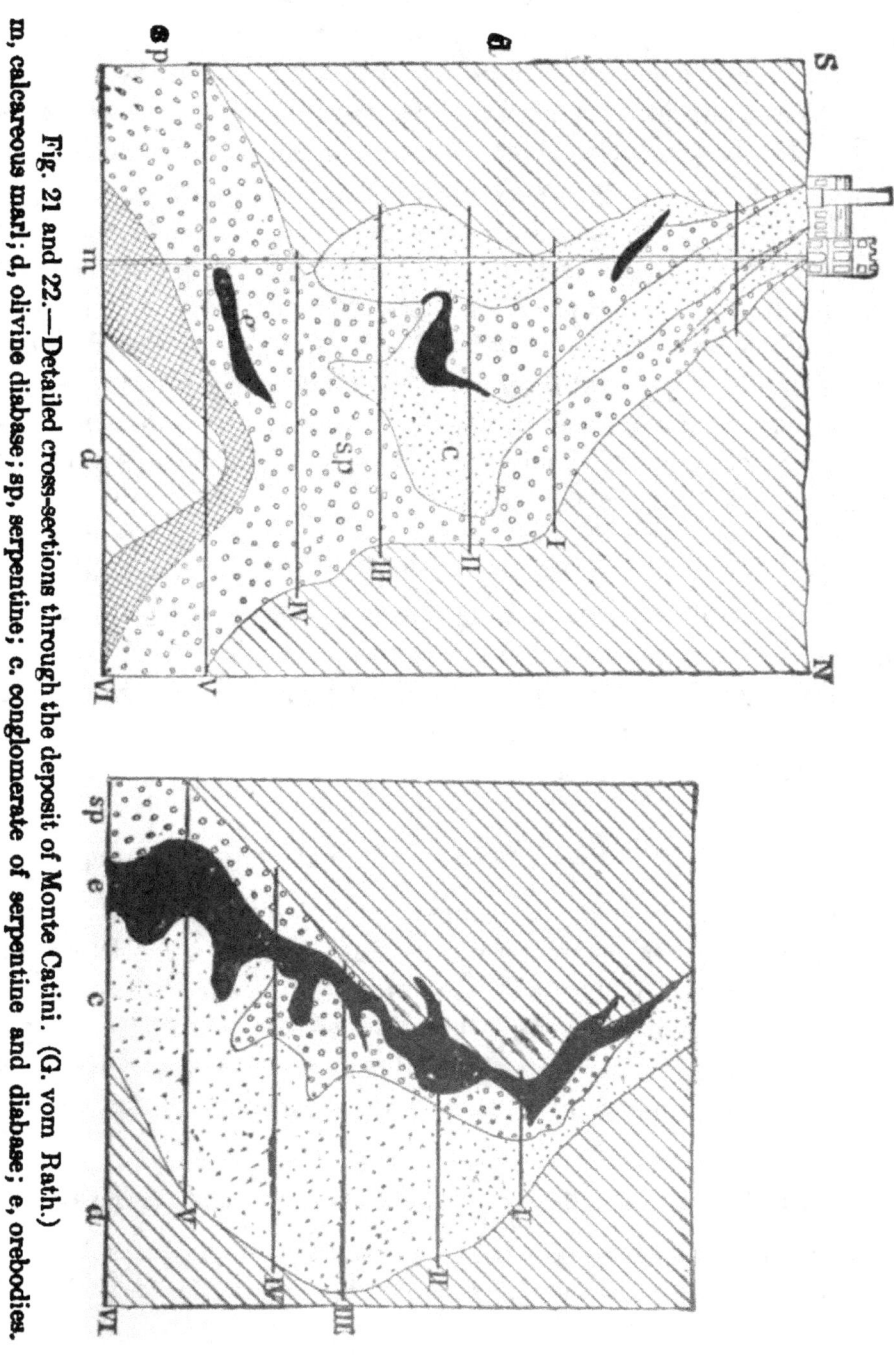

Fig. 21 and 22.—Detailed cross-sections through the deposit of Monte Catini. (G. vom Rath.) m, calcareous marl; d, olivine diabase; sp, serpentine; c, conglomerate of serpentine and diabase; e, orebodies.

quently, but not always, they show a concentric structure, with a core of copper pyrite surrounded by a shell of bornite, often covered by an outer crust of pyrite or of copper glance and native copper. Moreover, the serpentine which surrounds the rounded ore masses contains so much ore in a finely divided condition that it is worked. The ores of Monte Catini are classed as rich ores when the copper content is as high as 7%; as lean ores when it is but 1.25 to 1.50%. The irregular distribution of the ore masses, and their irregular form, together with the absence of continuous lodes or beds, has rendered mining very difficult; nevertheless, it had its periods of great prosperity. Some 400 meters east of the main stock and at a depth of 150 to 200 meters there is a second bed running east-southeast, the so-called Demetrio lode, with copper pyrite, bornite and copper glance.

According to G. vom Rath the orebodies of Monte Catini were formed by a concentration of the metallic sulphides of the original rock during their serpentinization. E. Reyer considers the deposit as a lode that subsequently had been strongly dislocated. B. Lotti, as already stated, considers the lumps of ore as original segregations. The mining industry of Monte Catini, abandoned during the Middle Ages, was resumed in the 15th century, again interrupted, and once more resumed in 1828. It attained its maximum of 3,000 tons of ore in 1860.

Similar, though less important copper ore deposits associated with serpentine are known on the island of Corsica (Ponte Alle Lecchia), in Servia[1], in Greece (Epidauros) and in Northern Norway (Hatfjelddal), and in Cuba[2].

The deposit of Cava Grande on the Temperino lode, in the Campiglia Marittima, which by some is placed in this category, will, in the present work, be treated under contact ore deposits.

3. *The Copper Ores of Ookiep in Little Namaqualand.*

According to A. Schenk[3], the copper ore deposits of Ookiep, formerly considered as lodes, are magmatic segregations in a rock consisting almost entirely of plagioclase with but little biotite, hornblende and augite, forming stocklike intrusion in gneiss. The ores contain bornite, chalcopyrite, some copper glance, pyrrhotite and molybdenite. The mines have yielded as high as 30,000 tons of ore a year, with an average content of 27.5% copper.

---

[1] R. Beck and W. Baron v. Fircks: 'Die Kupfererzlagerstätten' von Rebelj und Wis in Servia.' *Zeit. f. Prak. Geol.*, 1901, p. 321.

[2] W. H. Weed: *Engineering and Mining Journal*, Feb. 3, 1905.

[3] *Engineering and Mining Journal*, Feb. 10, 1905. See also *Zeit. d. D. G. G.*, 1902, Vol. LIII, pt. 4, p. 64.

SECTION II.

BEDDED ORE DEPOSITS.

Ore deposits of direct sedimentary origin, together with deposits of similar form resulting from impregnation and metasomatic replacement of sedimentary beds, are grouped together as bedded ore deposits.

GENERAL FEATURES OF BEDDED DEPOSITS.

According to the generally accepted definition of B. von Cotta[1], "Accumulations of ore, which lie parallel to the stratification or foliation of the rock enclosing them, consequently forming one or more subordinate layers between any stratified or foliated (schistose) rock, are called ore beds." We may further distinguish between *interbedded* deposits which are overlain by other strata and *surficial* deposits when there are no overlying beds, as, for example, a bog iron ore bed. If the stratified deposits are syngenetic, as has been assumed, the further tacit supposition follows that the ore beds and ore strata were deposited but little later than the underlying rock and but little earlier than their overlying bed. In the case of the metasomatic ore beds and impregnated layers this assumption refers only to the original layer, whose substance was subsequently displaced by ore or whose interspaces were subsequently filled by ore; they are then not true beds but bedded veins.

Like any other stratum of sedimentary rock, an ore bed may contain and be recognized by certain fossils (diagnostic fossils) which are usually mineralized. Thus, a characteristic fossil of the Oolitic iron ore beds of the Brown Jura is an ammonite (*Harpoceras Murchisonae*); and for the Eocene iron ore measures of upper Bavaria it is a sea urchin (*Conoclypeus Conoideus*).

Sometimes it is very difficult to distinguish the true stratified beds from bedded veins (see later on). The stratified beds are distinguished from

[1] B. von Cotta: 'Die Lehre von den Erzlagerstätten.' Freiberg, 1859, Vol. I, p. 85. Prime's translation, New York, 1870, p. 17.

the stratiform or bedded *lodes* mainly by negative marks; their boundaries above and below are for the most part not so sharp. They never cause displacements and never cut across another bed or a lode. They never form veinlike offshoots entering into the adjacent rocks, and do not enclose fragments of the country rock. In folded regions the beds follow all the windings of the surrounding strata.

Formerly ore beds when tilted at a high angle were considered as lodes, even in mining legislation, because mining such a deposit is like that of a lode.

The definition given for stratiform deposits involves, furthermore, a considerable extension of the deposit in length and breadth, combined with a relatively small thickness. The known exceptions to this law, such as the so-called ore 'pods' (Erzlineale) of the Norwegian pyrite deposits (see description) are, in point of fact, examples which are not true beds or are at least of doubtful genesis; as for instance, at the Mug mine, near Röros, in Norway, there is a low-dipping bed of phyllite and pyrite, which was found to be ten times longer than it was wide, the length coinciding with the dip.

It is customary in many places to apply the name 'seams' to those stratified ores which have a great horizontal distribution with a small and nearly constant thickness. The name 'lenticular' beds (*lager*) has been given to those which have a relatively small horizontal distribution with a relatively great thickness, and which may show great variation in thickness. As typical examples of these two extremes we may mention the copper shale measure of the German Zechstein formation, which maintains its thickness of 0.5 meter over about 4 square miles, though it does not everywhere pay; and the Norwegian pyrite beds, whose thickness varies between 0 and 26 meters and which have slight areal extent. Some writers call the *older* deposits *beds* (*lager*), the *younger* ones *seams* (*flötze*). In any case the distinction is not a sharp one.

The thickness of bedded deposits is never very great; the extreme thickness of the big iron ore lens of the Sjustjernberg, near Grängesberg, in Sweden, amounting to 90 meters, probably represents the greatest thickness known, and this really includes some intervening strata which are either barren or poor in ore.

Where the thickness of the orebody decreases until it disappears the ore bed is said to wedge out (as at *a* in Figure 23), or it may merely become much constricted with a narrow band of ore still remaining (at *b*), which increases in thickness some distance away, forming another orebody. This is termed "enlargement." After wedging out, a clay-filled or even entirely empty joint fissure sometimes indicates the horizon, and the miner must carefully follow this fissure if he wishes to find further orebodies.

Frequently this alternate wedging out and thickening cuts up a deposit into series of lenticular beds belonging to a common horizon, which may be termed a *chain*[1] (lagerzug) (Figure 24). The German term is also applied to a succession of beds which belong not exactly but at least approximately to the same horizon. An example of this is found in the series of 'minette' beds or measures of the Moselle region occurring in the *Harpoceras Murchisonae* formation; also the double series of magnetic iron ore deposits at Gellivare.

A bed which has a comparatively great thickness, but wedges out rapidly in all directions, is called a lens (see Figure 25). As examples we may mention many magnetic iron ore lenses in the crystalline schists. The flat, disc-shaped spherosiderite concretions in clay slates of the Carboniferous or

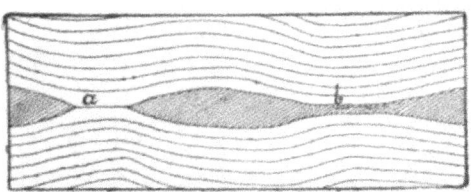

Fig. 23.—Wedging out (*a*) and constriction (*b*) of bed.

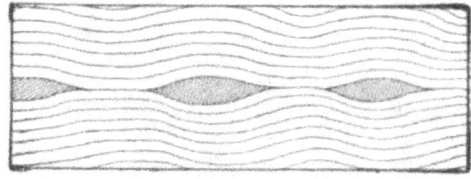

Fig. 24.—Series of ore lenses of bedded deposit.

of the Jurassic are miniature lenses, which are also called 'buttons' in some localities. Similar spheroidal formations constitute the 'muggeln' of the Austrian miners.

If the lens has an irregular outline it passes, according to its dimensions, into a bunch, pocket, or stock (Fig. 26). These irregular forms of deposit are less common in the stratified deposits, proper, than in the metasomatic orebodies, such as masses of spathic ironstone, while in still other cases they are the result of orographic pressure. One may further subdivide such deposits into flat and upright stocks, according as their longitudinal axis is horizontal or vertical.

Very frequently a stratified orebody contains barren or at least poor unworkable layers, intercalated parallel to the stratification. These are called 'partings' (zwischenmittel), or 'stone-band' (bergemittel), as, for example, the partings of gray gneiss in the iron orebodies of Sweden.

[1] *i. e.*, the lenses are the links of a chain of orebodies.

In certain cases a bed of ore does not terminate by simple wedging out, but by the intervention of thin bands of barren rock, which increase in number or thickness. In other cases there is simply a gradual increase in the barren constituents of the normal ore. In the latter case the bed is said to become *impoverished*. This is particularly frequent in metasomatic replacements and impregnations. Thus the pyrite and blende deposits near Schwarzenberg in the Erzgebirge pass very gradually along the strike into barren amphibole-garnet rocks.

The part of an orebody intersected by the earth's surface is called the outcrop. A distinction is made between a true or exposed outcrop and a blind or covered outcrop. In the latter case the ore bed is concealed by unconformable débris or deposits of alluvial strata.

To both the miner and the mining geologist the form of the outcrop of a bed is of much importance, since it indicates the direction for further exploitation.

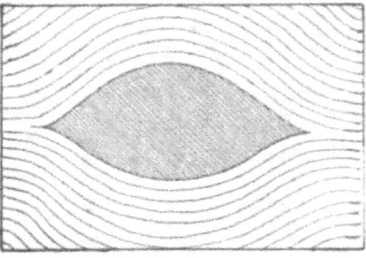
Fig. 25.—An ore lens.

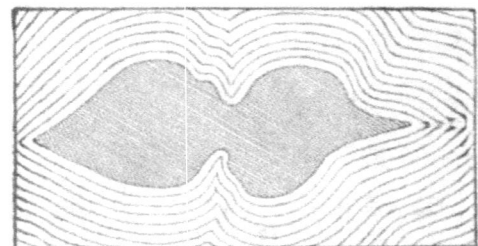
Fig. 26.—An ore stock.

The shape and extent of the outcrop depends on (1) the inclination of the bed, and (2) on the relief of the ground. A practiced eye will be able, from the form of the outcrop shown on an accurate geologic map, to infer with certainty the direction and angle of dip, and conversely, those factors will have to be known in order to plot a bed correctly on a map. According to K. Keilhack[1], the following typical cases occur.

1. Horizontal strata form lenticular outcrops running parallel with the contour lines of a good map. A horizontal bed outcropping on the slope of a mountain shows in a gorge as a crescent-shaped outcrop or bifurcated figure, in which one branch is seen on one side and the other branch on the opposite side of a valley. In a valley whose slopes are furrowed by lateral gulleys a horizontal bed will form a sinuous outcrop (Fig. 27-29).

[1] K. Keilhack: 'Lehrbuch der praktischen Geologie.' Stuttgart, 1896, p. 210, *et seq.*

2. Vertical beds give, under all circumstances, straight outcrops which run, undeviating, over mountain and valley, and at the same time represent the direction of strike (Fig. 30).

3. A more complicated form is assumed by an outcrop of inclined strata. In this case it is first to be noted that in general the boundary line or out-

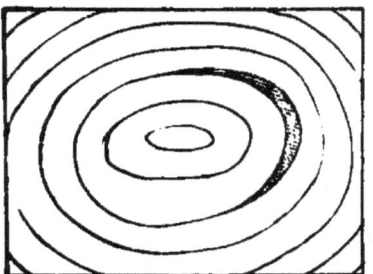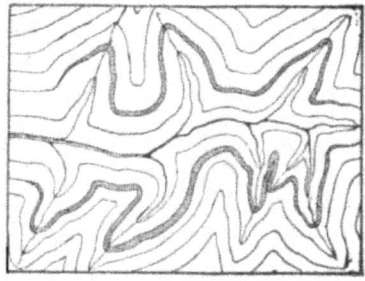

Fig. 27-29.—Behavior of the outcrops of horizontal strata in diverse types of topography.

crop of the bed runs parallel to the contour lines only when the strike of the bed agrees with the direction of the contour.

On the contrary, if the inclined lines of the outcrop run at right angles to a valley they form in it an acute angle; the water flows into this V when the dip of the bed is down the valley and steeper than the slope of its bottom; in all other cases it flows out of the Λ (Fig. 31 and 32). If out-

Fig. 30.—Course of the outcrop of a vertical bed in irregular topography.

crops of inclined strata run across a mountain ridge, they form arches which open toward the foot of the mountain when the dip runs in that direction and is steeper than that of the slope. In the opposite case the arches open

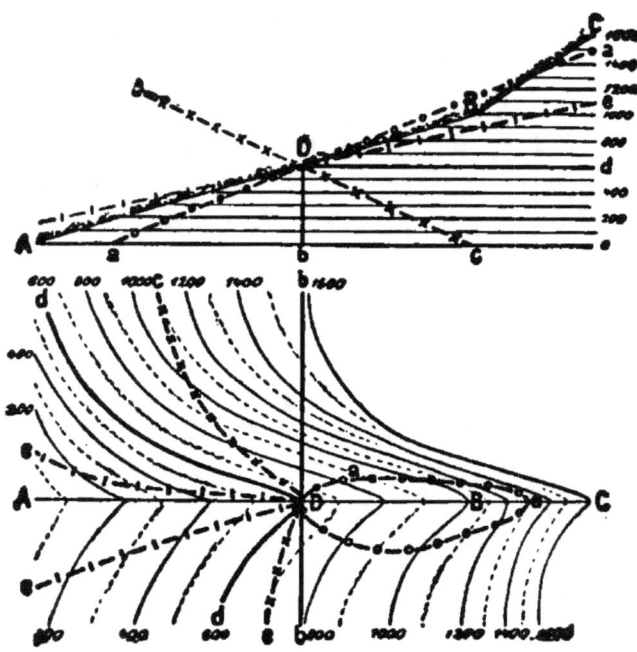

Fig. 31-32.—Outcrop of differently inclined beds in a valley.   (K. Keilhack.)

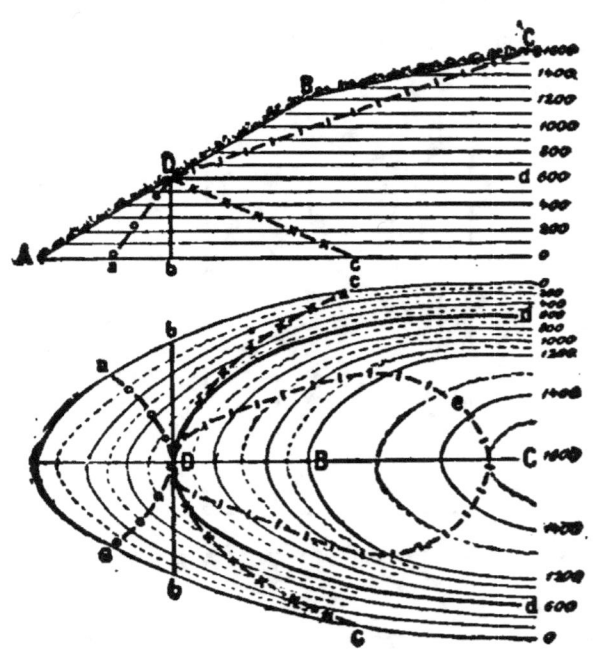

Fig. 33-34.—Outcrops of differently inclined beds on a ridge.   (K. Keilhack.)

toward the summit of the mountain or even form closed figures. (Fig. 33 and 34.)

In searching for ore beds, or for any other kind of beds, one should, in prospecting by tunneling, boring, or by sinking shafts, run as nearly as possible at *right angles to the general strike* of the stratified rocks of the region. If the beds form an isolated exposure, the first thing to be done is to obtain an exact section of the local series of strata to which it belongs, identifying individual beds by means of characteristic fossils or locating certain prominent strata by their particular lithological character, as the case may be. This section afterward often will be of the greatest practical use when the miner encounters dislocations in working the bed.

Disturbances or dislocations of strata show:

(1) In a bending and folding of the beds without interruption of their continuity.

(2) In a breaking and disruption of the strata, by flexure, pressure, thrusts, etc., accompanied by the formation of fissures or cracks. If displacements occur along these fissures we speak of them as *faults*.

1. Flexures.

In describing the various forms produced by folding we may begin with the closed syncline (Fig. 35). In this case the bed forms a 'canoe' or bowl, dipping downward from all sides to the bottom of the basin (center or axis of syncline). The structure is concentric and the outcrops form circles

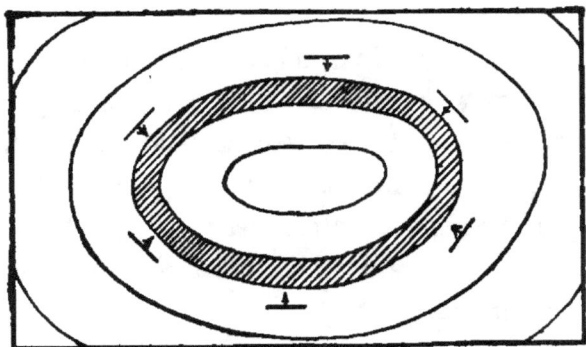

Fig. 35.—Plan of a Closed Syncline.

or ellipses. Thus, the Mansfeld copper shale, aside from small deviations, forms a trough 4 k. (2.4 miles) wide and open only toward the southeast.

The opposite of a basin or closed trough whose concavity is directed downward is the dome or closed saddle. The two forms are contrasted as synclines (trough) and anticlines (saddle). When the axis of a syncline or anticline has a certain length and is cut off at both ends by erosion or dislocation, *open synclines* and *open anticlines* result.   (Fig. 36.)

Every open form of this kind consists of two limbs or flanks connected at the axis of the syncline or the axis of the anticline.   The strike of the strata in these cases is not quaquaversal—*i. e.,* encircling—but rectilinear.

The axis of a fold may be horizontal or be inclined toward the horizon; in the latter case its inclination is called its pitch.

The transition from one limb of a fold to another is ordinarily effected by a gradual curve (open fold), more rarely by an abrupt, sharp angle (closed fold) or by a repeated plaiting of minor plications.   In large synclines, lesser flexures may produce subsidiary synclines and anticlines.   All these phenomena are well exhibited in the main bed of the Carboniferous spathic ironstone measures of Westphalia, and the anthracite coal beds of Pennsylvania.

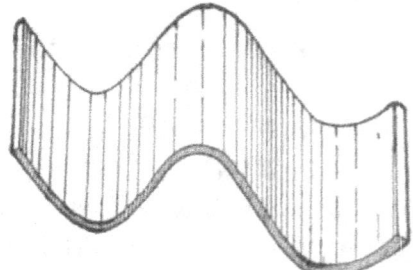

Fig 36.—Open syncline and open anticline, together forming a flat fold.

Both in anticlinal and synclinal folds a strong horizontal thrust may cause a complete inversion of the strata.   Examples may be found in the inverted, that is to say, reversed, succession of strata shown in the Lower Silurian iron ore deposits of the Steinach valley in the Franconian forest, and the inverted pyrite beds of Rammelsberg.

Anticlines whose upper parts have been removed by erosion and denudation are called aerial arches.

An anticline and a syncline that have been brought close together by horizontal thrust form a *recumbent fold* with a middle limb and two side limbs.

2. Displacements.   (Thrust Faults.)

Sometimes the limbs of an overturned fold have been subjected to such intense thrust or squeezing that they have finally yielded and been torn apart along a gliding plane. The discussion of such thrusts, or fold faults, is reserved for the general discussion of faulting, which, for various reasons, is not treated in this place, but later in connection with mineral veins.

DISTRIBUTION OF ORE WITHIN AN ORE BED.

When stratified deposits consist not of solid ore, but of ore mixed with barren, non-metallic minerals in one stratum, the distribution of the ore is an important factor. The determination of the distribution of the ore is especially important in those stratified deposits formed by impregnation, because in such cases its occurrence is apt to be very irregular, as for instance, in the gold-bearing conglomerates of South Africa. The particles of ore interspersed in the barren matrix may be of dust-like fineness, as in the Mansfeld copper shale; they may form small nodules, as in the galena-bearing variegated sandstones of Commern, or finally they may form larger concretions, as in many spherosiderite deposits.

If the rock in which the ore occurs in very fine particles or granules is a crystalline schist, it is customary to call the deposit in question a *fahlband.* The word was first used for the schist zones of Kongsberg, in Norway, containing iron pyrite, copper pyrite, magnetic pyrite, etc., in very fine particles, and though the beds are not sufficiently rich to be workable themselves, they are important because they enrich the silver veins which cut through them. German miners call these zones fahlbands because in the outcrop where the pyrites have been decomposed, the rock is gray. Typical fahlbands are found in the glance-cobalt deposits of Skuterud.

STRUCTURE OF BEDDED ORE DEPOSITS.

Like all stratified formations, bedded ore deposits quite frequently appear to be built up of different strata, as for example, with many iron ore deposits. If the stratum consists of layers of different mineralogic composition the true bedded deposits fail to show a symmetrical disposition of duplicate bands, in cross-section. This is an essential mark by which to distinguish them from the true veins which they often closely resemble. As a rule the true bedded deposits also lack druse cavities (vugs) extending parallel to the stratification. On the other hand, certain bedded ore deposits possess a feature lacking in lodes: they carry fossils or pebbles of true fluviatile or littoral origin.

MINERAL CONTENTS OF STRATIFIED DEPOSITS.

The only ore deposits thus far positively demonstrated to be of actual sedimentary origin (at least on a large scale) are those of iron and manganese, the ores assumed to be of this nature being described under the name of the metal. An important feature of ore beds is the presence of a matrix of certain non-metallic minerals which are entirely or almost wholly lacking in veins, viz.: green hornblende, light green pyroxene, pistazite and garnet. The primary ore content of a bed is, as a rule, apt to be much more constant than in a vein, and hence in upturned strata the value is not apt to change with increasing depth. The case may be different with epigenetic ore beds; for example, the beds of Meggen on the Lenne consist in one part of barite, in another part of iron pyrite. An exception is also found in beds of metasomatic origin, such as the Schwarzenberg beds, which in one place may be developed as copper, in another as a blende deposit.

As regards the differences in mineral content by reason of superficial alteration at and beneath the outcrop, the same results occur with strata as with vein outcrops, as will be shown later. Both have a gossan whose secondary minerals depend entirely on the primary contents. A bed of iron pyrite carries a gossan of brown hematite; a bed of galena forms a gossan with much cerussite, etc.

CLASSIFICATION OF SEDIMENTARY ORE DEPOSITS.

The primary ore deposits of this group may be divided into two classes: (a) Those in which the ore bed and its enclosing strata have undergone little or no change since their deposition (purely sedimentary deposits), and (b) those in which both ore and country rock have passed through a recrystallization or some other important structural transformation (metamorphic ore deposits). Typical examples of the first group are the lacustrine iron ores and most of the Oolitic iron ores; for the second group the Archean iron ores. In many cases, however, it is not possible to draw a sharp boundary between them; in general, the older the deposit the more likely it is to be metamorphosed. Hence the primary ore deposits of this kind will be arranged by geologic age, progressing from the older to the younger, and subdividing all into different groups according to the respectively predominant metals.

We include in sedimentary ore deposits those stratiform masses which contain the ores finely distributed, in the form of dust, granules, scales, or nodules, provided the assumption is justified that this ore content interspersed in particles was deposited simultaneously with the barren constituents. Here, too, a subsequent metamorphism has often occurred, as for example, in the specular iron schists. It will be difficult in individual cases to distinguish sedimentary ores whose finely disseminated metallic particles are of simultaneous origin, from those strata whose ores are the result of a secondary impregnation. The Mansfield copper shale, for example, is placed by various authors now in one group, now in the other. The discriminatior becomes even more difficult when such secondarily impregnated sediments as the cobalt fahlbands of Modum have undergone a subsequent metamorphism which has effaced the original character of the ore.

In a rigorously consistent classification we should have to add to the two categories just discussed, according to G. Gurich[1], a third category, a

Fig. 37.—Lens of spherosiderite from the clay slate of Bocksberg, split longitudinally. (One-fourth natural size.)

diagenetic group of stratified ore deposits, namely, those deposits in which the concentration of the ore took place in the muddy sediment of a water-laid deposit before it was hardened into rock, as for example, in the slate clays with kidneys of spherosiderite. A good instance of this process is seen in the clay ironstone lenses up to 1.5 feet across in size in the Lower Coal measures of western Pennsylvania. The concretions in the roofing slate of Bocksberg, Germany, show an outer layer impregnated with pyrite crystals, whose arrangement, as will appear by reference to Fig. 37, shows the stratification of the clay slate displaced by the spherosiderite. In the

[1] 'Sitz.-Ber. d. Schles. Ges. f. vaterl. Kultur.' February, 1899.

center is seen a network of quartz stringers.   However, for the sake of sim-
plicity we include this small group with the ore deposits regarded as normal
sediments.   We will now proceed to the special classification and sketch-
ing of the most important groups, laying special stress on certain examples.

1. SEDIMENTARY IRON ORE DEPOSITS.

(A) Sedimentary Iron Ores of the Crystalline Schists.

(a) CRYSTALLINE SCHISTS WITH DISSEMINATED IRON ORES.

Magnetite and hematite occur as accessory constituents disseminated
through portions of crystalline schists of almost all known areas.   At times,
however, they form the principal constituent of quartzites, mica schists,
and similar rocks, and thus attain importance as ore deposits.   The following occurrences illustrate this class:

Specular Hematite Schists.

Specular hematite schists are granular schistose mixtures of micaceous
hematite and quartz.   They are known to occur at several localities in
Germany[1], viz.: Soonwald and Marmarosch (Cotta), in many parts of Norway, and particularly in the ancient schistose rocks of Brazil[2].   In the latter
country at Itabira and Antonio Pereira they form a great series of strata
intercalated between clay slates and itacolumites.   They also occur in South
Carolina[3].   In both countries they are associated with gold deposits.

Itabirite.

Itabirite is a granular schistose mixture of quartz with hematite and
magnetite, which occurs at Itabira, Villarica, and other localities in Brazil,
associated with specular hematite schists.   It also occurs at Sutton in
Canada, included in metamorphosed Silurian strata.   In Brazil this rock
also contains gold, and Hussak[4] considers itabirite as the matrix of the
waterworn specimens of cinnabar found in the alluvial 'Cascalho' of

---

[1] Noggerath, in Karsten's *Archiv.*, Vol. XVI, 1842, p. 515.
[2] v. Eschwege: 'Beiträge zur Gebirgskunde Brasiliens.' Berlin, 1832.
[3] Lieber: *Rep.* Geol. Surv. S. Car., 1856.
[4] Hussak: *Zeit. f. Prak. Geol.*, 1897, p. 65.

Tripuhy. The same rock occurs in South Carolina associated with mica schist, itacolumite and talcose schist rich in magnetite grains, the so-called catawbirites. Here, too, gold deposits are known in this group. These examples are economically of less importance than certain Norwegian deposits of which a fuller description is given.

The Iron Ore Field of Naeverhaugen.

This district, situated on the west coast of Norway, 24 miles (40 kms.) east-northeast of the town of Bodö, is, according to A. W. Stelzner[1] and J H. L. Vogt[2], a region of metamorphosed Paleozoic rocks consisting of garnetiferous mica schists, quartz schists, amphibole schists, mica gneisses, pyroxene gneisses, epidote schists, and granular crystalline limestones. In certain layers both the mica and the quartz-schists carry numerous interspersed flakes of specular hematite, with subordinate granules of magnetite. Thin, highly ferriferous layers alternate with barren layers. These finely banded layers of ore attain a thickness of 16 to 22 ft. (5 to 7 m.), sometimes 25 to 30 ft. (8 to 9 m.); greater thicknesses, up to 50 ft. (16 m.), occur, but are due to folding. The foliated schists contain, moreover, layers of rich ore sometimes 0.3 feet to 0.6 ft. (10 to 20 cm.) rarely a foot thick, in which the amount of iron, ordinarily about 50%, may rise to 58%, with a phosphoric acid content of 0.2 to 0.4%. The barren layers of the banded ores consist of quartz, with some hornblende, pyroxene, epidote, and garnet. Accordingly these ores approach the Swedish Striberg type.

These occurrences form a transition from the deposits of disseminated ore to the beds of the compact iron ore in the crystalline schists. This last named class includes a variety of ores. Those of spathic ore will be described first, then those of magnetite and iron glance:

Compact Iron Ore Beds of the Crystalline Schists.

THE CARBONATE IRON ORE BEDS OF THE CRYSTALLINE SCHISTS.

Owing to the ready solubility of iron carbonate where it is associated with limestone or dolomite, and the frequent migration of material, the original sedimentary character of all the deposits of this class has been greatly obscured by secondary metasomatic processes. This is particularly true of the most important example, described below.

[1] A. W. Stelzner: 'Das Eisenerzfeld von Naeverhaugen.' Berlin, 1891.
[2] J. H. L. Vogt: 'Salten og Ranen.' Christiania, 1891. Cited in *Zeit. f. Prak. Geol.,* 1894, p. 30.

1. *The Iron Ore Deposits of Huttenberg, in Carinthia*[1].

In the eastern Alps crystalline schists and granites make up the central east to west range of the system. To the north and south the rocks of other ages occur as steep-dipping and often dislocated strata lying against this zone; first a low area of essentially Paleozoic rocks, then the Mesozoic limestone of the steep and high ranges of the Alps. The spathic ironstone beds of Huttenberg, northeast of Klagenfurt, belong to the central zone. The beds and the associated granular crystalline limestones form a series of strata which extend from the vicinity of Friedsach on the Olsabach ranges, across Huttenberg and Lölling, and farther across Wolfsberg into the Lavant valley. The limestone in which the several beds of these series are intercalated is itself interstratified between mica schists and gneisses.

At the Erzberg, near Huttenberg, which in popular parlance is called the "main iron root" (Haupteisenwurze) of the country, as shown by a section (Fig. 38, after Seeland), the lower member of the crystalline schist series is the same gneiss that contains the well known mineral deposits of the Sualpe. Its strata dip southwest. It is followed by mica schist with ore-bearing limestone intercalations, as well as with intervening layers of tourmaline rock (probably a pegmatite rich in tourmaline), eclogite, and amphibolite. These are finally followed by phyllites.

The main limestone bed at Erzberg has a length along the strike of more than one and one-third miles (2,400 m.), and a thickness of 1,312 feet to 2,296 feet (400 to 700 m.). Lenses of gneiss and mica schist, together with small quartz streaks are intercalated in it, and it often carries muscovite, passing into calcareous-mica schists. Disseminated grains of pyrite, arseno-pyrite, and rarely chrome-mica and realgar, occur here and there. Sometimes the limestone is ankeritic (that is to say, it contains much iron carbonate).

The spathic ironstones, which when undecomposed are called white ores on account of their light gray color, contain considerable manganese which, when weathering begins, shows on the surface as a violet sheen (blue ores) and eventually forms a large amount of wad and pyrolusite. These manganese minerals form a gray coating called mold (schimmel) over ore fragments. The orebodies have been largely transformed into limonite, and as this alteration is accompanied (despite the absorption of oxygen and water) by a loss of volume amounting to one-fifth of the original mass, the

---

[1] F. Münichsdorfer: 'Geologisches Vorkommen am Hüttenberger Erzberg.' *Jahrb.* d. k. k. geol. Reichsanst, 1856. F. Seeland: 'Der Hüttenberger Erzberg und seine nächste Umgebung.' With plates I–IV. *Jahrb.* d. k. k. geol. Reichsanst., Vol. XXVI, 1876, pp. 49-112. A. Brunlechner : 'Die Form der Eisenerzlagerstätten in Hüttenberg.' (Kärnthen.) *Zeit. f. Prak. Geol.*, 1893, pp. 301-306.

limonite has a spongy structure. Their cavities often contain **hematite stalactites,** and sometimes pseudomorphs of göthite after blende.

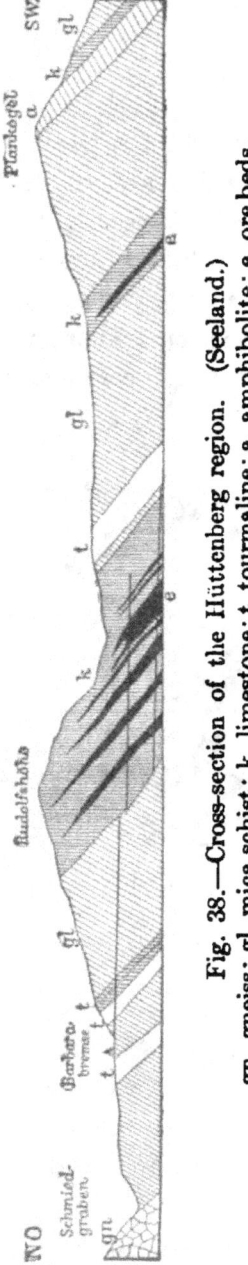

Fig. 38.—Cross-section of the Hüttenberg region. (Seeland.)
gn, gneiss; gl, mica schist; k, limestone; t, tourmaline; a, amphibolite; e, ore beds.

According to F. Seeland the spathic ironstone of Hüttenberg contains in **fresher samples:**

Iron.	41.28 to 44.33 per cent
or Ferrous oxide	47.62 to 56.11 "
Manganese oxide	up to 5.02 "
Lime	0.79 to 1.33 "
Magnesia	3.05 to 4.35 "
Carbonic acid.	32.79 to 37.70 "
Silicic acid	up to 2.47 "
Water	0.43 to 2.47 "

The different orebodies retain a partly lenticular form, but as a whole they form a rather irregular system of deposits since they overlie one another variously, both in the direction of the strike and in the direction of the thickness, or are connected by spurs. Divisions of beds recur rapidly in one and the same orebody. On a small scale, too, the outlines of the lenses next to the limestone wind about in a most irregular way, and sack-shaped spurs 6 to 10 feet long occasionally project into the underlying or

Fig. 39.—Cross-section through a part of the Hüttenberg field. (Brunlechner.)
The shaded patches are ore.

overlying limestone. At the end the masses either wedge out or break up into several wedges or strings. All this is shown in the section, Figure 39. Sometimes the ore encloses also boulders of slate or limestone.

In isolated cases distinctly lode-shaped ore masses were also observed; in the Seeland shaft a schistose, well stratified limonite was traversed by a vein of younger limonite.

In the Glücklager the conditions of stratification seem to clearly indicate that the ore has been formed by a progressive replacement of the limestone. In a mass of brown ankerite [ankerite=(Ca, Fe) CO_3] obtuse-angled fragments of crystalline limestone, with an etched appearance, are found entirely isolated.

According to Brunlechner there has been a partial transposition of the ore, as .follows: The surface water seeping downward absorbed the carbonic acid which had been liberated by a superficial alteration of the siderite into brown hematite. Charged with this solvent the water was now able to decompose the ironspar below the zone of oxidation. The redeposition of the ferric carbonate taken up by these waters had to occur where the water flowed away from the siderite mass and encountered $Ca\ CO_3$, that is to say, at the margins of the lenses, at the contact with the lime, through the replacement of the limestone and the exchange of the iron carbonate with the calcium carbonate. Hence it occurs that the ores of such sacklike offshoots are particularly pure. Similar replacements seem to have occurred very soon after the first layers of the limestone were deposited. However, that the iron carbonate has in the main been directly deposited as a sediment, is shown by the distinct stratification of many orebodies as well as by the manner of the connection with the series of which it forms but a member.

The Hüttenberg deposits were known to the ancient Romans. In the seventies of the last century an average of 146,000 tons of ore per year were produced, averaging 68.8%. In the nineties the annual output was, in round numbers, 100,000 tons of spathic ironstone and limonite, the output for 1898 being 7,244.1. tons of spathic ironstone, and 58,558.8 tons of limonite.

2. Gyalar in Transylvania (Hungary).

An analogous iron ore deposit exists in Transylvania, in the mountainous country south of the Maros, near Vajda-Hunyad. Crystalline limestones extend for miles through mica schist rocks and hold vast intercalations of spathic ironstone, which, however, in the levels thus far worked has in almost every case been altered into limonite. In the nineties there were produced near Vajda-Hunyad annually about 180,000 tons of brown hematite, with a content of 52 to 56% of iron.[1]

(b) BEDS OF MAGNETITE AND HEMATITE.

1. *The Iron Ore Beds of the Archean Rocks of Sweden.*

Sweden is very rich in iron deposits, the total iron ore production of 1903 being 3,677,841 tons. There are in the main two great areas containing

[1] F. Beyschlag: 'Das Montanwesen auf der Milleniumausstellung.' *Zeit. f. Prak. Geol.,* 1896, p. 465.

these underground treasures. One lies in central Sweden on both sides of the 60th degree of latitude; it is the long famous Jernbäraland, with the well-known mines of Persberg, Taberg-Nordmark, Striberg, Grängesberg, Norberg and Dannemora. The other field lies far to the north, immediately north of the Polar Circle; it consists of the much discussed fields of Gellivara and those of Luossavaara and Kiirunavaara and Svappavare in the province of Norrbotten, described previously. With the exception of the last named, all the deposits belong to the crystalline schists whose classification must, for that reason, be briefly described in order to define the geologic position of the occurrences.

According to A. E. Törnebohm[1] the primitive rocks of Sweden are classified as follows:

Upper division ..	Granites; in part also gneisses (Vermland).
	Phyllitic schists[2], dark hälleflintas[3] at bottom, with a diorite (Grytthyttan) sheet.
	Porphyroid and hälleflintas.
	Fine-grained, scaly gneisses (the so-called granulites or eurites).
	Banded gneissoid granulites.[4]
Lower division. .	Red and gray granites and granitic gneisses.
	Banded gneisses as well as cordierite gneisses (eastern Sweden) and epidote gneisses (western Sweden).
	Iron gneisses (with finely interspersed magnetite).

Iron ores and crystalline limestones—the two almost always found together—occur but rarely in central Sweden in the lower series of gneisses, but are very abundant in the upper, in the group composed of granulites, porphyroids and hälleflintas. Both magnetite and hematite ores occur; the former are closely associated with either limestone or with a rock consisting essentially of pyroxene and hornblende, often carrying garnet and epidote, rock called *skarn* by Swedish authors. The hematite ores, on the contrary, are usually directly interbanded with the granular gneisses.

In the description of examples from the mining fields it is advisable to begin with Norberg, because it contains a great variety of ore types crowded

[1] Törnebohm: 'Ofverblick öfver Mellesta Sveriges Urformation.' *Geol. Fören. Förh.*, Vol. VI, part 12.

[2] Phyllite is a semi-crystalline somewhat metamorphosed clay slate containing sericite, ottrelite, ilmanite and garnet.

[3] Hälleflinta, an exceedingly compact hornstone-like rock formed of microscopic particles of feldspar and quartz with scales of mica and chlorite. Porphyroid is a fibrous schistose siliceous micaceous rock with scattered porphyritic crystals of feldspar and quartz.

[4] Granulite, a schistose rock (German and English usage), with peculiar granular structure resulting from crushing and re-crystallization. Not a granite as understood in French. Composed of feldspar, garnet, etc.

into a small space. Among the many others we select Persberg, Dannemora and Grängesberg.

I. Norberg.

"Norberg Bergslag," in Westmanland, contains several hundred iron
mines distributed over a narrow zone 12 miles (20 k.) long and 2 miles
(3 k.) wide. All these mines lie in a zone of fine-grained, scaly micaceous
gneisses (granulites of Swedish authors) and of hälleflintas, bounded on
both sides by granite and gneiss, and containing intercalations of mica
schist as well as of crystalline limestone and dolomite. Three kinds of ore
occur there: (1) Red hematite ores, consisting of crystalline hematite with
numerous exceedingly thin quartzite lamellae, often crumpled and hand-
somely folded by rock pressure, to which is given the name dry ores, *i. e.,*
needing a flux; (2) finely crystalline magnetic iron ore intermingled with
a garnet-pyroxene-skarn; they need no flux and hence are called self-fluxing
ores; (3) magnetic iron ores, mostly highly manganiferous (at Klackberg
up to 7% of Mn_2O_3) in lenses in the midst of the dolomite and limestone.
and hence highly calcareous, being called fluxing ores because as a rule they
are added to others to make a smelting mixture.

The dry quartzose hematite ores which occur under precisely similar
conditions at Striberg[1] in the Örebro area, and hence are called by Swedish
authors "iron ores of the Striberg type," are, in the main, confined to tne
lower geologic horizon of the ore zone. Above them come the ores
associated with pyroxene skarn, and in the uppermost horizon the calcareous fluxing ores of the dolomite intercalations predominate[2]. The iron
contents of the ore vary between 43 and 60%, with a phosphorus content
of only 0.004 to 0.035%. During the years 1891 to 1895 the average annual output was 172,516 tons of iron ore[3].

Sometimes the magnetic iron ores associated with limestone contain an
admixture of pyrite, and hence have to be roasted before being put in the
furnace, as for example at Klackgrufva. In some mines the pyrite is accompanied by such an abundance of other sulphide ores, especially galena

[1] Birger Santesson: 'Beskrifning till Karta öfver Berggrunden af Orebro Län
II. De vigtigare Grufvefälten.' Stockholm, 1889.

[2] A. E. Törnebohm: 'Om lagerföljden inom Norbergs malmfällt med karta.'
Geol. Fören Förh., Vol. II, 1874-75, p. 329. Also see G. A. Granström: 'Nagra
underrättelser om grufvorna och grufdriftn inom Norbergs berslag.' *Jern-Kontorets
Annaler*, 1876, p. 1.

[3] Törnebohm: 'Geologiska Öfversigtskarta öfver Mellersta Sveriges Berslag
Blad. 1 med Beskrifning,' 1880. Nordedström's 'Katalog Mellersta Sveriges Grufut-
ställning.' Stockholm, 1897.

and chalcopyrite, as to pay for their working.   (See further account under Epigenetic Ore Beds.)

II. Persberg[1].

The iron ore mines of Persberg, which according to legend were worked as early as 1390, lie in Vermland on the peninsula projecting into the Yngen Lake, in a low, hilly country.   Although the production has declined, the annual average from 1891 to 1895 was 31,884 tons of ore, in which the iron content was 53 to 60% in the better ores.   Magnetite (svartmalm) is the only ore produced.   The best grades contain only slight admixtures of pyroxene in their fine crystalline mass.   The poorer grades enclose also garnet and talc.   The phosphorus content amounts on an aver-

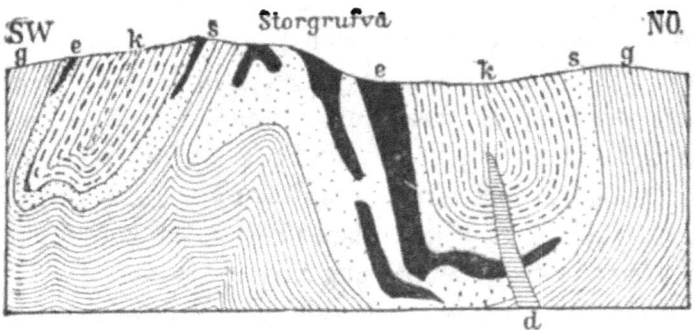

Fig. 40.—Section through the Storgrufva at Persberg.

g, Micrograular scaly biotite gneiss; k, limestone and dolomite; s, skarn; e, orebodies; d, diorite.

age only to 0.002, at most 0.013%.   The manganese content varies between 0.20 to 0.35%.

The surrounding rock also consists here of a so-called granulite, an exceedingly fine-grained gneiss which in many cases passes into dense looking hälleflinta-like rocks and adjoins extensive granite areas.   The orebodies form lenses or often very irregular and wholly shapeless masses, imbedded in the garnet and epidote-salite skarn.   The skarn occurs either as an independent intercalation in the granulite gneiss or at the boundary of several large beds of crystalline limestone and dolomite, inserted in the granulites.

The association of these strata is apparent from the accompanying profile through the Storgrufva (Figure 40).   The skarn, together with the ore beds, takes part in all the numerous foldings of the country rock.   The skarn

[1] A. E. Törnebohm: 'Karta öfver Berggrunden inom Filipstads Berslag,' 1874. The same author: 'Geognostisk Beskrifning öfver Persbergets Grufvefalte,' 1875. Walfr. Petersson: 'Högbergs faltet vid Persberg,' 1897.   Nordenström's 'Katalog.'

beds in the Högsberg field are especially strongly folded and crumpled. Walfr. Petersson states that he ascertained that at the flexure points of the arches, *i. e.*, anticlinal folds, the largest orebodies are to be found. A special kind of skarn, consisting mainly of talc, prevails in the area of the Alabama mine. Sometimes the skarn shows a bedlike stratification of alternating layers of garnet and pyroxene, but this is mostly confined to its barren portions. The ore beds are either sharply defined or pass gradually into the skarn by transitions containing more and more of its mineral constituents. On the other hand, pyroxene-bearing bands may at times be seen occurring in the granulite and thus effecting a transition to the skarn. In some parts of the mine magnetic iron ore is filled with stringers of calcspar.

In many places so-called *skölar* are seen to run through the ore beds. By a *sköl* the Swedes understand a zone of slipping or movement, mostly vertical, and generally nearly parallel to the strike, its filling consisting of crushed rock, mostly altered and either chloritic or serpentinized. They may be several meters in thickness, and in mining sometimes cause the loosening of large portions of the walls.

III. Dannemora[1].

The famous Dannemora mine lies near the railway line from Upsala to Gefle, in a low, hilly country, on the shore of the reed-bordered Grufva lake. Directly east of the mines are the furnaces of Osterby, where the choice Dannemora steel is obtained from the ore of this mining field. The mines are mentioned as early as 1481, although the real industry began only in 1532. From 1891 to 1895 the average annual output was 55,440 tons of iron ore, and 390 tons of zinc-blende.

The Dannemora ore is a very dense magnetic iron ore, with 20 to 65% of iron, averaging 50%, the percentage of iron depending upon the amount of the intermingled actinolite (a variety rich in manganese called dannemorite) as well as on the calcspar.

The geologic conditions are as follows: A broad mass of granite, the Upsala granite, partly gneissoid, occurs intruded in a very thick series of bands of hälleflinta, fine-grained gneiss (granulite, eurite of Swedish authors) and crystalline limestones, carrying considerable manganese. The hälleflinta of the Dannemora has for a long time attracted the attention of geologists. It consists of eruptive rocks, in part clearly recognizable as quartz porhpyry, in part of strongly altered porphyry tuffs, which may alternate with the limestone in layers, sometimes as thin as a sheet of paper: thus distinctly showing their sedimentary origin. The orebodies occur in

[1] A. E. Törnebohm: 'Geologisk Atlas öfver Dannemora Grufvor vid Beskrifning.' Stockholm, 1878.

three main bands, which are found intercalated in the limestone, or between it and the hälleflinta. These ore belts strike north-northeast for a distance of over a mile (2 km.) and dip northwest at 75 to 80°.

The largest of the ore lenses, situated in the middle field, has been exposed by a vast open-cut, whose almost vertical walls descend to a depth of 475 feet (145 m.). (See Fig. 41.) The present workings are all underground, below the floor of this gigantic open-cut. This orebody has a thickness of 98 feet (30 m.) and has been produced by the convergence of three closely adjoining parallel lenses.

At the boundary between the limestone and the ore masses, which are often distinctly stratified, a layer of skarn is often developed; this is an

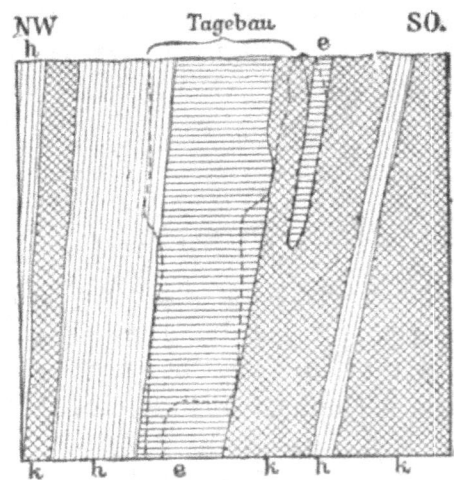

Fig. 41.—Cross-section through Storrymningen at Dannemora.
h, hälleflinta; k, limestone; e, iron ore; Tagebau-open-cut.

amphibole salite rock, frequently garnetiferous, and with streaks of magnetite. At many points in the field dikes of felsite-porphyry and diorite cut across the beds.

A peculiar feature of the Dannemora field is seen in the Svafvel mine in the southern field. At this place the ore beds and accompanying rocks are obliquely traversed in strike by a broad lodelike zone of impregnation extending from the surface to a depth of about 213 feet (65 m.). This often contains good-sized masses of pyrite and zinc-blende, with accessory galena, magnetic pyrite, copper pyrite and arsenopyrite. A similar impregnation with secondary sulphide ores, but following the strike of the strata, is observed at a depth of about 754 feet (230 m.). The Dannemora south field formerly produced up to 2,000 tons of blende per year.

Finally mention may be made of the remarkable occurrence of asphalt in small calcspar veins traversing the iron ore beds, asphalt pellets being enclosed by calcite skalenohedrons.

IV. Grängesberg[1].

Grängesberg is at present the largest iron ore district of central Sweden.

The geologic mode of occurrence is as follows: The field is about 3 miles (5 km.) long and 0.6 mile (1 km.) wide, lying in the granulite zone of the primitive rocks, whose prevailing type is a fine-grained scaly biotite gneiss. The principal ore beds occur in the lower strata, in which there is intercalated at Grängesberg a vast bed of a coarse, fibrous, reddish granite gneiss. All the schists strike north-northeast and dip steeply east-southeast. The district is divided into four parts, the Lomberg field at the south, the Ormberg and Risberg field at the west, the Export field at the east and the Norra-Hammer field at the north. The first two contain a very great number of small beds of hematite carrying an admixture of magnetite and a phosphorus content of 0.02-0.8%. These ores are in part associated with the pyroxene skarn, in part intergrown with calcspar.

The conditions are totally different in the Export field. In the first place, we are here dealing with far greater deposits. They consist in the main of two great lenticular masses. In the southernmost mass the greatest thickness in the Bredsjö mine is 212 feet (65 m.), from there northward it has for a great distance a thickness of 100 to 130 feet (30 to 40 m.), until finally the lens wedges out abruptly. This, however, includes a number of intervening beds 10 to 13 feet (3 to 4 m.) thick, as well as intercalated pegmatite masses. The orebody proper consists exclusively of fine-grained crystalline magnetite, with 60 to 62% of iron and 0.7 to 1.2% phosphorus. Quartz, feldspar, fluorspar and actinolite are accessory constituents. The great northern ore lens of Sjustjernsberg, whose exposures form the Bergsbogrufva, attains a length of 1,312 feet (400 m.) along the strike, and a thickness of 295 feet (90 m.). The upper part consists of magnetite, which, however, contains so much apatite that the phosphorus content may rise to 2.8%; moreover, layers of almost pure apatite rock up to 0.06 to 0.1 feet (2 to 3 cm.) in thickness are not infrequently intercalated in the ore. The lower half of the lens, on the other hand, consists of hematite ore with 0.5 to 2% of phosphorus. At times large octahedrons of magnetite are found sparsely scattered through the fine-grained crystalline hematite. The Bergsbo lens also carries some intervening beds of granulite, which may attain a thickness of 8 meters. From their course it is seen that the entire ore mass must be regarded as a complex of several closely crowded lenses. Furthermore, the ore is frequently traversed by vast intrusive stocks

[1] N. Hedberg: 'The Grängesberg Iron Mines in 1898.' Falun, 1898.

and dikes of pegmatite. This rock, poor in mica, contains apatite, beryl, and, curiously enough, asphalt in kidney-shaped or drop-shaped pieces in the midst of druse-like portions, also as inclusions in feldspar and quartz. Where the larger pegmatite dikes traverse the hematite ore they have transformed it into magnetite, sometimes for a distance of 6.5 feet (2 m.) from the contact. It is also affirmed that the iron ores in the vicinity of the pegmatite have a higher phosphorus content, up to 2.8%.

The northernmost mining field, Norra-Hammargrufva, yields a magnetic iron ore exceedingly rich in apatite, and holding 6 to 8% of phosphorus. The immediate country rock is a hornblende gneiss with layers rich in mica. This particular occurrence at Grängesberg shows very great petrographic similarity to the Gellivare ores. The ores of Norra-Hammargrufva are characterized by pegmatitic segregations with large individuals of titanite, with hornblende and asphalt, sometimes also with scheelite and zeolitic minerals.

Nowhere in the Swedish iron ore fields are there more distinct indications of the genetic conditions of these ore masses than just here at Grängesberg. When we see how, for example, in the Mor mine an alternation of extremely thin layers of normal granulite and magnetite-bearing rock is repeated a thousand times, there can be no doubt that the magnetite and hematite were crystallized out simultaneously with the ingredients of the country rock. This is confirmed by microscopic investigation. In the rock of the Mor mine just mentioned the magnetite is clearly of the same age, and is syngenetic with the quartz, orthoclase, plagioclase and mica, of which the striped granulite of that mine is chiefly composed. It occurs partly in grains and partly in concentric layers about other minerals; also as inclusions in the midst of quartz and feldspar, while conversely it may enclose the minerals named. The same is true of the apatite-rich ores of the Bergsbö lens. Here, too, oft-repeated alternations of ore and rock and the microscopic conditions of mutual intergrowth show a simultaneous crystallization of ore- and rock-forming minerals. In the apatite-rich ores of Norra-Hammargrufva the microscopic sections suggest, it is true, a certain succession in the segregation of the several ingredients, which is absent in the examples above mentioned. The apatite of the latter mine is always free from inclusions, and this, together with the abundant brown titanite, seems in each case to have been the earliest formed mineral in the various layers of ore. Next in age follows the hornblende, then the infrequent quartz and finally the fluorspar. The magnetite existed before the hornblende only in small individuals, for only in this form is it found enclosed in the hornblende. Subsequently it filled the gaps between the hornblendes in the form of larger grains and aggregates.

The question as to the form in which the iron compounds were present in the rock before they were altered by regional metamorphism cannot be definitely answered. Their present mineralogic nature, however, was certainly imparted to them simultaneously with the general alteration of the rock. The mineral wealth of Grängesberg seems destined to last for many years. Inasmuch as on Vestra Ormberg an orebody has been followed to a depth of 984 feet, it is fair to assume that the others also extend far down.

The total production from Grängesberg (629,802 tons) in 1897 exceeded that of Norberg, and by a small amount that of the Gellivare field of northern Sweden, which in 1897 produced only 623,110 tons. This great development has, however, taken place only in recent time; though the ore beds of Grängesberg have been known since the beginning of the 17th century, they have been worked but little. The very largest deposits were practically untouched because the high phosphorus content of the ores rendered them unsuitable for furnace treatment. It is only since the invention of the Thomas basic Bessemer process that the revolution took place here, and all at once production increased tremendously. The many companies which formerly operated small mines are now consolidated into four large companies. By far the larger part of the phosphoric ores is at present shipped from the port of Oxelosund, in eastern Sweden. The bulk of it goes via Stettin and Rotterdam to upper Silesia and Westphalia.

In the following table we give some typical analyses of different varieties of iron ores from central Sweden taken from four localities:

I. Pyroxene-bearing magnetic iron ore of the Storgrufva at Persberg. (Average sample taken by W. Petersson.)

II. Quartzose hematite ore from Striberg (Kärrgrufva). (C. J. Särnström, analyst.)

III. Apatite-bearing hematite ore . . . . {From Grängesberg, Bergsbö mine,

IV. Apatite-bearing magnetic iron ore. . . . {   after N. Hedberg.

	I.	II.	III.	IV.
Magnetic oxide...........................		26.78	14.81	79.04
Ferric oxide 55.24		47.32	73.50	9.13
Ferrous oxide.......................23.80		. . . .	......	....
Iron.............................57.18		52 52	62.18	63.63
Manganese oxide.................... 0.21		0.19	0.04	0.10
Lime............................... 5.17		0.95	5.09	3.61
Magnesia.......................... 1.85		0.40	0.82	2.72
Alumina........................... 2.38		0 35	0.63	1.78
Silica............................10.45		23.80	1.62	2.00
Phosphoric acid..................... 0.007		0.034	3.53	2.42
Phosphorus 0.0033		0.015	1.54	1.06
Carbon dioxide 0.51		......	......	....
Sulphur........................... 0.02		0.020	Trace.	0.013
Copper .... Trace.		......	......	....
	99.637	99.844	100.04	100.813

V. The Iron Ore Deposits in the Province of Norrbotten, Gellivare and Svappavara.

The famous ore mountain of Gellivare,[1] 80 kilometers north of the Polar Circle, in Swedish Lapland, province of Norrbotten, which is now connected by a railway line of 132 miles (220 k.) with the port of Luleå, and is to be connected in the near future by the Ofoten railway with the Norwegian coast, is the largest iron ore deposit of Scandinavia. About one-twentieth of the slopes of the mountain, in all about 65 hectares, are occupied by the outcrops of the iron orebodies, arranged in two series. The prevailing country rock, in which the beds are intercalated, is a coarse-grained, reddish gneiss, which is seen outcropping especially on the slope of the 'Välkomann' peak. Farther south it becomes more fine-grained, and contains a good deal of hornblende. It is always garnetiferous. Toward the west it passes over into hälleflinta[2]. At the Capitan and Frederika grants, and thence to the Tingfall peak, it is banded, owing to the development of layers especially rich in hornblende, and in such case shows intense folding. The gneisses and ore are often broken through by granite, as at Välkomman and Kungsryggen.

The series of ore lenses take part in all the changes of strike of the granite, even the finer plications.

The ore consists of magnetite and hematite, and is usually rich in apatite, sometimes even passing over into apatite rock proper. Out of 41 samples 26 showed an iron content of 70 to 74.3%; 13 a content of 60 to 70%; 2 a content of 50 to 60%. The phosphorus content varies from 0.05 to 1.5%. The sulphur content is quite insignificant. The manganese content is only about 0.15%, and titanic acid 0.45 to 1.9%.

The outcrop of the orebodies is in part concealed by glacial drift.

Gellivare has been officially known as a mineral deposit since 1704. It was only in 1797 that Hermelein, to whom the development of Lapland is so largely due, established an industry of note in this locality. Transportation was effected solely in winter by means of reindeer. The Koskull and Capitan mines were then the largest. In 1837 that first mining period attained its greatest annual production. with 540 tons. After 1870 the work was almost entirely abandoned, but of late years there has been a remarkable development, as noted previously.

---

[1] 'Berättelse om Malmfyndigheter inom Gellivare och Jukkasjärvi Socknar,' etc. afgiven af Chefen för Sveriges Geol. Undersökning, 1877. Contains the work of D. Hummel, O. Gumaelius, O. Trysén and C. A. Dellwik. Wedding: 'Die Eisenerzvorkommen bei Gellivara,' etc. *Preuss. Z. f. B., H. u. S. Wesen,* Vol. XLVI, 1898, pp. 69-78. For further literature consult *Förh. Geol. Fören.,* Stockholm.

[2] The Norwegian hälleflintas are exceedingly compact hornstone-like rocks produced by metamorphism from fine sediment. (Geikie.)

Svappavara lies between the rivers Tornio and Kalis, 43 kilometers southeast of Luossavaara, already described. The deposits are intercalated between mica schists and fine-grained quartzites. They were discovered as early as 1654 and were worked at that early date, but became important only in recent time. The ore resembles that of Gellivare—but that it has a higher sulphur content, which is a disadvantage. Copper ore deposits are also known in the same locality.

2. *The Iron Ore Deposits of the Arendal Region in Norway.*

In southern Norway there are iron ore deposits, formerly extensively worked, which appear to be related to those of Persberg, although Th. Kjerulf and Telleff Dahl,[1] in accordance with the state of knowledge at that time, were inclined to regard them as lode-like occurrences. The deposits are near the coast and form an ore zone about 15 miles (25 km.) long. The most important mines lay along the Langsev Vand (Langsev and Vas mines) at Näskillen and on the Hellesund on Langö and Gamö. The country rock is biotite gneiss with intercalations of mica schist, quartzite, hornblende schist and crystalline limestone. The ore consists of magnetite, with some intermingled garnet and augite, the deposits being enveloped in a mantle of skarn composed of garnet, light green augite, together with some epidote and calcite. The various ore masses are irregularly lenticular in shape, and attain a thickness of 6.5 to 65 ft. (2 to 20 m.), with a length along the strike of 300 to 650 feet (90 to 200 m.). While the boundary between garnet skarn and country rock is very sharply defined, stringers of ore frequently penetrate the skarn, the ore occurring also in smaller lenses by itself in the surrounding gneiss. The fragments of the country rock described by Kjerulf and Dahl in these deposits may indicate local disturbances. The ore beds themselves and the pegmatite veins by which they are traversed are rich in various minerals, including some rare species.

3. *The Iron Ore Deposits of Krivoi-Rog, in Southern Russia.*

These ore deposits[2], ordinarily grouped as those of the Saxagan basin, lie on the Inguletz, a western tributary with north-south course entering into the Dnieper river above Cherson. They are of exceedingly great importance to Russia, because they lie close to the Donetz coal basin extending east of

---

[1] Th. Kjerulf and Tellef Dahl. 'Ueber das Vorkommen der Eisenerze bei Arendal,' *Neues Jahrb. f. Min.*, 1862, p. 557.

[2] See reviews of the works of Kontkiewitsch, Piatnitzky, Domherr, Monkowsky, Karpinsky, and others: *Zeit. f. Prak. Geol.*, 1896, p. 271; 1897, pp. 182, 186, 278, 374; 1898, p. 139 (Macco). See also, L. Strippelmann: 'Süd-Russlands Magneteisenstein u. Eisenglanzlagerstätten in den Gouvernements Jekaterinoslaw u. Cherson,' 1873. J. Cordeweener: 'Geologie de Krivoï-Rog et de Kertsch.' Paris, 1902.

the Dnieper. The ores occur in a strongly folded crystalline schist striking north-south, whose geologic age is as yet uncertain. In its upper part this rock consists of carbonaceous slate with but few layers of ore; next below come the ore-bearing quartzite schists, underlaid by clay slate, actinolite schist, quartz chlorite schist, talc schists, arkose and itacolumite-like mica schists, and finally by gneiss (probably dynamo-metamorphic granites) and true granites. The ore-bearing strata form a long-extended fold, and show a close minor plication by which quartzite beds have, according to Macco, changed into a succession of quartzite nodules, like conglomerate cobbles in a clay slate. In the double row of deposits extending parallel to the Saxagan river the two most important orebodies are those near Krivoï-Rog; the lower one about 98 feet (30 m.) and the uppermost about 262 feet (80 m.) thick. The ores consist of magnetite, for the most part altered to red hematite, with 45 to 70% of iron and 0.01 to 0.02% P_2O_5. The very irregular ore masses lie in a finely banded yellowish white, red or brown ferruginous quartz schist whose crystalline quartz grains enclose numerous magnetite particles. In 1894 the production of Krivoï-Rog had already risen to 880,000 tons, and by 1900 to about 2,700,000 tons. According to J. Cordeweneer there are about 73,000,000 tons of ore yet available.

4. Iron Deposits of Spain (El Pedroso) and the Bukowina.

Spain, also, possesses important iron ore deposits in crystalline schists, which recently have been worked on a large scale. At El Pedroso,[1] on the south slope of the Sierra Morena, famous for its mines, there is in the mining field of Juanteniente, northeast of Seville, a hematite deposit 4 to 5 meters thick, interbedded with upturned mica schist; it has been followed over 600 meters. Near Navalazaro another magnetite deposit 19 to 26 ft. (6 to 8 meters) thick, containing some granite and pistazite, has been disclosed in the midst of the gneiss. Further discoveries have been made lately in this region by means of magnetic prospecting.

A brief mention should be made of the magnetic iron ore deposits in the mica schist area of southern Bukowina (Rusaja mine and others)[2].

5. Archean (Laurentian) Magnetic Iron Deposits of North America.

Deposits of magnetic iron are found in various places in North America in Archean gneisses and crystalline limestones. The most important are

[1] F. Römer: 'Ueber die Eisenerzlagerstätten von El Pedroso in der Provinz Sevilla.' Z. d. D. G. G., 1875, pp. 63-69.

[2] B. v. Cotta: 'Lehre vom den Lagerstätten der Erze,' II, p. 260. B. Walther: Jahrb. d. k. k. geol. Reichsanst., 1876, pp. 391, 415.

found in the hills of southeastern New York and northern New Jersey (Tilly Foster mines, Forest of Dean mine), and in western North Carolina (Cranberry).[1]

6. *Pre-Cambrian (Algonkian) Iron Ore Deposits of North America.*

Marquette Iron Ore District in the State of Michigan.

Deposits belonging only in part to the class of syngenetic stratified deposits, probably more properly to the metasomatic formations, are those of the Marquette district, in the State of Michigan, which have recently been the subject of painstaking study. According to Van Hise and Bayley[2] the geologic conditions there are as follows:

This district is about 29 miles (65 km.) long and 1.2 to 3 miles (2 to 5 km.) wide, extending along the south shore of Lake Superior between Mar-

Fig. 42.—Profile through the area south of Negaunee. (Van Hise.)

gn, gneiss; g. granite; q, quartzite; s, clay slate; e, iron-bearing Negaunee strata; d, diorite and diabase.

quette and Michigamme. The oldest bedrock consists of Archean mica schists, hornblende schists, gneisses, granite gneisses and intrusive granite masses. Above these, as shown in profile (Fig. 42) come the unconformable Algonkian (Huronian) formations, the lower and the upper Marquette. The first consists of quartzites, dolomites, schists and the iron-ore-bearing Negaunee strata; the latter of quartzites, schists, graywackes and conglomerates, in part marked by strong regional metamorphism as well as basic eruptive sheets

The iron-bearing Negaunee formation, 984 to 1,476 feet (300 to 450 m.) thick, is in its lower level, where it is found unchanged, mainly formed of finely banded sideritic slate, consisting of alternating lamellae of spathic iron ore and quartz, and containing about 30 to 40% of ferrous oxide. These slates, however, are very commonly altered in the lower levels by metasomatic processes into grunerite-magnetite slate. These are composed

[1] J. Kemp: 'Ore Deposits of the United States,' 1900, p. 166. H. B. Nitze: 'Iron Ores of North Carolina,' N. Car. Geol. Surv., *Bull.* 1, 1893. Arthur Keith: Cranberry folio, Geologic Atlas of the U. S., folio No. 59. Washington, 1903.

[2] Van Hise and Bayley: 'The Marquette Iron-Bearing District of Michigan.' With Atlas. Washington, 1897, U. S. Geol. Survey, *Monograph* XXVIII. J. E. Jopling: 'The Marquette Range, its discovery, development and resources.' *Trans.* Amer. Inst. Min. Eng., Vol. XXVII, 1898, pp. 541-555.

of thin layers of quartz, magnetite and grunerite, the almost pure ferric silicate in the form of hornblende. In many cases they also carry a garnet of secondary origin. These metasomatic processes are supposed, by the above-named authors, to be connected with the intrusion of the diabase characteristic of the entire Negaunee group. These diabases have been intruded in vast stocks and penetrate through the ore-bearing strata in numerous dikes.

Subsequent alkaline solutions are supposed to have passed through the sideritic slates and occasioned the exchange of the iron carbonate with alkaline silicates. At the higher levels, on the contrary, the sideritic slates by recrystallization under the influence of atmospheric seepage water, entering the deeper strata with its charge of oxygen, have been formed into hand-

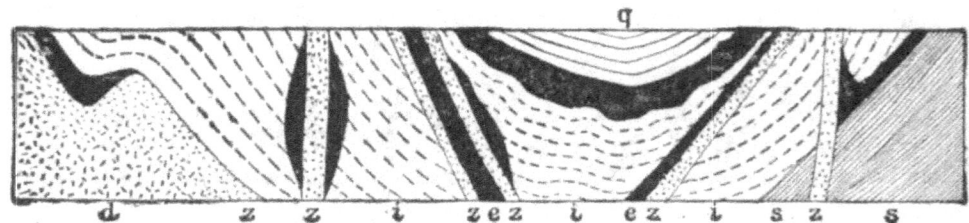

Fig. 43.—Diagram showing the occurrence of the iron ores of the Marquette district. (Van Hise.)

s, schist; i, Jasper schist; q, quartzite; d, diabase and diorite; z, decomposed diabase and diorite; e, orebodies.

somely banded hematite-limonite-quartzites. In these, strata of specular ore and hematite are seen to alternate with bands of ferruginous quartz or jasper, which in many areas have been subjected to great pressure and curiously folded or often comminuted into a breccia. Crystalline hematite has at the same time penetrated into the cracks between the individual jasper fragments. From the chemical mobility of the iron compounds of the Negaunee strata the authors infer the origin of the orebodies forming the actual object of the mining industry. These orebodies consist partly of distinctly granular-crystalline to dense hematite mixed with magnetite (hard ores) partly of red hematite, with transitions into brown hematite (soft ores).

Though the orebodies in general are associated with a definite stratigraphic horizon, yet their position in detail, as appears by investigation, is dependent on structural conditions. The ore masses are always found in synclines of folded, impermeable rocks, or in troughs formed of such strata, or in basins formed by impermeable chloritic and talcose eruptive rocks, as shown in the accompanying section (Fig. 43). These solid ore masses, therefore, cannot have been formed by direct sedimentation, although they may have originated genetically from the undoubtedly sedimentary sideritic slates. Accordingly they are supposed to be rather enrichment formations deposited by iron-bearing water penetrating downward, which received their charge from the upper strata greatly disintegrated by the folding process.

They would thus constitute an excellent example of ore deposits which have to be explained by the aid of the descension theory (see later). The greatest objection to this theory is its failure to show how the silicic acid was removed from the quartzose rock masses in order to leave the spaces free for the invading iron ore, thus finally forming compact orebodies. The authors of the monograph ascribe this action to alkaline compounds derived from the diabasic rocks.

Other deposits similar in character and genesis occur in the Menominee district, on the boundary between Michigan and Wisconsin, as well as of Penokee-Gogebic district in the latter State, and the Vermilion and Mesabi districts north of Lake Superior in Minnesota[1].

In 1903 the annual production of all the iron ore fields in the vicinity of Lake Superior was 26,573,271 long tons. The composition of the ores is shown by the following analyses, the material being dried at 100° C.:

	Marquette Range. (Barnum)	Menominee. (Appleton)	Gogebic. (Anvil)	Vermilion. (Chandler)	Mesabi Range. (Adams)
Iron	65.30	63.30	62.74	64.70	64.18
Silicic acid	3.49	4.61	4.09	4.26	2.80
Phosphorus	0.075	0.018	0.055	0.036	0.035
Manganese	0.36	0.27	0.82	0.13	0.40
Alumina	1.79	1.30	1.10	1.37	0.80
Lime	0.33	0.52	0.47	0.33	0.21
Magnesia	0.26	0.47	0.11	0.10	0.10
Sulphur	0.026	0.019	0.018	Trace.	0.007

The subject of the origin of the Lake Superior iron ores has been investigated by a number of geologists during a long term of years. Prof. R. D. Irving at first adopted the theory that the ores were due to the replacement of the limestone, but he afterwards abandoned it, considering that the original rock must have been an iron carbonate and not a carbonate of lime. For the ores of the Vermilion Range of Minnesota, N. H. and H. V. Winchell propounded the explanation that they were due to direct precipitation from the waters of a hot primordial ocean. Alternating conditions produced from these waters alternating precipitations of iron and silica, forming the characteristic banded ores.

[1] H. V. Winchell: 'The Lake Superior Iron Ore Region,' *Inst.* of Mining Engineers (Eng.), 1897. Irving and Van Hise: *Monograph* XIX, U. S. Geol. Survey, 1892.

It was held by J. D. Whitney as early as 1854 that the iron-bearing rocks of the Lake Superior region were volcanic rocks, erupted in practically their present condition, and this view was held by subsequent geologists. To this supposed volcanic rock the name jaspillite was given. In their monograph on the Penokee-Gogebic iron range, Irving and Van Hise came to the conclusion that the iron was originally precipitated as carbonate or oxide, and that the accompanying chert was simultaneously precipitated. It is supposed that the atmosphere was more highly charged with carbonic oxide than at present, and that the general temperature of the earth's crust was somewhat higher. J. E. Spurr[1] studied in 1893 the iron ores of the Mesabi Range in Minnesota and found the rocks substantially like those of the Penokee-Gogebic Range. However, the iron carbonate, which, according to the researches of Irving and Van Hise was the oldest type that could be found, and was therefore held to represent nearly the condition of the original rock, was found on the Mesabi to be in nearly all cases of undoubtedly secondary origin and to have been formed by the carbonization of iron oxides. It was found, moreover, that all of the constituents of the rocks, including the iron carbonate and oxides and the silica, could be shown to have been derived from an original green ferrous silicate which Spurr classed as glauconite, although the analysis showed very little potash. This explanation of the origin of the Mesabi ores has been accepted generally. Recently the work of Leith on the Mesabi Range[2] confirms Spurr's observations concerning the nature of the original mineral, its manner of formation and the manner in which iron carbonates and oxides are derived from it. On account of the scarcity or lack of potash, the writer objects to the name glauconite for the original ferrous silicate and coins a new name, greenalite. Spurr[3], however, after a review of the evidence, is still in favor of retaining the name glauconite for the Mesabi mineral, although acknowledging, as before, that it constitutes a somewhat unusual variety.

7. *African Iron Ore Deposits in Crystalline Schists.*

Among these the best known are the numerous iron ore deposits which France possesses in Algiers. The most important of them have hitherto been those of Mokta-El-Hadid (or Ain Mokra) in the Department of Constantine, at the foot of the south slope of the Coast Range, between Cape de Fer and Bône, near Lake Fezzara.

The ores consist of magnetite and red iron ore, and occur interbedded with limestones, forming part of a series of garnetiferous-mica schists, in-

---

[1] *Bull.* Minn. Geol. Surv., No. 10, 'The Mesabi Iron-Bearing Rocks.'

[2] *Monograph* XLIII, U. S. Geol. Surv.

[3] *Amer. Geologist,* June, 1902.

tercalated in crystalline gneisses. The main ore deposit has a thickness of 40 meters (131 ft.). The production in some years rose to 430,000 tons[1] (1874). Other important occurrences have recently been exploited.

Mention may also be made of the red hematite intercalated in the quartzite of the crystalline schists in the German possession of Togo at Banyeri, Kabu and Basari.[2]

(c) Origin of the Iron Ore Deposits of Crystalline Schists.

To the remarkable works of Hj. Sjögren and J. H. L. Vogt[3], on which the following remarks are in the main based, we owe some deeper insight into the genesis of the iron ores of the crystalline schists. It cannot be said, however, that the question of their genesis has as yet been conclusively answered.

The theory of a direct sedimentary origin of the magnetite and hematite ore deposits is sustained, according to Vogt, by the following facts:

1. Their complete conformity with the country rock.
2. Their very pronounced stratification.
3. Their association with definite stratigraphic horizons.
4. Their frequent association with limestone.
5. The occurrence of chemically similar formations in younger nonmetamorphic formations.

Sjögren is inclined to regard the older Swedish iron ore deposits as lacustrine or terrestrial, at most as littoral, which he supposes to have been segregated from very dilute iron solutions in the manner of bog ores and lake ores (see p. 101), that is, with the aid of vegetable material. In point of fact this hypothesis is supported by the occurrence in many deposits of substances probably of vegetable origin, such as petroleum, asphalt, bitumen and anthracite. The stock-like or lenticular form of many old iron ore deposits is also readily explained if we imagine that they were deposited on a somewhat uneven surface under lacustrine conditions. On the other hand, it must be remarked that we know also of younger genuine marine iron ore deposits even containing mineralized ammonites and sea urchins, such as those of the Brown Jura and the Eocene. While this is true in the individual cases, it must be assumed that the iron ores were primarily deposited at the bottom of some body of water from highly dilute solutions. The formation of such solutions is to be explained by the decomposition of the finely interspersed iron ores and ferrous silicates contained in so many

---

[1] E. Fuchs and L. de Launay: 'Traité des Gîtes Minéraux,' 1893, I, p. 721.

[2] Fr. Hupfeld: 'Die Eisenindustrie in Togo.' *Mitth.* a. d. deutschen Schutzgebieten, Vol. XII, 1899, part 4, pp. 175-193.

[3] Hj. Sjögren: 'Om de svenska jernmalmlarerens genesis,' *Geol.* Fören i. Stockholm, Förh. 13, 1891, p. 373. J. H. L. Vogt: 'Salten och Ranen,' Christiania, 1891.

older rocks, through the agency of terrestrial water charged with carbonic acid, sulphuric acid or organic acids. (On this mode of precipitation of iron, see p. 102). The decomposition undoubtedly must have originally taken place in many cases in the form of iron carbonate. This was in most cases at once further oxidized into ferric hydrate. Only with an abundance of reducing organic substances could the carbonate remain as such; in most cases ferric hydrate was at once formed. The silicic acid in the quartzose iron ores may in part be derived from alkaline iron solutions. From such a solution the silicic acid is precipitated simultaneously with the segregation of ferric hydrate. The hematite ores are in general rich in silicic acid, while the magnetic iron ores are poor in that respect. The latter often carry sprinklings of sulphur ores. These facts point to reducing processes in the formation of magnetic iron ores, by which the iron was probably precipitated from carbonate solutions as carbonate, or from humic acid solutions at first as iron crenate, while during the formation of the hematite ores rich in silicic acid these reductions did not take place.

The phosphoric acid content is derived originally from the apatite of th: rocks. Phosphoric acid is accumulated by plants; when they decay it passes into solution as ammonium phosphate. During the sedimentation of the iron ores the phosphoric acid was also precipitated as iron or lime phosphate.

Manganese is precipitated from solutions in the same way as iron, but since the iron is more rapidly oxidized, the manganese is precipitated later when the solution contains less carbonic acid. In this way we may explain the fact that manganese ores are ordinarily found above the iron ores and associated with limy ores, often connected with limestone and dolomite.

It was only by means of regional metamorphism, more particularly the factors which are probably most active in that process, pressure, moisture and heat, that the final alteration of the iron carbonate and ferric hydrate into magnetite and hematite was effected. Not only the ore beds, but also the country rock, undergo a complete recrystallization in this process (see remarks on p. 71 on Grängesberg).

(B) Iron Ores as Original Intercalations in Normal Sediments.

(a) SILURIAN IRON ORES.

1. *The Iron Ore Deposits in the Lower Silurian of Central Bohemia.*[1]

The Bohemian iron ores are intercalated in Lower Silurian strata (Bar-

---

[1] M. V. Lipold: 'Die Eisensteinlager der silurischen Grauwackenformation in Böhmen.' *Jahrb. d. k. k. geol. Reichsanst.*, Vol. XIII, 1863, pp. 339-448. With 40 Fig. Jos. Vala and R. Helmhacker: 'Das Eisensteinvorkommen in der Gegend von Prag und Beraun.' Prage, 1873. C. Feistmantel: 'Die Eisensteine in der

rande's stage D) made up in the main of quartzites, associated with slates, graywackes, conglomerates, diabases, amygdaloids and diabase tuffs.

The ores occur within these stages at different horizons:

The lowest horizon, D_1, consisting of conglomerates, graywackes, graywacke slates, diabase tuffs and tuff slates, contains the iron ore deposits of the Sárka, Svárov, Libecov and Chynava. Two vast beds of diabase tuffs and tuffaceous slate, separated and overlain by gritty slate, are here the ore-bearing rock proper. This bed here consists of oolitic red hematite up to 5 meters in thickness, but is elsewhere a poor blackish gray oolitic chamosite (a water-bearing aluminous iron silicate with slight magnesia content). These chamosite beds attain a thickness of up to 20 meters. The several ore beds in the tuff slates form extended lenses or short beds which either wedge out or pass along the strike into tuff slate poor in iron.

An overlying horizon, D_4, consisting essentially of an alternation of graywacke slates and quartzites, encloses the ores of Jinocan, Nucic, Chrustenic and Vraz.

The most important is the ore bed of Nucic west of Prague.

The Nucic ore possesses a pronounced oolitic structure. It consists of a groundmass of light brownish gray spathic iron ore, or more frequently of dark gray chamosite, in which the ooliths are interspersed as concentric-shelled ellipsoids of chamosite. The size of these small concretions ordinarily varies between 1 and 2 millimeters in diameter. A very firm and brittle ore called sklenenka (glass ore) also occurring there is especially rich in ironspar and calcspar, which completely impregnate the groundmass of chamosite.

Owing to the high percentage of phosphorus, most of the ores are treated by the Thomas process.

The bed attains its greatest thickness of 16 in. at Nucic itself; the length in the strike has been ascertained to be 15 kilometers. It shows distinct stratification. Overlying and underlying it are graywacke slates, farther on also is quartzite with lower Silurian fossils. The ore bed itself has furnished some fossils, among them *Trinucleus Ornatus* Barr., *Asaphus Nobilis* Barr., *Orthis Macrostoma* Barr. At times the bed has been influenced and bleached by decomposing solutions. In the outcrop to a depth of 6 to 12 m. (19 to 38 ft.) the ore has been transformed into brown hematite.

The Nucic beds especially, which are either flat or but gently inclined, are cut obliquely by numerous faults. At the present time the greatest output is from the open cuts and underground workings of the two companies

Ftage D des böhmischen Silurgebirges.' *Abh.* d. k. Böhm. Ges. d. Wiss. VI, Vol. VIII, 1875 and 1876. Also, 'Ueber die Lagerungsverhältnisse der Eisensteine,' etc. *Sitzungsber.* d. k. Bohm. Ges. d. Wiss., 1878, pp. 120-132.

operating the Nucic mines. The Zdic mines are the only others of importance. In 1898 Bohemia produced a total of 632,183.7 tons of iron ore. In 1899 the Prager Eisenindustrie at Nucic produced 296,734 tons; the Montangesellschaft 344,718 tons of ore, with an average iron content of 37.25%.

2. *The Lower Silurian Iron Ores of the Thüringerwald and Vicinity.*

Oolitic iron ores similar to those of Bohemia[1] occur in the lowest horizon of the lower Silurian in the Thüringerwald and the Frankenwald, as well as in the Vogtland. In recent years the deposit at Schmiedefeld (Sachsen-Meiningen) near Gräfenthal in Thüringia has become especially important.

The petrographic character of the Schmiedefeld iron ore has been studied by H. Loretz. According to this author it consists of thuringite and chamosite. The thuringite is a compact olive-green mineral with a finely foliated

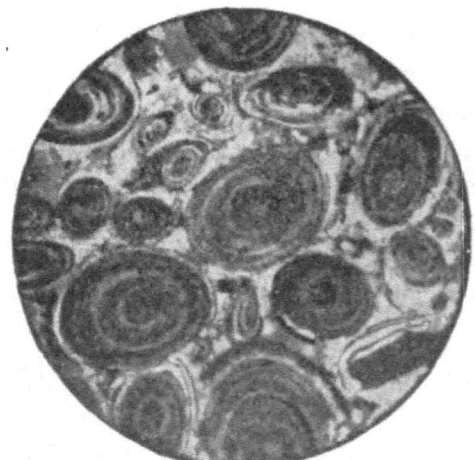

Fig. 44.—Thin section through a chamosite oolite from Schmiedefeld. (Enlarged fifty times.)

crystalline texture, which on weathering often becomes oolitic. Its average chemical composition as given by Loretz is as follows:

SiO$_2$	22.61	per cent.
Al$_2$O$_3$	16.80	"
Fe$_2$O$_3$	15.43	"
FeO	33.10	"
MgO	1.20	"
H$_2$O	10.60	"
Total	99.74	"

[1] C. W. von Gümbel: 'Fichtelgebirge,' 1879, pp. 235, 236, 420-428. Th. Liebe: 'Uebersicht über den Schichtenbau Ostthüringens,' *Abh. z. geol. Spezialk. v. Preussen.* Vol. V, part 4, 1884. H. Loretz: 'Zur Kenntniss der untersilurischen Eisensteine im Thüringer Walde,' *Jahrb.* d. k. preuss. Landesanst, 1884, pp. 120-147.

The thuringite often forms small thin layers or oolitic granules in an ordinary soft clay slate, in which case it is called thuringite slate. The chamosite, which is far more abundant than the thuringite, consists of a dark gray compact aggregate of small concentric pellets, of about the size of millet, lying close together and cemented by siderite.

We give the figure of a thin section through a typical chamosite oolite of Schmiedefeld. (Figure 44.)

The chemical composition of typical chamosite ore from Schmiedefeld (W. Böttcher) is given under (a), while column (b) shows the calculated theoretical composition of true chamosite, as calculated by H. Loretz:

(a)		(b)	
SiO_2	18.63%	SiO_2	29%
Al_2O_3	8.48	Al_2O_3	13
Fe_2O_3	3.73	Fe_2O_3	6
FeO	45.13	FeO	42
MgO	1.68	H_2O	10
CaO	0.84		
P_2O_5	0.44	Total	100
CO_2	13.00		
TiO_2	1.63		
H_2O	6.44		
Total	100.00		

The structural conditions as outlined by Loretz are as follows: The Cambrian rocks, near Schmiedefeld, are overlaid by quartzite similar to

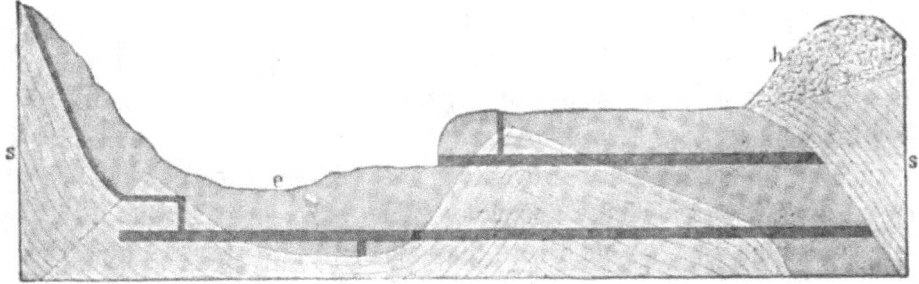

Fig. 45.—Cross-section through the main bed at Schmiedefeld.
s, clay slate; e, iron ore; h, dump.

the phycod slates and by Griffel slates, these beds forming transition strata between Cambrian and lower Silurian beds. These are overlaid by iron-stone beds, followed by typical lower Silurian slates. The main iron ore bed is sometimes as much as 18 meters (59 ft.) thick and is now worked in the Maxhütte at Unterwellenborn. Its position is shown in the cross-section (Fig. 45) kindly furnished by the mine manager.

The production at Schmiedefeld for 1899 was 140,000 tons of iron ore.

A similar deposit of chamosite ore and thuringite slate occurs northwest of Hof in the valley of the Saale, and in inverted beds in the Steinach Valley of the Franconian forest. The thuringite rocks of Leuchtholz, near Hirchberg, contain numerous casts of *orthis*. (Gümbel.)

The following examples are, because of their structure, classed in the same category as the lower Silurian iron ores of central Europe.

3. *The Clinton Hematite Ores.*

The Clinton[1] iron ore received its name from the town of Clinton, near Utica, New York, where it was extensively mined. The ore is an almost constant accompaniment of the Clinton series of the Upper Silurian throughout its exposure in New York, Pennsylvania, Wisconsin, Virginia, Kentucky, Tennessee and Alabama. The ore forms one or more beds, interbedded with clay slates, sandstones and impure limestones, which are overlain by the Niagara slate and underlain by the Medina sandstone. The ore beds vary greatly in thickness and position. At Clinton three beds are known, the uppermost, 4 to 6 ft. thick, not being workable. The middle bed, 2 ft. thick, is the only one now worked. At Birmingham, Alabama, the ore is rich in fossil remains and occurs in several beds which may reach a total thickness of 9 meters (29.5 ft.).

The Clinton ore is often oolitic, and in that case is then called flaxseed ore. In this ore, from which the calcareous cement has often been leached out, the grains consist of concentric shells of iron oxide and some amorphous silica, about a nucleal grain of quartz.

Some of the deposits often consist of an aggregate of countless fragments of various calcareous marine organisms, such as bryozoa, corals, crinoids, brachiopods, etc. (fossil ore) all more or less replaced by iron oxide, and cemented by carbonate of lime. The soft ores are mostly red in color, of earthy appearance and loose texture. The hard ore is a highly ferruginous limestone. The hard ore has been found to average between 35 and 40% of iron with 0.5% to 1% of phosphorus, and is seldom a bessemer ore[2]. According to C. H. Smyth and others, these hematite beds are original deposits in an inland sea that received the drainage of an extensive area of crystalline rocks. The ferruginous and somewhat siliceous waters of this shallow body of water deposited the iron about nucleal grains of sand and replaced the calcium carbonate of the fossil fragments before the sediment forming the overlying strata was deposited. The alternative explanation of these red hematite beds by a subsequent replacement of beds of

---

[1] C. H. Smyth, Jr: 'On the Clinton Iron Ore.' *Am. Jour. Sci.*, June, 1892, p. 487. *Zeit. f. Prak. Geol.*, 1894, p. 304.

[2] W. B. Phillips: 'Iron Making in Alabama,' 1898, Alabama Geol. Survey.

limestone by iron-bearing seepage waters seems to us to have been completely disproved by Smyth.

The oolitic hematites of Belle Island, near Newfoundland, seem to belong to the Clinton group. They form two beds intercalated in slates and sandstones, one 3 m. (9.8 ft.) and the other 1.8 m. (5.9 ft.) thick, the ore fracturing in remarkably regular parallelopipeds.

The precipitation of amorphous silica and the formation of iron oolite is a very common phenomenon, as has been proved by Ch. Bleicher's[1] investigations upon samples of the most diverse origin.

(b) Ironstone Deposits of the Carboniferous Rocks.

Wherever the carboniferous rocks contain coal seams, there are almost always beds of iron ore of more or less economic importance. One of the most carefully investigated European examples is the following:

1. *The Iron Ores of the Ruhr Coal Basin.*

The iron ores of the Ruhr coal basin[2] are associated with the lowest and leanest coal measure of that locality. Baumler distinguishes the following occurrence:

I. A Spathic Iron Ore Measure.

This bed, which is 0.24 m. (0.7 ft.) to 0.48 m. (1.5 ft.), more rarely 1.4 m. (4.6 ft.) thick, has been traced for a distance of several miles, and has here and there been worked. It is not, however, a single bed throughout, but consists of thin lenticular beds as much as 1 km. (0.6 mile) in extent, with intervening barren layers. This ore is a yellowish gray granular crystalline spathic ironstone ore without lamination or cleavage structure and carrying 45.66% of iron (65.3% when roasted) with some carbonaceous material. Underlying the iron ore measures there is usually a layer of coal (Kohlenpacken) varying up to 30 cm. (1 ft.) thick, while a foot above the barren overlying rock often contains a layer of spherosiderite concretions. The spathic iron ore horizon lies 80 to 100 m. (260 to 328 ft.) below the Mausegatt, the leading bed of the lean measures. Of course the ore bed

---

[1] Ch. Bleicher: 'Sur la structure mikroscopique du mineral de fer de Lorraine.' *Compte Rendu*, 114, p. 590.

[2] R. Peters: 'Der Spathcisenstein der westfälischen Steinkohlenformation.' Z. d. V. Deutsch. Ing. I. 1857, p. 155. Baumler: 'Ueber das Vorkommen der Eisensteine im Westfälischen Steinkohlengebirge. Z. f. B. H. u. S.-Wesen im preuss. St., 1868, Vol. XVII, p. 426. W. Runge: 'Das Ruhr-Steinkohlenbecken.' Berlin. 1892, pp. 70-73.

participates in all the flexures and faults to which the entire formation has been subjected.

II. Clay Ironstone Measures.

These ores accompany the coal beds, especially in the lowest measures, lying either above or below or forming partings, which may replace the coal seam altogether. This clay ironstone is very rich in carbon, and contains as high as 39% of iron (up to 60% roasted). In the fifties and sixties large quantities of this black band ore were produced. About fourteen different layers, varying up to a meter in thickness, have yielded workable ores in the past. The ores are sometimes so rich in phosphorus that they are used for the production of superphosphates.

III. Spherosiderites.

Spherosiderite concretions containing up to 45% of iron are not infrequently found in the clay slates and coal seams of the lower measures, and sometimes unite and form extensive beds of clay iron ore.

The production of iron ores in the Ruhr coal district reached, according to Baumler, 167,609 tons in 1890.

2. *Carbonaceous Iron Ores of Upper Silesia and Saxony.*

Workable ore beds also occur in the coal basin of upper Silesia.[1] Spherosiderites within the clay slates are especially common, and in rare cases, as in the upper coal bed of the Saara measure, and at Czernitz, bituminous iron ores (black band ores) have also been developed. The spherosiderites form large lenticular and nodular masses, which may extend along the strike for a few meters. Fig. 46 illustrates this occurrence.

The largest production in the sixties took place at Antonienhutte, Friedenshütte and Ruda, at Zalenze, Janow, Orzesze, Dubensko and Ornontowitz.

At many points, as at Zalenze, it has been noted that the spherosiderites occur especially above the flat depressions in the coal measures.

In 1900 upper Silesia produced 7,147 tons of carbonaceous iron ore. In the Zwickau area district[2] of Saxony spherosiderites and carbonaceous iron ore were for a while produced along with the coal.

Deposits of the first named ore were mined from especial pits in the region

[1] F. Römer: 'Geologie von Oberschlesien.' 1870. Appendix by W. Runge, p. 533.

[2] H. Mietzsch: 'Erläuterungen zu Section Zwickau der geol. Spez.-K.,' 1877. p. 11.

of the Russkohlen measures and also from the Segen Gottes measures (lower reef) within the clay slates. Carbonaceous iron ore, found, for example, in the lower part of the measure at the Bahnhof shaft at Zwickau, formed a stratum 0.2 m. (0.6 ft.) in thickness, containing veins of zincblende, but possessing little extent.

At the coal mines of Kastner & Co., in the Reinsdorff, near Zwickau, a spathic iron ore deposit up to 0.8 m. (2.6 ft.) thick has recently[1] been discovered under the earthy coal measures. The spherosiderite found between the third and fourth measures of the Hilfe Gottes shaft, near Zwickau, contained the rare mineral *whewellite,* besides zincblende and iron pyrite in its fissures.

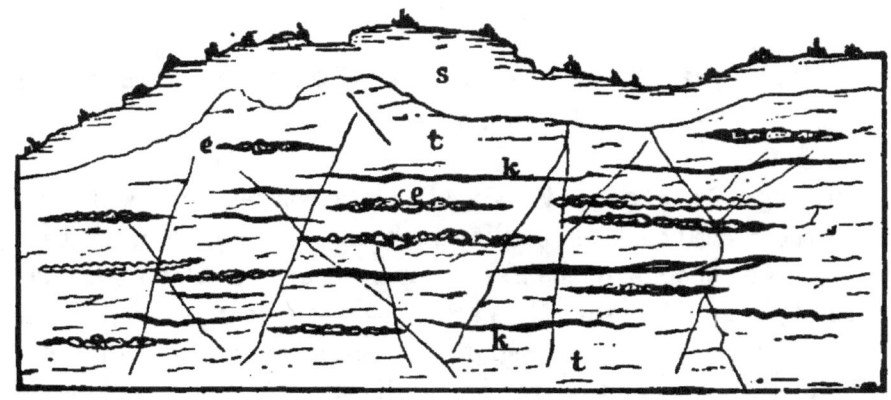

Fig. 46.—Section through the carbonaceous clay slate at Janow.   (F. Römer.)
s, sand; t, clay slate; e, spherosiderite lenses; k, coal pockets and carbonized sigillaria trunks.

Spathic iron ore and brown hematite are also found in the Carboniferous formations of the eastern Austrian Alps, as at Turrach, southwest of Murau, where 5,327.3 tons of brown hematite were produced in 1898.

3. *Carbonaceous Iron Ores of Great Britain.*

Great Britain is rich in deposits of this kind. In southern Wales[2] most of the ironstones are found in the lower divisions of the productive coal measures. Many of the numerous blackband beds of that region seem to be continuous throughout the entire district, as, for example, the Three-quarter Balls measure. The main production in Wales comes from Ebbw Vale,

[1] *Jahrb.* f. B. u. H. in Sachsen, 1899, p. 141.
[2] Phillips-Louis. 'Ore Deposits.' 1896, pp. 285, 323.

Blaenafon, Pontypool, Abercarn and Dowlais. The ore contains 21 to 38% of iron. The production of South Wales in 1880 was 170,000 tons. Important deposits of blackband ironstone also occur in Scotland. The iron ore measures of that country occur, according to H. Louis, in both the upper and the lower Carboniferous formations. The upper Carboniferous contains seven beds of this ore, one of which reaches a thickness of 0.9 m. (2.9 ft.). The lower Carboniferous holds three iron ore beds 0.3 m. thick. These blackband ores contain such an abundance of carbonaceous matter that they burn in the roasting furnaces without the addition of fuel, the residue containing 50 to 70% of iron. Besides these blackband ores, clay ironstone occurs at Bauton, Denny and elsewhere. In 1881 Scotland produced 1,402,-700 tons of blackband ore and 1,192,675 tons of clay ironstone. In 1894, however, the total output had dropped to 631,304 tons.

(c) The Iron Ore Formations of the Northern Alps (Probably Permian).

Iron ore deposits occur in the so-called graywacke[1] zone, which lies between the central range of crystalline rocks and the limestone peaks of the Tyrolean Alps on the north, the Salzkammergut of Styria and Austria. Recently, through the labors of M. Vacek, the geologic age and structure of the rocks of this complicated region have been determined, and the true stratigraphic position of these iron ores, hitherto in dispute, has been clearly established. In the area between Enns and Mur the iron ores consist of spathic ironstones and brown hematites with associated slaty, conglomeratic and breccia-like rocks, especially limestone breccias with a sericitic cement, whose fragments are derived from the upper Silurian. Ordinarily the iron ores form the lowest beds, resting unconformably on lower Devonian limestone, as at Eisenerz, or sometimes on still older gneisses. The slates and breccias, which are well developed at Admont, but are quite subordinate in the Eisenerz measures, are often unconformably overlain by lower Triassic Werfener slates. The exact age of the formation is still unknown, but is either Permian or between that and upper Devonian.

A continuous series of such iron ore deposits is always found on the unconformity at the base of the lower Triassic formation, from Schwatz and Pillersee in Tyrol, across Dienten, Flachau and Werfen in Salzburg, Lietzen, Admont in the Radmer, Eisenerz, at Feistereck, on the Veitsch and Neu-

[1] "Graywacke is an old name of loose signification, but chiefly applied to metamorphosed shaly sandstones that yield a tough irregularly breaking rock different from slate on the one hand and from quartzite on the other."—J. F. Kemp, 'Handbook of Rocks,' pp. 140.

berg in Styria as far as Reichenau in lower Austria. The strata unquestionably attain their greatest development at Erzberg, near Eisenerz.

The Erzberg Near Eisenerz.

Eisenerz[1] lies on the northwest side of the Prebichel Pass, which connects the valley leading to the Enns near Hieflau with the valley discharging into the Mur at Leoben. The district is in a depression of the high mountain range lying between the lower Tauern on the west and the Hochschwab on the east.

The Erzberg itself is an almost isolated cone 1,537 m. (5,131 ft.) high, whose summit affords a magnificent panorama. The upper part of the cone is formed of an enormous thickening of an iron ore bed, which, according to M. Vacek, rests unconformably on lower Devonian limestone, carrying *Bronteus palifer* Beyr, etc. This unconformity is well exposed in the limestone cliffs, which show the underlying rock projecting into the ore deposit. The accompanying cross-section after M. Vacek (Fig. 47) gives a clear representation of the structural conditions. It shows that the lower Devonian Sauberg limestones underlying the ore beds rest unconformably on the so-called granular graywackes of Eisenerz, rocks which, according to Vacek, are really gneisses. In the northeastern part of the Erzberg, in the Söbberhaggen, the ore lies immediately above this graywacke-like gneiss. It should be noted that in the eastern part of the district, near the Barbara chapel, the Erzberg ore deposit is unconformably overlain by the Werfener slates which dip north. These slates ordinarily begin with a breccia formation, seen in the Peter Tunner adit. In the western part of the Erzberg, however, through which the profile is drawn, these lower Triassic Werfener strata have been removed by denudation, and the ores outcrop.

According to F. v. Hauers the ore bed is sometimes as much as 125 m. (410 ft.) in thickness. A part of the deposit shows, however, a large proportion of ankerite, and all transitions from siderite and ankerite to ordinary limestone, and in several cases the deposits consist entirely of "Rohwande," as the ankerite ore is called.

The ore averages 40% of iron. The spathic iron ore is not entirely free

[1] F. Ritter von Ferro: Innerberger Hauptgewerkschaft. *Tunners mont. Jahrb.*, Vol. III, 1845, p. 197. A. von Schouppe: 'Erzberg bei Eisenerz.' *Jahrb. d k. k. geol. Reichsanst.* 1854, p. 396. A. Miller von Hauenfels: 'Die steiermärkischen Bergbaue,' in 'Ein treues Bild des Herzogth, Steiermark,' Vienna, 1859, p. 16. D. Stur: 'Vorkommen obersilur. Petrefacten am Erzberg.' *Jahrb. d. k. k. geol. Reichsanst.*, 1865, p. 267. F. von Hauer: 'Geologie der österr. Monarchie.' 1875, p. 223. M. Vacek: 'Ueber den geologischen Bau der Centralalpen zwischen Enns und Mur.' *Verh. d. k. k. geol. Reichsanst.*, 1886, p. 71. Also 'Skizze eines geologischen Profiles durch den steierischen Erzberg.' *Jahrb. d. k. k. geol. Reichsanst.*, 1900, Vol. L., p. 23. (The foregoing represent the most important publications about Eisenerz.)

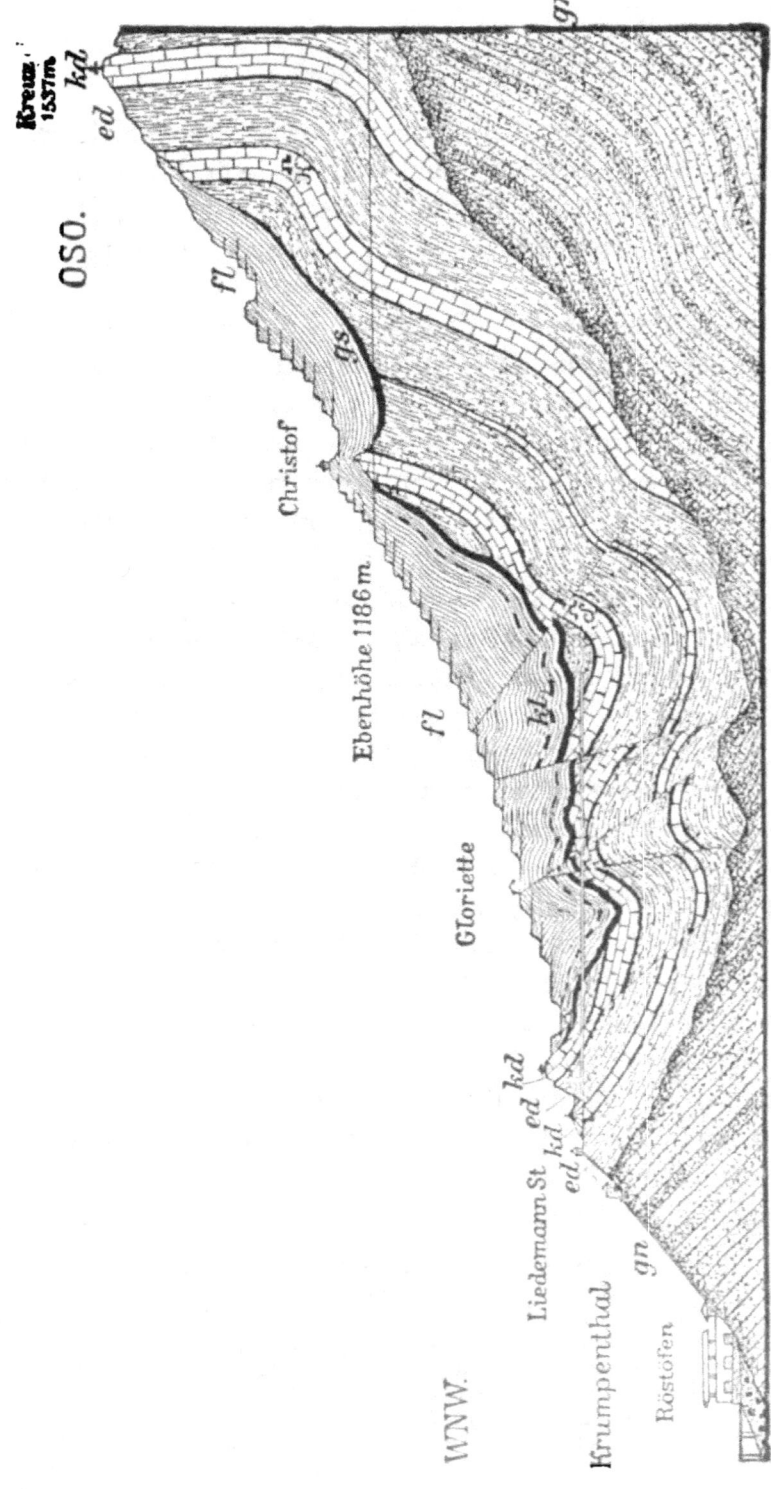

Fig. 47.—Profile through the Erzberg at Eisenerz. (M. Vacek.)

gn, gneissoid rocks ("Eiseners graywacke"); kd, lower Devonian limestone; ed, lower Devonian iron ores and "Rohwande"; fl, "Flinze" spathic iron ore of the Perman (?); kl, impure calcareous layer; gs, basal shale of the Perman (?).

from sulphides, as pyrite and chalcopyrite occur in scattered grains, and rarely it holds small streaks of cinnabar. The phosphorus content hardly reaches 0.01%.

The mining industry of Eisenerz began in Roman time. Documents extending back to the year 712 A. D. are said to have been formerly in existence. At any rate Tacitus and other classic writers mention "Noric iron." The mining is carried on both by open cuts rising in terraces along the slope, and by underground work.

The production in 1898 was as follows:

Innerberger Gewerkschaft, near Eisenerz, 923,454.2 tons of spathic iron ore; Vordernberger Gewerkschaft (Oesterreichische Alpine-Montangesellschaft), 75,038 tons; Veitsch mine, 152.5 tons; Olimie mine, near Windisch-Landsberg, 965 tons.

(d) PERMIAN SPHEROSIDERITES.

Clay ironstone concretions occur in the *Acanthodes* slates of the Lebach formation of the lower Rothliegende at Saarbrucken. These ellipsoidal concretions often enclose organic remains, such as *Archegosaurus* and *Walchia*. They are called "Knopfstrich."

At Goldlauter, near Suhl, in Thuringia, the dark clay slates of the Rothliegende carry nodular concretions consisting of alternate shells of slaty brown spar, arsenopyrite and pyrite, and often containing a central core of copper pyrite, gray copper and native silver.[1]

(e) SEDIMENTARY IRONSTONES OF JURASSIC AGE.

1. *Liassic Iron Ores.*

Germany possesses true ironstone measures in many of the Jurassic formations[2]. The Lias, for example, contains four beds of finely oolitic, usually porous limonite, with an aggregate thickness at Bundheim, near Harzburg, of 4 m. (13.1 ft.) (zone of *Amm. Bucklandi*). The output of the Friederike mine at that place averaged 44% of iron. The ironstone measures of Willershausen, Calefeld and Oldershausen in Hanover also belong to the Lias (zone of *Amm. Jamesoni* and *Amm. Ibex*).

The Middle Lias of England also contains iron ores mined in the Cleveland district. The principal bed extends over an area of 350 square miles,

[1] B. von Cotta: 'Erzlagerstätten,' II, 1861, p. 72.
[2] J. Haniel: 'Ueber das Auftreten und die Verbreitung des Eisensteins in den Jura-Ablagerungen Deutschlands.' *Z. d. Deutsch. geol. Ges.*, Vol. XXVI, 1874, pp. 59–118.

and is workable for one-fifth of this area[1]. The ores average 30% of iron and 1.5% of phosphoric acid.

2. *Iron Ores of the Dogger Formation of the Jura.*

The most widely distributed iron ore beds of Jurassic age are those of the Dogger formation. The Minette ore beds belong to this formation.

Oolitic Iron Ores (Minettes) of Luxemberg and Lorraine[2].

The Minette beds belong to the lower Dogger formation, which is characterized by the fossils *Harpoceras Murchisonae* and *Harpoceras Opalinum*. The accompanying profile (Fig. 48), after W. Branco, gives the detailed classification of the various beds of the Minette horizon.

The Minette measures occur mainly throughout the frontier region between Luxemberg, German Lorraine and French Lorraine, in a strip of 100 km. (60 miles) long, 18 km. (10.8 miles) wide, west of the Moselle, the portion belonging to Germany being 60 km. (36 miles) long and 12 km. (7.2 miles) wide.

The most important mines of Luxemberg are at Beles, Esch and Rümlingen; of Lorraine (German) Ottingen, Tillots, Neufchef, Hayingen, Moyeuvre, Rosslingen, Maringen, Vaux, Chabonniere, Varraines, Novéant and Arry; in France, Longwy and Briey. This plateau-like area is traversed by long cliffs running north and south, which show at the foot a gently sloping area of Liassic beds, and at the summit an abrupt slope cut in oolitic limestone (Dogger). At the boundary between the gentle slope and the steep escarpment the outcrops of the Minette measures are usually found. In Luxemberg, the valley of the Alzette divides the formation into two flat basins. The different beds do not always extend uniformly throughout both basins, but show the distribution indicated in the diagram (Fig. 49, after Van Werveke), which also applies to other areas.

[1] J. D. Kendall: 'The Iron Ores of Great Britain,' 1893. Cited by Phillips-Louis, 'Ore Deposits,' p. 44.

[2] E. Giesler: 'Das oolithische Eisensteinvorkommen in Deutsch-Lothringen.' *Z. f. B. H. u. S.-Wesen im preuss. St.* Vol. XXIII, 1875, pp. 9–41. W. Branco: 'Der untere Dogger Deutsch-Lothringens.' Strassburg, 1879. Van Werveke: 'Erläuterungen zur geologischen Uebersichtskarte des westlichen Deutsch-Lothringen.' Strassburg, 1887, pp. 83–99. L. Hoffmann: *Stahl u. Eisen*, 1896, Nos. 23 and 24. Schrödter: *Zeit. f. Prak. Geol.*, 1897, p. 296. W. Kohlmann: 'Die Minetteformation nördlich der Fentsch.' *Stahl u. Eisen*, July, 1898. *Zeit. f. Prak. Geol.*, 1898, p. 363. 'Uebersichts-karte der Eisenerzfelder des westlichen Deutsch-Lothringen.' 3d edition, Strassburg, 1899. F. Greven: 'Das Vorkom. der Oolith Eisenerz im südlichen Deutsch-Lothringen,' *Stahl u. Eisen*, 1898, No. 1. H. Ansel: 'Die Oolith. Eisenerzform. Deutsch-Lothringens.' *Zeit. f. Prak. Geol.*, 1901, pp. 81–94. L. von Werveke: 'Ueber die Zusammensetzung und Entstehung der Minetten.' Reviewed *Zeit. f. Prak. Geol.*, 1901, pp. 396–403.

In German Lorraine the following strata are distinguished from below upward:

The *black* bed, the most extensive, but so siliceous that it is only mined

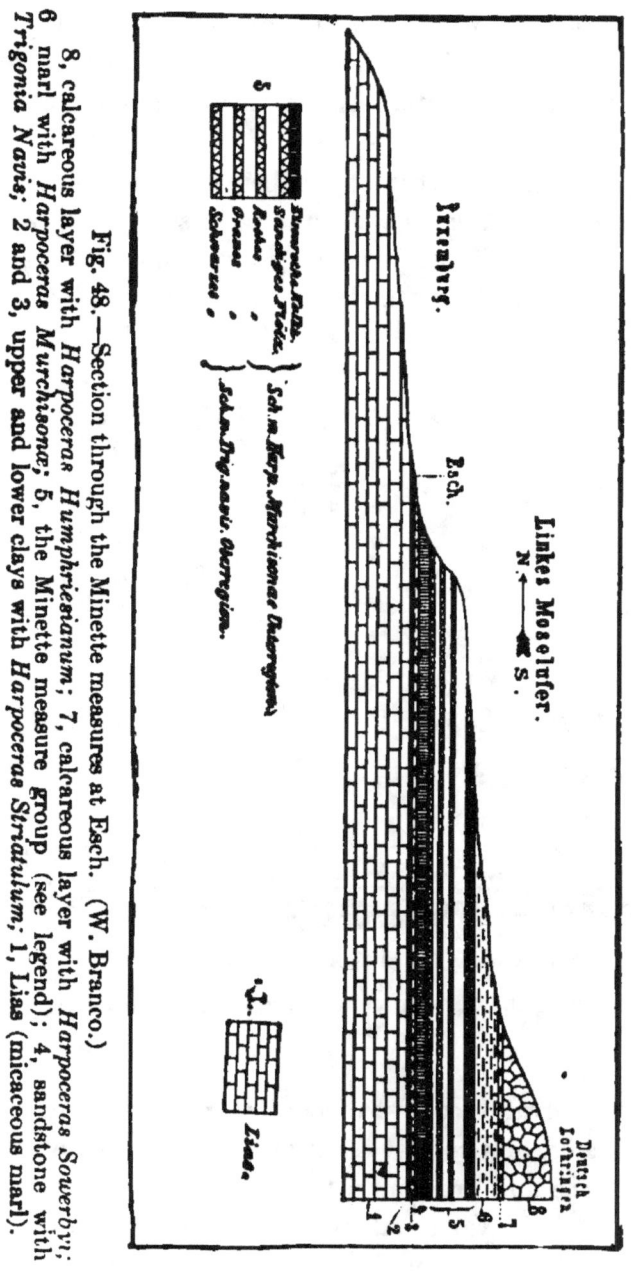

Fig. 48.—Section through the Minette measures at Esch. (W. Branco.) 8, calcareous layer with *Harpoceras Humphriesianum*; 7, calcareous layer with *Harpoceras Sowerbyi*; 6, marl with *Harpoceras Murchisonæ*; 5, the Minette measure group (see legend); 4, sandstone with *Trigonia Navis*; 2 and 3, upper and lower clays with *Harpoceras Striatulum*; 1, Lias (micaceous marl).

for a flux for the calcareous ores; 1.2 to 3.9 m. (4 ft. to 13 ft.) thick.

The *brown* bed, whose ore is also siliceous, is separated from the preceding by a marly parting 0 to 4 m. (0 to 13.1 ft.) thick.

The *gray* bed, whose nearly uniform thickness of 5 to 6 m. (16.4 ft. to 19.6 ft.) and excellent quality make it the most important, is accompanied by a yellow bed and carries as fossils *Gryphaera ferruginea, Amm. radians,* etc.

The *red limy bed,* the most valuable of the Luxemberg beds, has a maximum thickness of 8 m. (26.2 ft.), though seldom workable in German Lorraine; its lower layers contain *Amm. Murchisonae* and *Pholadomya reticulata.*

The *red sandy bed,* the thickest of all, being sometimes 13 m. (42 ft.) through; but it contains a great deal of quartz in sandy grains.

The following table, by H. Ansel, gives a summary of the average chemical composition of the different beds, only the technically important elements being given:

Bed.	Fe	C_aO	SiO_2	Al_2O_3	P_2O_5	MgO
Black.....................	30	6	24.5	10	1.4	1.5
Brown.	34.3	8.6	16.6	6.5	....	2
Gray	39	8.0	7.5	6	1.7	1.6
Yellow....................	36	12.3	8.5	3	1.3	1.4
Red (limy).................	40	9.5	7.5	5	1.8	1.2
Red (siliceous).	31	5.3	33.6	4.2	1.6	9.5

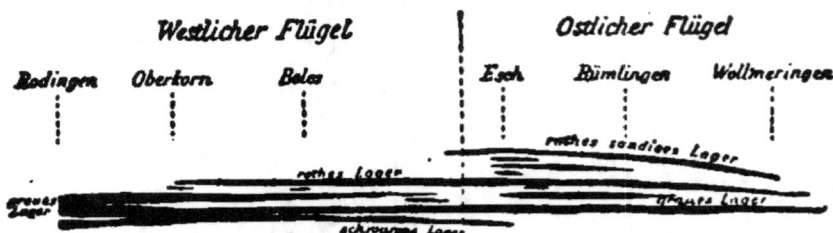

Fig. 49.—Diagram showing the development of the Minette measures. (Van Werveke.)

The Minettes of Lorraine consist of oolitic grains, mostly less than 0.25 millimeter in diameter, and consist essentially of ferric hydrate. The grains have a smooth surface and are bound together by a cement that is usually calcareous, but is rarely clayey or siliceous from the presence of interspersed quartz grains. In thin section under the microscope the oolites show very distinct concentric shells about one or sometimes two cores. In rare cases organic remains are found to have served as a nucleus, consisting of fragments of echinoderms, segments of crinoid stems or small gastropods. On the other hand, the grains sometimes contain scales of an indeterminate green iron silicate, and the presence of a microscopically fine siliceous skeleton enclosed in the limonite oolites indicates a former greater abundance of the iron silicates. More frequently scales are found of a chamosite-like mineral in the cementing material. In some beds iron spar is present in considerable amount, constituting as much as 60% of the ore of the lowest bed. Pyrite is also occasionally present, particularly in the black bed ores. In some Minettes the green mineral of the cement and .the limonite of the oolites are both replaced by magnetite. Not infrequently the ores contain calcspar casts of large fossils, and sometimes carbonized wood remains.

The beds of the ore-carrying Dogger formation are inclined gently westward. The structural feature of the greatest importance to the mining industry is the faulting, and there are numerous displacements along northeast fissures such as the fault of Deutsch-Oth, with a throw of 120 m. (393 ft.) and the fault of Gorze-Metz, known for a distance of 85 km.. (51 miles).

A valuable summary of our present knowledge concerning the genesis of the Minettes is given by L. van Werveke (1901). According to him most authors agree that the oolitic iron ores of Lorraine were deposited on the bottom of a shallow coastal sea. "The iron was brought from the land to the sea by brooks and rivers, and was precipitated in very diverse forms as silicate (looking like glauconite), also as carbonate, sulphide and sesquioxide in the upper strata, possibly also as ferric hydrate." (L. Van Werveke.) The cement in part is a mechanical sediment.

Another theory less generally accepted is advanced by Villain. According to him, the iron-bearing solutions were carried directly to the sea by thermal springs issuing from fissures in the ocean's bed. For the deposits of German Lorraine at any rate the agency of such "failles hourricieres" (feeding faults) seems to be untenable. A metasomatic origin of the Minettes seems to be altogether out of the question.

The production of Minette ores in the French department of Meurthe-et-Moselle was 2,630,311 tons in 1890. In German Lorraine 5,955,351 tons of iron ore were produced in 1899. The stock of ore still available is very large, being estimated at two thousand million tons in the German areas.

3. *The Iron Ores of the Dogger Formation of Wurtemburg, Upper Silesia and Switzerland.*

The iron ore measures of Wurtemburg belong to the Jura (zone of *Amm. Murchisonae*), especially in the region of Wasseralfingen and Aalen.

The sedimentary sequence at that locality is, according to Schuler,[1] as follows:

Sandstones, sandy slates and clay sandstones overlying the measure; upper ironstone bed 1.1 m. (3.6 ft.); sandstones and sandy slates 10 m. (32.8 ft.); lower ironstone bed, 1.6 m. (5.2 ft.); so-called stalhstein and sandy hard limestone .02 m. (4-5 in.); sandstones and clay sandstones underlying the above.

The ores of Wasseralfingen and Aalen are oolitic red and brown hematites, containing about 40% of iron. They not infrequently enclose fossils, especially *Amm. Murchisonae, Avicula elegans* and *laevigata, Venulites aalensis, Pecten demissus.* The production at the present day is inconsiderable.

The Dogger formation of upper Silesia also encloses iron ore in a zone characterized by fossils of *Ammonites parkinsoni,* according to F. Romer[2]. At Bodzanowitz, Wichrow and Sternalitz, southeast of Landsberg, a sandy upper bed and a purer lower bed of spherosiderite have long been mined for the Malapane furnaces. Similar occurrences are mined in Poland at Kostrzyn and Przystayn, as well as at Stara Kuznica.

In the iron mines of the Kleine Windgalle, in the canton of Uri (Switzerland) oolitic iron ores belonging to the Dogger are seen altered by dynamo-metamorphism; the hematite concretions have been pressed into flat lenses and iron silicates and magnetite crystals have been formed[3].

(f) The Eocene Iron Oolites of Kessenberg and Sonthofen, Bavaria.

In the Nummulitic sandstone[4] of Kressenberg and Sonthofen in upper Bavaria three groups of oolitic limonite beds are intercalated, which have been steeply upraised along with the other strata and dislocated by many faults. The local miner gives the name lodes (gange) to these measures, the principal one of which attains a thickness of two meters. Besides glauconite, the ore contains numerous quartz grains, and thus gradually passes into an iron sandstone. The most abundant fossil of the measure is *Conoclypeus conoideus* Ag., whose shell is entirely filled with oolitic brown hematite.

(g) Bog Iron Ores and Lake Ores.

This class of the iron ore deposits is of especial interest because the beds

[1] J. Haniel: 'Ueber das Auftreten und die Verbreitung des Eisensteins in den Jura-Ablagerungen Deutschlands.' Z. d. Deutsch. geol. Ges., Vol. XXVI, 1874, p. 97.

[2] F. Römer: 'Geologie von Oberschlesien,' 1870, p. 210.

[3] A. Heim: 'Mechanismus der Gebirgsbildung,' Vol. I, p. 62, Vol. II, p. 98.

[4] C. W. Gümbel: 'Geogn. Beschr. des bayerischen Alpengebirges,' 1861, p. 647. O. M. Reis: 'Zur Geologie der Eisenoolithe führenden Eocänschichten am Kressenberge in Bayern.' *Geogn. Jahresh.,* Munich, 1897.

are now forming, and the process of ore formation is taking place, as it were, before our eyes; so that we may obtain important information bearing upon the genesis of the older formations. In addition to this geologic importance, they have recently acquired an economic value, since it is now possible to smelt these high phosphorus ores, and at the same time to recover the phosphoric acid in them, and return it to Nature as the fertilizer known in Europe as Thomas meal.

Character and Mode of Deposition of Bog and of Lacustrine Iron Ores.[1]

Bog iron ore, also called swamp ore, meadow ore and bog ore, is yellowish, brownish or blackish brown limonite, with resinous luster on fresh fractures, always highly porous and cavernous, often slag-like and hard, sometimes ochrous, loose, earthy and mingled with many other substances. The ores contain hydrated iron silicate (a gelatinizing basic iron-silicate), also iron phosphates, crenates, ulmates and humates. The ores contain between 20 and 60% of Fe_2O_3. The phosphoric acid content rises as high as 10%. There is also a mechanical admixture of sand grains and clayey particles.

Chemical Analyses of Lacustrine and Bog Ores[2].

	I.	II.	III.	IV.
Iron oxide	62.92	67.46 ⎫	51.10	62.57
Manganese oxide	4.13	3.19 ⎭		5.58
Silica	8.12	7.00	9.20	12.64
Phosphoric acid	3.44	0.67	10.99	0.48
Sulphuric acid	.....	3.07	.....	0.07
Alumina	4.60	.....	0.41	3.58
Lime	.....	0.90	.....	1.37
Magnesia	.....	.....	.....	0.19
Water	18.40	17.00	28.80	13.53
	101.66	99.29	100.50	99.82

I. Bog ore from Schleswig (Pfaff).

II. From Auer near Moritzburg in Saxony (Bischof).

III. From Leipzig (Erdmann).

IV. Average of thirty analyses of Swedish lake ores (Svanberg).

Deposits of bog iron ore are found where surface water stagnates in the shallow depressions of flat lands, especially in the vicinity of sluggish streams whose waters are colored brown by dissolved humous acids or humic salts,

---

[1] F. Senft: 'Die Torf und Limonitbildungen,' etc. Leipzig, 1862. F. M. Stapff: 'Ueber die Entstehung der Seeerze.' *Z. d. Deutsch. geol. Ges.*, Vol. XVIII, 1866, pp. 86-173. A. W. Cronquist: 'Om sjömalmsfyndigheten,' etc. *Geol. För. Förh.* No. 65, Vol. V, 1880, p. 402. Hj. Sjögren: 'Om de svenska jernmalmens genesis.' *Geol. För. Förh.* Vol. XIII, 1891, p. 373. R. Klebs: 'Das Sumpferz.' Vortrag, Königsberg, 1896.

[2] F. Zirkel. 'Petrographie.' III, p. 574.

and in the moor and meadow bottoms of the lowlands of northern Europe, Asia and North America. The Saxon and Prussian parts of lower Lusatia, Brandenburg, Mecklenburg, Pomerania, Prussia, Masuria, Poland, European Russia, Holland, Finland and Sweden are rich in bog iron ores. In North America the Three River district, Province of Quebec, contains a typical deposit, in which mining was begun as early as 1730. Sometimes ores of this kind are also found on high plateaus of the mountains of central Germany.

These deposits rarely attain a thickness of over a meter. For the most part they show no stratification. Often they form isolated lumps, cakes or slabs, which exert an unfavorable influence on the immediately overlying field or meadow land, because they shut off water and air from the deeper layers of soil and hence are disliked by the farmers.

These iron ores have been worked since the earliest period known; hence Linné called them Tophus Tubalcaini, because Tubal Cain, the first iron worker, is supposed to have manufactured iron out of them.

The Lake ores (Sjömalmer of the Swedes) are somewhat different in quality and quite different in their deposition. They are found at the bottom of innumerable lakes in the Swedish Province of Småland, Ostergötland, Dalarne, Herjeådalen, Jämtland and Norrland; in Finland, European Russia and in Canada. They are mostly found on a sandy bottom at a distance of about ten meters from the shore and up to a depth of about 10 m. (32.8 ft.). The deposits are usually thin, rarely reaching 0.5 m. (1.6 ft.) in thickness, but as they may be obtained by simple dredging, they are worked even if but 10 cm. to 15 cm. (4 to 6 inches) thick. The supply is renewed in about fifteen to thirty years. "Estque thesaurus hic perennis et inexhaustus" (this is a perpetual and inexhaustible treasure), says Swedenborg of these lake ores of his home. The ore in the lakes does not form a continuous sheet, but occurs in round or elongated patches, whose direction and arrangement is evidently determined by the currents due to streams entering the lakes, since the ore beds are in shallows covered by an abundant growth of water plants, while the currents supply sand and mud. Lake ores sometimes occur in rivers, as, for example, in the channels connecting Swedish lakes, but the deposits only occur in quiet water found on the convex side of curves and not in the rapid current. Thus we have an example, showing how streaks or linear masses of ore of fairly uniform character may be formed with and as part of a series of sediments.

The formation of these lake ores is accomplished in several stages, each characterized by different material. In the first stage the iron oxide settling on the bottom, at first as a light ochrous mud, gradually hardens into crusts, having the luster, color and hardness of true ore. This mud has a blackish gray, brownish or greenish color, and is filled with vegetal débris. Exposed to the air, it dries to a gray or yellow powder. It is rich in gelatinous silica and contains numerous algae. On hardening, the masses of mud form either compact lumps (rusor), small or large discs and balls, or else they encrust roots, portions of trunks and branches of plants and animal remains, such as beetles and worm tubes, Phryganid quivers and the like. All of these deposits consist partly of hard, brown resinous-lustrous and partly loose, yellowish or brownish ochrous ore. In the spherical masses concentric shells of hard and loose ore alternate, often about a nucleal grain of sand or a vegetal remnant. The Swedish miners class the ores according to size and shape of these concretions as krutmalm (powder ore), aertmalm (pea ore), bonmalm (bean ore), penningmalm (penny ore) and skraggmalm (fragment ore). The fresh and soft unhardened ores often contain the phosphoric acid in the form of earthy vivianite, a blue iron ore, which may be greatly concentrated at certain spots. A good deal of manganese ore, probably in the form of wad, is also frequently mixed with the pulverulent lake ores of Sweden.

General Remarks on the Origin of Lake and Bog Ores.

It is certain that the deposition of all these ores took place from very dilute iron solutions, belonging to either the groundwater or that of lakes and rivers. The origin of the iron is apparent, for almost all rocks contain iron compounds, which, under certain circumstances, are soluble. If rare metals are present in the ores, one must look for older primary deposits of sulphide ores, whose decomposition furnished the material for such solutions. Thus in the Lake ores of Sweden there are found traces of copper, nickel, cobalt and zinc, which unquestionably are derived from decomposed pyrites of older deposits, which are abundant in the vicinity. Thus a bog ore from the Tertiary trough between Grochau and Briesnitz, which lies southwest of the nickel ore deposit of Frankenstein in Silesia, was found to contain:

Nickel	3.5%
Cobalt	1.3
Copper	0.1

Next follows the question, What was the nature of the solutions? The following are the chief solvents:

1. Sulphuric acid formed by the decomposition of iron-bearing sulphides.

2. Carbonic acid supplied by the air and by decaying organisms, and to some extent by the living animals. This enables it to attack various silicates.

3. Organic acids also play a part. These are, moreover, transformed into carbonic acid by oxidation, when vegetable masses decompose. In the presence of decaying vegetable matter, deprived of an adequate oxygen supply, iron sesquioxide is reduced to ferrous oxide, which forms soluble double salts, with humous acids and ammonium.

The precipitation of iron from these dilute solutions may take place in various ways.

In solutions of iron sulphate, the mere addition of ammonium humate, which is always present in the brown waters of peaty areas, effects a precipitation of iron oxide and later on of ferric hydrate.

From carbonated solutions the iron is precipitated as ferric hydrate by the escape of carbonic acid into the air, or by its absorption by plant cells. The deposition of iron carbonate is only possible when the air is excluded or in the presence of organic matter, which seems to harmonize with the known facts concerning spherosiderite and blackband ores.

From humates and other organic compounds the ferric hydrate is precipitated by the oxidation of the humous acids and their decomposition into carbonic acid and water. Here, too, the plant cell accelerates this process by furnishing oxygen. Lastly, by the mingling of iron humates and sulphates, the sulphuric acid, which kept the iron sesquioxide in solution, unites with ammonium, and iron is precipitated as hydroxide or as ferric humate.

In this action, the life processes of plants take a part, entirely independent of any products of plant decay. According to Ehrenberg, the algæ. especially the so-called iron algæ, *Galionella ferruginea*, Ehrenb., are active ore precipitants, coating their cell walls with ferric hydrate and opaline silica. This alga is abundant on the sea bottoms. According to the recent works of Molisch and Winogradsky, these and most other supposed algæ are ciliated bacteria of different kinds, especially *Leptothrix ochracea*.[1]

The silica of these ores may originally have been held in solution as alkaline silicates, which are supposed to be decomposed by carbonic acid. This silica is precipitated simultaneously with the ferric hydrate. The phosphoric acid was certainly present as ammonium phosphate and is precipitated at first as iron phosphate and as calcium phosphate in calcareous ores.

The further speculations which these processes suggest as to the genesis of the older stratified iron ore beds are given on page 81.

We saw that in the case of lake ores the deposition took place quite slowly. This process is more rapid where the drainage from the gossan (which see) of a large pyrite deposit is carried into a lake basin, or into the sea, or

[1] Weed: 'Geological Work of Plants,' *Am. Geol.*, June, 1894. Walther: 'Einleitung in die Geologie.' Jena, 1893-4, p. 655.

where mining operations produce an inflow of great quantities of iron-bearing mine waters. Thus the bottom of Lake Tisken, near Falun, is covered with a layer of ocher-mud several meters thick that has been furnished by the neighboring pyrite stock. The bed of the Rio Tinto carries ocher-mud and diatoms derived from the waters of the copper mines as far as Palos in Huelva Bay. That this was the case even before mining began at that locality is proved by the deposit of iron ore on the Mesa de los Pinos and the Cerro de las Vacas (see Figure 50). These limonite deposits were formed in a bog which was afterwards dissected by the river. The iron-stones contain plant remains of the same character as the present flora. Slabs of this ore were used by the Romans for tombstones.[1]

The deposition of iron ore is also rapid in places where iron-bearing mineral waters come to the surface in swampy hollows. An interesting case is

Fig. 50.—Profile through a part of the Rio Tinto ore field. (A. Phillips.)
s, clay slate; p, porphyry; k, copper ore bed; e, Pleistocene iron ores.

the extensive mineral marsh in the Soos near Franzensbad, described by O. Bieber.[2] The marshy beds of that locality are strongly impregnated by the waters of mineral springs, carrying sodium sulphate, magnesium sulphate, iron sulphate and other salts, and are overlain in many cases by whole beds of bog iron ore, blue iron ore and iron ocher. East of the sour spring called Polterer or Kaiserquelle, for example, we find the following section:

(1) Mineral bog 3 to 5 meters; (2) bog iron ore 0.3 m.; (3) blue iron ore (vivianite) 0.5; (4) iron ocher 0.3 to 0.5 meters.

The occasional occurrence of pyrite and marcasite in the mineral moor is noted later on.

(h) Recent Iron Ores of Marine Origin.

Aside from the deposits of iron ocher above mentioned, which must be forming at present in Huelva Bay, and perhaps to a less extent on some other portions of the coast, we do not know of any really recent iron ores found on the sea bottom. However, deep sea investigations have proved the very wide distribution of a highly ferruginous sediment, the glauconite

---

[1] H. Louis: 'Ore Deposits.' Second edition. 1896, p. 41.
[2] O. Bieber: 'Das mineralmoor der Soos,' Marburg, 1887, p. 29.

ooze.[1] It is probable that certain older iron ore deposits may be sediments of this kind altered by subsequent metamorphosism and modified perhaps by a subsequent concentration of the iron compounds.

According to J. Walther[2] the S. S. Tuscarora found off the coast of California, at a depth of 180 to 730 meters, black sands consisting almost entirely of dark green glauconite grains 0.6 millimeter in size. Glauconite sand of such purity is rare, while glauconitic sediments have been shown by the Challenger and other expeditions to be very widely distributed at depths of 590 to 5,395 ft. Phosphate concretions also occur associated with the glauconite. The amount of lime may reach 56%, and increases with the distance from the land. These masses of green mud contain an insoluble residue of 44% or more, consisting of siliceous shells and portions of skeletons of organisms, as well as of granules of quartz and very diverse rock silicates. Towards the middle of the ocean basin these formations are wanting.

The following analyses of recent glauconite sediments are taken from the work, 'The Voyage of H. M. S. Challenger; Report on the Deep-Sea Deposits,' p. 387:

Station.	Depth in fathoms.	SiO$_2$	Al$_2$O$_3$	Fe$_2$O$_3$	FeO	MnO	CaO	MgO	K$_2$O	Na$_2$O	H$_2$O	Total.
164 B	410	56.62	12.54	15.63	1.18	Trace	1.69	2.49	2.52	0.90	6.84	100.41
164 B	410	50.85	8.92	24.40	1.66	Trace	1.26	3.13	4.21	0.25	5.55	100.23
164 B	410	51.80	8.67	24.21	1.54	Trace	1.27	3.04	3.86	0.25	5.68	100.32
164 B	410	55.17	8.12	21.59	1.95	Trace	1.34	2.83	3.36	0.27	5.76	100.39
185 B	155	27.74	13.02	39.93	1.76	Trace	1.19	4.62	0.95	0.62	10.85	100.68

II. SEDIMENTARY DEPOSITS OF MANGANESE ORE.

A. WITHIN THE CRYSTALLINE SCHISTS.

1. *Manganese Ore Deposits of Långban, Sweden.*[3]

This mining district lies north of Philipstad, in Vermland, between Lake Långban on the east and Hytt Lake on the west. The deposits are associated with a band of dolomite, 4 km. (2.6 miles), with north-south course and west dip. This dolomite is intercalated in granulite,[4] that is,

[1] The manganiferous ironstone concretions of the deep sea are spoken of later. See Gümbel 'Ueber die Natur und Bildungsweise des Glaukonits.' *Sitz der K. Ak.* Munich, 1886.

[2] J. Walther: 'Einleitung in die Geologie.' Jena, 1893-4, pp. 880-882.

[3] *Zeit. f. Prak. Geol.*, 1899, pt. I.

[4] Törnebohm: 'Ofverblick öfver Bergbyggnaden inom Filipstads Bergslag,' 1877. Nordenström's *Katalog*, p. 38.

a fine-grained biotite gneiss poor in mica, forming an island-like mass in a great area of granite. Dioritic rocks occur on the east.

There are six main deposits, besides many small ones of very irregular shape and having an approximately lineal distribution. The deposits are ordinarily considerably enlarged near their base and thus approach close to one another. They do not consist of workable ore throughout, but sometimes wedge out and are replaced by dolomite or a pyroxene skarn containing scattered streaks of ore.

The Långban dolomite contains about 20% of magnesia and is of granular-crystalline structure. While it is white when fresh and unaltered, it usually turns brown upon exposure to the air, owing to the decomposition of finely interspersed manganese minerals. The ores are divided into iron ores and manganese ores. The former consist mainly of specular hematite and, to a smaller extent, of magnetic iron ores. The manganese ores are mainly braunite and hausmannite in dolomitic matrix. Furthermore, a great number of other manganese minerals appear as rock constituents, the commonest being rhodonite, tephroite, schefferite (a lime-manganese-pyroxene with 8 to 10% of MnO), and richterite (a soda hornblende with 8 to 11% MnO). Not infrequently large bunches of rose quartz and of red, ferruginous quartz are found, the latter material being formerly cut up into cups, paper weights, etc. Certain crush zones or shear zones (skölar), traversing the beds, are frequently coated with manganophyllite (a reddish magnesia mica, with up to 20% MnO).

According to H. Tiberg[1] the strata are arranged in the following manner from above downward. (See Fig. 51, which shows too gentle a dip. The numbers below correspond to those in the figure.)

1. Dolomite with lenticular layers of fine-grained gneiss, mostly poor in mica.

2. Thin layer of hornblende-pyroxene-garnet skarn.

3. Magnetic iron ore with melanite (up to 2 meters).

4. Iron ore with nests of ferruginous quartz (up to about 20 meters).

5. Hausmannite in dolomite (to 3 meters).

6. Braunite and hausmannite (up to 20 meters).

7. Thin layer of skarn (schefferite, richterite, tephroite and rhodonite).

8. Footwall dolomite with streaks and layers of fine-grained gneiss.

The most important layer is the manganese deposit, mainly braunite, forming the lower part of the iron bed of the Kollegii mine, a bed which reaches a thickness of 40 m. (131 ft.) and has been followed for a distance of 65 m. (213 ft.) along the strike. The braunite was not recognized until

---

[1] *Wermlandska Bugmannafören*, 1901. Lately Tiberg ascribes the origin of the deposit to replacement and not sedimentation.

1878, the hausmannite somewhat earlier. The braunite ore contains as high as 45% of manganese, the hausmannite ore up to 47%. The manganese ore is usually sorted into three grades, with 40, 30 and 20% of manganese. The two last named are concentrated to a 54 to 56% product. Most of the ores are used in the production of bessemer steel, as well as in the glass industry.

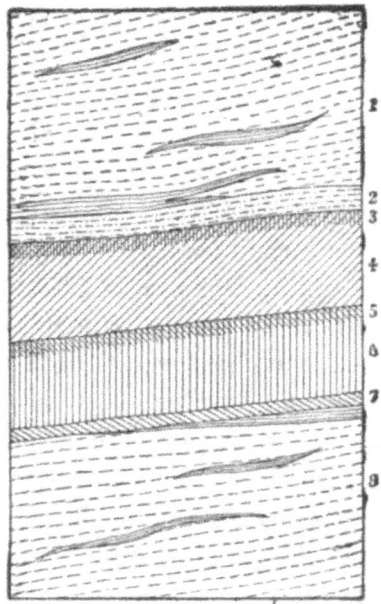

Fig. 51.—Diagrammatic cross-section of Langban ore bed. (H. Tiberg.) (Explanation of figure in text.)

Similar deposits, also associated with dolomite, are found at Pajsberg near Nordmarken, at Jakobsberg and in the Sjö mine in the Oerebro district. It may here be remarked that to the north of Långban, in the Stora Getberg mine, lead, zinc and copper ores occur in the same dolomite intercalation. At Långban itself these ores occur only in subordinate amounts. A remarkable feature of this mine is the association of native lead[1] with hausmannite in fissures in the dolomite. The manganese ore produced in Sweden in 1896 was 2,056 tons.

2. *The Manganese Deposits of Southern Bukowina, Austria.*[2]

The manganese mines of Oberarschitza and Arschitza-Anna occur high up on the slopes of the Eisenthal, 4 km. (2.4 miles) from Jakobeny. The

[1] Igelström in *B. u. H. Z.*, 1816, p. 26.
[2] B. v. Cotta: 'Einlagerungen im Glimmerschiefer der südl. Bukowina,' *B. u. H. Z.*, 1855, No. 39, p. 319. B. Walter: 'Die Erzlagerstätten der südl Bukowina,' *Jahrb. d. k. k. geol. Reichsanst.*, Vienna, 1876, pp. 372-382.

ore deposit is 40 m. (133 feet) thick, being enclosed in mica schists that are either horizontal or dip northeast at 30°. Directly beneath is a bed of siliceous schist 6 to 20 m. (19 to 65 ft.) thick, lying upon a hornblende-bearing mica schist. Above the ore is a much decomposed brownish yellow hornblende schist. The ore deposit consists of a mixture of pyrolusite, some hausmannite, brown hematite and quartz. The whole is locally called Schwarzeisenstein (black ironstone). This mass shows distinct bedding, even when decomposed, and in places its primary composition is disclosed, especially at another mine, the Oitza, near the Transylvania boundary. The deposit is in such cases seen to be built up of beds 0.1 to 2 m. thick of manganese silicate (rhodonite), of grayish green to flesh color, and with interspersed portions of rose-colored manganese-spar and quartz, and layers of a yellowish green mica hornblende schist. The transformation of the rhodonite into the more highly oxidized manganese ores and into quartz is traceable step by step through every gradation, the alteration starting along cracks and fissures. The hornblende schist when decomposed furnishes the intermingled brown hematite.

The mines belonging to the Greek Oriental Church yielded 2,063 tons of manganese ore in 1898.

3. *The Manganese-Zinc Deposits of New Jersey.*[1]

Although there is considerable doubt as to the sedimentary origin of this deposit, we include in this section the remarkable ore deposits of Franklin Furnace and Sterling hill in New Jersey, deposits whose exact genesis is still uncertain. It is possible that they do not belong in this group, but to that of contact metamorphic formations.

These deposits occur in a zone of white crystalline limestone, lying upon gneiss and regarded as a metamorphosed Cambrian rock, since it becomes dense away from the deposit, where, according to Nason, it contains Cambrian fossils. If this is correct, the gneiss is probably an altered granite. This hypothesis is further supported by the presence of intrusive masses and veins of a pegmatitic granite, traversing the limestone and carrying augite, hornblende, allanite, zirkon and orangite (Groth). The limestone is exceedingly rich in Mn CO_3 (16.57%) as shown by the brown color of the surface of the weathered rock owing to the formation of MnO_2. In the limestone at its contact with the gneiss, layers of magnetic ore occur, which were formerly worked. At a somewhat higher horizon, however, this marble con-

[1] H. Credner: 'Franklinit und Rothzinkerz im krystallinen Kalkstein,' *B. u. H. Z.*, 1866, p. 29. F. L. Nason: Geol. Survey, New Jersey, 1890, XIV; and 'Franklinite Deposits of Mine Hill,' *Trans. Am. Inst. Min. Eng.*, February, 1894. P. Groth: 'Zinkerzlagerstätten von New Jersey,' *Zeit. f. Prak. Geol.*, 1894, p. 230. J. F. Kemp: 'Ore Deposits of U. S.,' 1900, p. 251. Also *Bull.* 210, U. S. Geol. Survey, 1903.

tains two stratiform orebodies, composed of black crystals of franklinite (ZnFeMn)O,(Fe$_2$Mn$_2$)O$_3$, zincite (ZnO with up to 8% MnO), willemite and calcite. One of the beds outcrops on Mine hill, near Franklin Furnace; the other on Sterling hill, near Ogdenburg. Between the two hills the valley shows no outcrops. The Franklin Furnace deposit, 12 m. (39 ft. thick), has been the most carefully investigated. As shown by the accompanying section (Fig. 52, after P. Groth), it forms a syncline followed to the southeast by a strongly compressed anticline. After the folding the strata were cut obliquely across by a later dike of diabase.

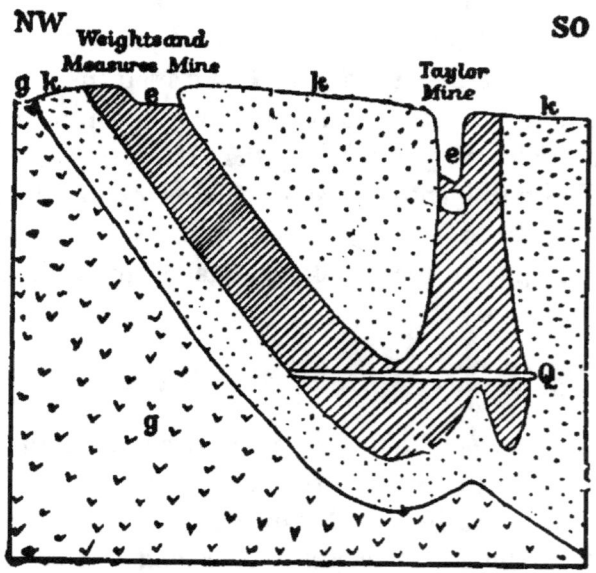

Fig. 52.—Profile through the ore bed of Franklin Furnace.   (P. Groth.)
g, gneiss; k, crystalline limestone; e, ore bed; q, cross-cut.

Besides the minerals already mentioned, jeffersonite (a zinc manganese pyroxene), hornblende, tephroite, troostite, fluorite and chloanthite occur in the mine. Near the granite intrusions there are certain other minerals not found elsewhere, namely, garnet, rhodonite, cleiophane (colorless zincblende) and axinite.

4. *Manganese Ore Deposits in Minas Geraes, Brazil.*

These deposits of manganite and pyrolusite occur as beds up to 2 m. thick, traceable for over 2 km. (1.2 miles) along the strike. The beds occur intercalated in crystalline schists in the region, between the railway stations of Queluz and Miguel Burnier of the Brazilian Central railway.

According to Lisboa[1] the steeply upraised strata at Miguel Burnier show the following succession:

1. Mica schist.
2. White limestone with 1.5% manganese content (over 10 meters).
3. Impure earthy iron and manganese ore (24 meters).
4. Pure manganese ore, mostly hard (over 3 meters).
5. Itabirite (quart-schist rich in specular hematite).
6. Gray limestone with 1.5% manganese, 11% iron, 13.8% SiO_2.
7. Overlying mica schists.

The production reached 100,000 tons in 1900. The ores contain between 45 and 55% of manganese, and very little silica (up to 1.3%) and phosphorus (up to 0.07%). They are shipped from Rio de Janeiro to North America.

Similar deposits are exploited, according to J. C. Branner,[2] in the Pedras Pretas mine, 26 kilometers west of Nazareth in the State of Bahia.

B. Manganese Ores in Normal Sediments.

(a) *Manganese Ore Deposits in the Carboniferous.*

The ores obtained from the Kaiser Franz mine near Elbingerode, in the Harz, occur, according to W. Holzberger and C. Zerrenner,[3] as pocket-shaped intercalations a meter or so thick in the siliceous shales of the Culm.[4] The ore consists of psilomelane in dense and botryoidal masses, some pyrolusite and coatings of wad, with rhodonite, rhodochrosite and quartz present as accessories. The ore formerly worked contained on an average 60 to 63% of manganese peroxide, sometimes rising to 67%. Zerrenner considers these maganese ores as later material separated out of the siliceous shales, a theory which needs further investigation The bed-like deposits of rhodonite and manganspar near Lautenthal in the Harz, mentioned by Klockmann,[5] are found also in the siliceous shales of the Culm formation.

The manganese ore deposits of Alosno, north of Huelva, in Spain, which contain pyrolusite and psilomelane, are also said to belong to the lower Carboniferous (Culm).

Beds of psilomelane, pyrolusite and brown hematite, as well as those of arsenious black oxide of copper and malachite, are found in the horizontal

---

[1] R. Ribeiro Lisboa: *Jornal do Commercio*, June, 1898, and March, 1899. Review *Zeit. f. Prak. Geol.*, 1899, p. 256. H. K. Scott: 'Manganese ores of Brazil,' *Jour. Iron and Steel Inst.*, London, 1900, No. 1.

[2] *Trans. Am. Inst., Min. Eng.*, 1900 (Meeting September, 1899).

[3] W. Holzberger: 'Neues Vorkommen von Manganerzen bei Elbingerode am Harze,' *B. u. H. Z.*, 1859, p. 383. C. Zerrenner: 'Die Manganerz-Bergbaue in Deutschland, Frankreich und Spanien.' Freiberg, 1861, p. 103.

[4] The Culm is the basal formation of the Carboniferous of Eastern Europe.

[5] *B. u. H. d. Oberharzes*, p. 65.

upper Carboniferous (lower Nubian) sandstones of Wadi Nasb and Wadi Chalig of Mount Sinai.[1]

(b) *Mesozoic Manganese Deposits of Chile.*

The extensive manganese deposits in the districts of Coquimbo and Carrizal, in Chile, worked by the Chilean Manganese Mining Company, have received but little study as yet. According to H. Louis[2] they occur interbedded with Jurassic and Cretaceous sandstones, clay slates, limestones and gypsum that rest on eruptive rocks. The ores consist of manganese oxides, peroxides and silicates in a gangue of silica, calc and heavy spar, the ores averaging about 50% manganese and having very little phosphorus. Chile produces some 30,000 tons of manganese ore annually.

(c) *The Eocene Oolitic Manganese Ores of Transcaucasia.*

The principal manganese deposit of Transcaucasia lies near Tschiatura on the Kwirila river in the province of Kutais, 42 kilometers from Kwirila railway station on the Tiflis-Poti line. The ores form a nearly horizontal bed in the Eocene, underlain by limestones and marls of the Turonian, which, in their turn, are underlain by granite below Tschinopoli. The Eocene series begins with a red or greenish sand 0.4 to 4 meters thick, containing teeth of *Lamma elegans*. Above this follows the manganese ore bed, averaging 2 meters and reaching 5 meters thick. The bed is composed of 5 to 12 layers of hard oolitic pyrolusite, with a cement of pulverulent ore. It outcrops underneath the bluffs of the low plateaus into which the Tertiary has been dissected by the Kwirila river and its tributaries. The overlying strata are sandstones and limestones of later Tertiary epochs. The ore bed can be traced 120 kilometers, so that the bed will last a long time.

The average amount of manganese in the ores, as determined by F. Drake, is 40 to 45%, while within certain areas it rises to 50%. When dressed for shipment, the ore averages 51 to 52%, reaching as much as 61%. The ore averages 0.16% of phosphorus and not over 8% of silica.

According to the same authority, a complete analysis of concentrates exported from Tschiatura gave:

MnO_2	86.25%	$Na_2O + K_2O$	0.22%
Mn_3O_4	0.47 "	SiO_2	3.85 "
Fe_2O_3	0.61 "	CO_2	0.63 "
CuO	0.01 "	S	0.23 "
NiO	0.30 "	P_2O_5 (P = 0.14)	0.32 "
Al_2O_3	1.74 "	H_2O	1.85 "
CaO	1.73 "		
MgO	0.20 "	Total	99.95%
BaO	1.54 "	Mn	54.90 "

[1] M. Blankenhorn: 'Neues zur Geol. u. Paleon. Aegypten·,' *Zeit. f. Prak. Geol.*, 1899, p. 392.

[2] 'Ore Deposits,' 1896, p. 878.

The brittle nature of the ore is a serious detriment in shipping.

The deposit was discovered in 1848 by the geologist Abich, and has been worked since 1879. In 1897 it produced 231,868 tons. The total production has already reached one and one-half million tons. The deposits of Samtredie and Novo-Senaki are similar.

European Russia, also, contains manganese ore deposits in the Eocene formation, at Nicopolis on the lower Dnieper. The manganese ore bed is 0.5 to 3 meters thick, and where worked it contains as high as 50% of manganese. The production in 1894 was about 58,000 tons.[1]

(d) *Recently Formed Deposits of Manganese Ore.*

The discovery that manganese-iron concretions of irregular flat forms are of frequent occurrence on the sea bottom at depths of 1,800 to 3,000 meters is of much genetic importance. In the region of the Gulf Stream the Albatross party found the sea bottom covered with them. Among them were pieces 2 to 15 centimeters thick and 10 kilograms in weight, the lower side of which often consists of tough, blue clay.

In a finely divided condition manganic hydrates in combination with ferric hydrates are, according to J. Walther[2], among the most widely distributed substances in marine sediments; all the rocks, shells, calcareous algæ and limestone fragments covering the sea bottom at Millport, Scotland, at a depth of 90 meters, show thin, black coatings of manganese compounds. The Challenger observed similar manganese coatings on pteropod and globigerina shells obtained from depths of 2,560 to 2,743 meters, and the dredgings corroborate previous observations showing the great distribution of the manganese concretions previously noted in the deep sea deposits of the Indian and Pacific oceans and in the vicinity of volcanic islands in the Atlantic Ocean[3].

[1] A. Macco: 'Reisebericht,' *Zeit. f. Prak. Geol.*, 1898, p. 203. F. Drake: 'The Manganese Ore Industry of the Caucasus,' *Trans. Am. Inst. Min. Eng.*, Vol. XXVIII, 1899, pp. 191-208.

[2] 'Einleitung in die Geologie,' Jena, 1893-4, p. 700.

[3] Gümbel: 'Ueber die Manganknollen in Stillen Ocean.' *Sitz.* der K. Ak., Munich, 1878. Murray-Irvine: 'On the Manganese Oxides and Nodules in Marine Deposits,' *Trans.* Roy. Soc.. Edinburgh, 1894, Vol. XXXVIII.

SECTION III.

EPIGENETIC DEPOSITS.

I. MINERAL VEINS.

A. General Description of Mineral Veins.

(a) DEFINITION OF THE TERM 'MINERAL VEIN.'

The material filling a fissure in the earth's crust is called a vein; if the material is igneous the vein is called a dike, if not, it is a true vein; if the material contains minerals of economic importance it is a *mineral vein*. More briefly, veins are fissure fillings; mineral veins are those containing ores. Their general form resembles that of ore-beds, since they are more or less irregular sheets wedging out in two directions, and showing great variations in thickness and great departures from rectilinear strike and dip. The enclosing rock is called *country* or *country rock* and the rock walls of the fissure are *vein walls*. The outer surface of these slab-shaped formations are often called *selvages*, but this term is only appropriate for the layer or sheet of clayey material, often soft and unctuous, which frequently lies between the body of the vein and the country rock. It is also known as *gouge* or fluccan.

As veins exhibit a great diversity in character, the definition of the term given above is not sufficient and needs amplification. According to Lindgren[1] a *fissure vein* is a mineral mass tabular in form as a whole, though often irregular in detail, filling or accompanying a fracture or set of fractures in the enclosing rock. This mineral mass has been formed later than both the country rock and the fracture, either through the filling of open spaces along the fracture or through chemical alteration of the adjoining rock. Many veins are not single fissure fillings, but zones of rock with numerous parallel, very narrow fissure fillings or stringers. Such lines of stringers practically form a *lode,* being, strictly speaking, composed of very small veins. According to Emmons,[2] "the term *vein* should be confined as far as possible to a single mineralized fissure, or the orebody formed along a single fissure, while the term *lode* is applied to an assemblage of ore-bearing

[1] W. Lindgren: 'Metasomatic Processes in Fissure-Veins'; 'Genesis of Ore Deposits,' 1902, p. 500.

[2] Text of Folio 39, U. S. Geol. Survey, 1899.

fissures so closely spaced that the ore that has been formed along and between them may in places be considered to form a single orebody. The principal fissures making up a lode are nearly parallel, but there are also smaller cross fissures connecting them, through which the intermediate country (granite) becomes impregnated with ore until it may constitute vein material. A lode may be only ten feet wide, or it may be a hundred or more; and within it are included zones of more or less altered country rock, sometimes carrying enough values to pay for mining; in others too barren to pay, when they are generally called *horses*. When there is an assemblage of veins too widely spaced to be included in a single lode, and

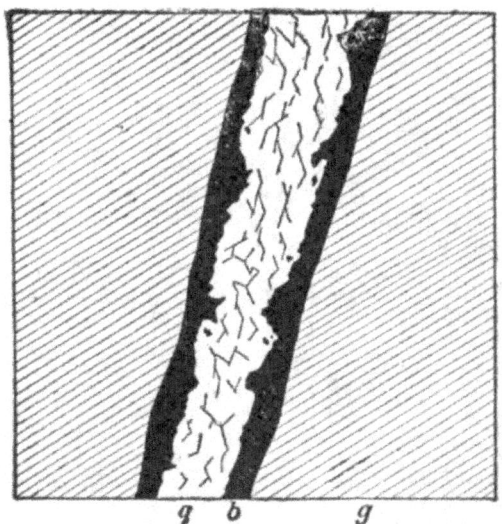

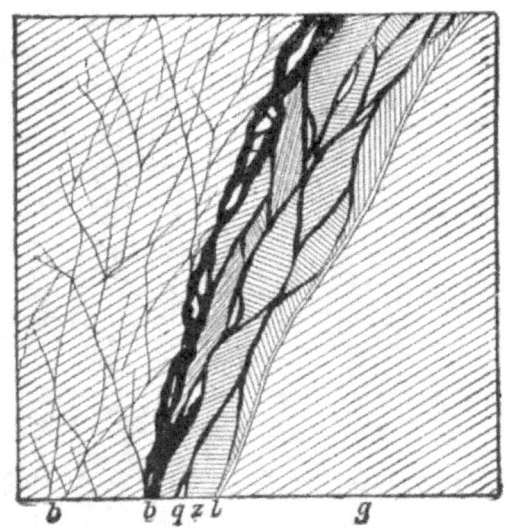

Fig. 53.—Simple vein (Krieg and Frieden
　Stehender of the Himmelfahrt, near
　Freiberg).
q, quartz; b, zinc-blende; g, gray gneiss.

Fig. 54.—Compound vein.
　b, lead glance; q, quartz; z, blocks
of decomposed gneiss; l, clay selvage;
g, gray gneiss.

yet to a certain extent coördinated into one system, they may be designated as a *vein system.*" The various stringers of such lodes may, indeed, have a transverse course. Parts of the country rock, either in great blocks or little bits, by means of such transverse and parallel stringers, may be entirely detached from connection with the country rock. B. von Cotta[1] has given the name *compound veins* to such lodes. They consist essentially of country rock traversed by ore stringers, thus differing from the simple fissure veins which, according to the general definition, are formed by the filling of a single fissure. This distinction was accepted by A. von Groddeck and is now much used. Compound veins or lodes often lack a sharp wall, at least

[1] 'Ueber den. sog. Gangthonschiefer von Clausthal,' *B. u. H. Z.*, 1864, p. 393.

on one side, and it is then left to the judgment of the observer to locate the boundary between the lode and the country rock.

The two figures, 53 and 54, represent these two types of lodes.

A part of a lode or compound vein is shown in Fig. 55, reproduced from a photograph.[1]

A good example of a so-called lode not embraced in any of the above definitions, but, on the contrary, representing a great, often very thick, zone of veins and lodes, is the famous mother-lode in California, containing numer-

Fig. 55.—Part of a lode or compound vein (of the Traugott Spat of the Gesegnete Bergmann's Hoffnung Mine, near Obergruna) showing gray gneiss traversed by narrow stringers of galena and some quartz.

ous gold quartz veins. Which one of the numerous parallel fissures of this zone are to be regarded as its bounding surfaces is a matter of arbitrary judgment and is of economic rather than of geologic interest.

Compound veins are common at Freiberg, Clausthal in the Harz and at Kremnitz in Hungary.

The term 'filling' may sometimes be misused in this connection. Certain veins, especially gold veins, are nothing less than zones of rock, traversed by innumerable practically barren fissures, from which they have been impregnated with gold-bearing pyrites or other ores to a workable

[1] Rickard: *Trans.* Am. Inst. Min. Eng., Vol. XXVI, p. 216.

degree. Such zones of disintegrated and impregnated rocks, called lodes by miners, usually coincide with lines of displacement. Finally, the definition of a filled fissure is unsuitable when the barren fissure, almost devoid of minerals, has been a conduit for solutions penetrating a zone of country rock on both sides and entirely replacing the rock by new-formed ores and other minerals, as seen in many galena deposits in limestone. The author (R. B.) does not entirely agree with S. F. Emmons,[1] who attributes an important rôle in the formation of many veins to metasomatic processes, as such processes are regarded by him as always subordinate phenomena in vein formation.

(b) DIMENSIONAL RELATIONS OF MINERAL VEINS.

"The width of the filled fissure, measured at right angles from wall to wall, is called the *thickness.* The rock in which the fissure has been opened is called the *country rock.* If the vein is not vertical the rock above is called the *hanging-wall* rock and that below the *footwall* rock. The horizontal direction of the vein is called the *strike,* that which comes nearest to the vertical is called its *dip.* If the body of the vein has undulations these may, by several observations, be plotted and reduced to a plane corresponding to the average attitude of the vein, on which the strike and dip is called the average or principal strike and the principal dip of the lode." (Von Cotta.)

The strike of a lode was usually determined in European mines by means of a miner's compass divided into twice twelve hours. The dip is ordinarily on a graduated circle connected with the compass. This form of compass is, however, not used outside of Europe, and even there is being more and more displaced, even among miners, by one with a graduation into twice 180°, and in most scientific monographs we find the strike of veins given in degrees, as has long been done in general geology.

The old division of veins into different categories, according to their strike and dip, long in use in the Erzgebirge of the Harz and other German mountains, is still used in Europe. The diagram (Fig. 56, page 116) gives the designations of this kind based on the strike.

According to this scheme, 'low' strike veins are those situated at the beginning of a division; 'high' strike veins are those situated near the end of it. A morgengang striking 4 hours is thus a 'low,' while one striking 5 hours is a 'high' lode. As to the veins striking almost centrally, 3, 6, 9 and 12 hours, they are said to strike in the alternating hours (Wechselstunden).

It is to be noted that in the German mines, in giving a name to a newly discovered vein, its insertion in one of the above named divisions was made

[1] 'Structural Relations of Ore Deposits.' *Trans.* Am. Inst. Min. Eng., February, 1888.

according to the observed magnetic strike, not the true strike, that is to say, without regard to the declination of the magnetic needle at the time. Furthermore, on the plats the name once given was retained, although greater development work might show that the portion of the vein first worked had a somewhat different strike from that of the vein as a whole. Thus it may happen that a vein mentioned on official plats and in diagrams as high-strike Spatgang really belongs to the group of Flache Gänge. Such contradictions have also arisen from the fact that when, at a much later time, a survey of the mine is made, the declination may have changed materially from that

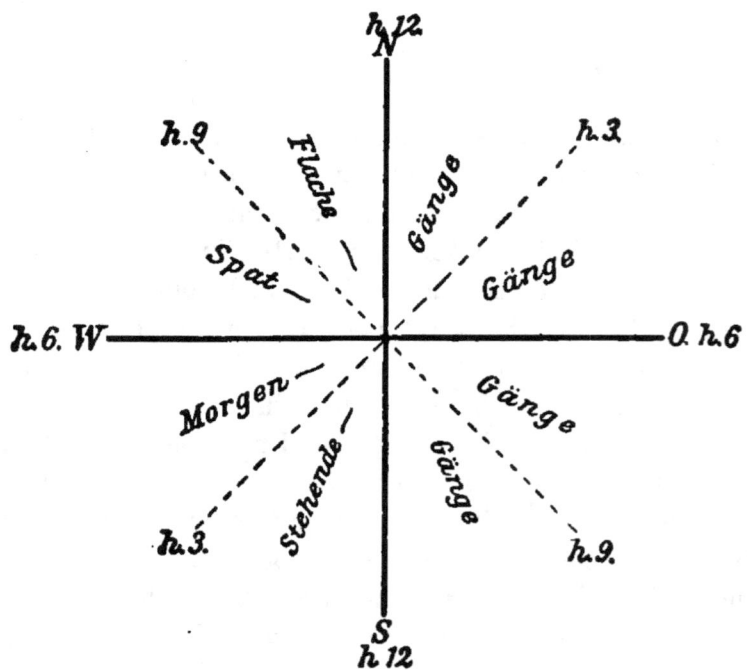

Fig. 56.—Division of veins according to their strike.

officially entered on older maps, and thus the observed strike may no longer agree with the older observations. Hence it is now the custom to plot the lodes with reduced strike, that is to say, referred to the true meridian.

In the Austrian mines the terminology is somewhat different. A compass divided into 24 hours is there generally used, and lodes striking between 21 and 3, or 9 and 15 hours, are called midnight lodes; those between 3 and 9, or 15 and 21, are called morning lodes.

The divisions according to the difference in dip are shown in the diagram (Fig. 57, page 117):

Most of the veins belong to groups between 90 and 75° or 75 and 45°; the other two divisions occur much more rarely. German terms rechtsinnig (right-wise) and windersinnig (contrary-wise) are not used in America and only have a local significance. In the Freiberg area, for example, most of the lodes dip west, and hence those members of the network of lodes that by way of exception dip east are said to dip contrary-wise. In another area the reverse may occur.

If a lode suddenly changes its dip it is said to 'get out of the hour'; to return to the old direction a short distance farther on it is said to 'strike or throw a hook.' The vein 'straightens up' if a prevailing low dip suddenly changes to a high dip; it 'flattens out' if the reverse takes place.

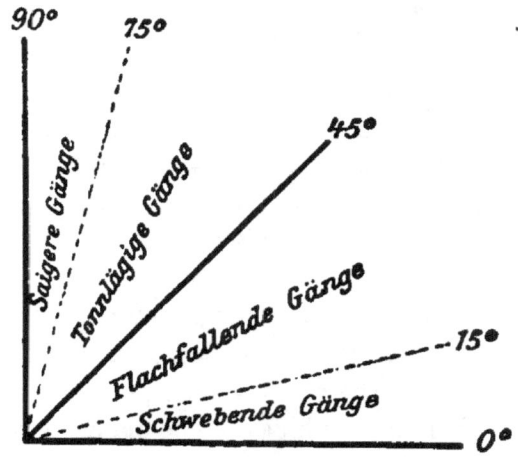

Fig. 57.—Division of veins according to their dip.

The thickness often shows great and frequent changes. It may be reduced to nothing, when the lode is said to pinch out (at "a," Fig. 58); to 'open up' perhaps a short distance farther on (at "b," Fig. 58).

In the great majority of veins the thickness varies between 0.5 to 1 meter. In the Freiberg area the greatest thickness shown by the barytic lead veins is 6 meters, while the others ordinarily vary between 0.15 to 0.5 meters. In exceptional cases lodes may be 50 meters and more thick, as, for example, the famous Comstock lode in Nevada and Veta Madre of Guanajuato,[1] Mexico. In such cases we are always dealing with combined lodes, made up cf closely spaced parallel lodes, or broad zones of impregnated rock.

Veins of slight thickness are called stringers, veinlets, etc.

A vein showing frequent pinching out and widening up, repeated at short intervals, is called a lenticular vein (Playfair) (Fig. 59). When the lenses are approximately perpendicular, it may be defined as a series of pyrite

[1] Carl Henrich: 'Mines of Guanajuato, Mexico,' *Mining Magazine*, 1904, Vol. X, pp. 63-75, 101-108.

stocks. Such lenticular lodes, according to K. Schmeisser, are quite frequent, for example, in the region of Coolgardie, in western Australia, in the auriferous quartz lodes, especially the Edjudina lode group.

By 'leaves' the Austrian miner means 'fissures,' sometimes no thicker than a sheet of paper, which though barren themselves are sometimes accom-

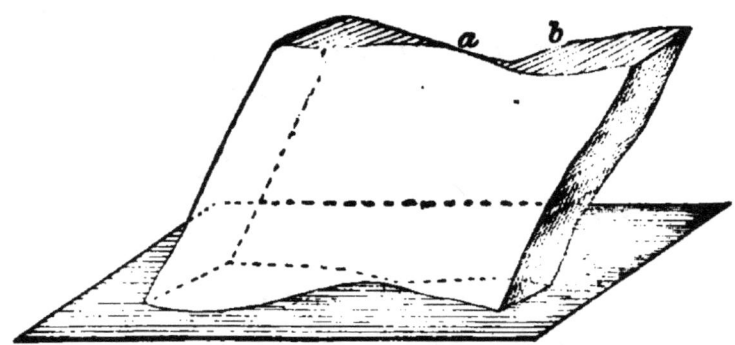

Fig. 58.—A lode which pinches out (a) and comes in or 'opens up' (b).   (A. von Groddeck.)

panied by a zone of country rock impregnated with ore. At the Rathausberg, near Gastein, a whole system of such leaves exists in the gneiss or mica schist, between which the country rock proves to be impregnated with native gold and auriferous sulphides. The name leaves is also given to the fissures connected with the ores in the dolomite of Raibl.

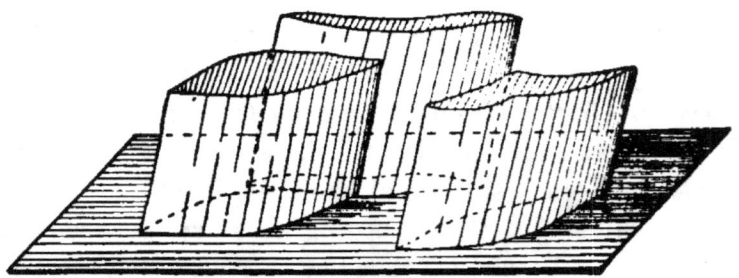

Fig. 59.—Lenticular lodes.

Somewhat similar fissures, called Zwitterbander, *i. e.,* narrow fissures, are often no thicker than a knife-blade, filled with quartz, topaz, lepidolite, fluorspar and a little tin ore, from which the country rock, granite or schist has been 'impregnated,' as it is expressed by the Saxon miners, with tinstone and its accompaniments, on both sides to a distance of 1 to 10 centimeters from the fissure.

'Verticals' are similar to the 'leaves' just noted. The term is used in the Black Hills of Dakota for barren fissures that have been channels for mineralizing solutions forming ores in favorable strata.

(c) The Termination of a Vein.

The vein 'wedges out' if its sides gradually converge and finally join (Fig. 60). Often, however, a joint plane, at least, continues on through the rock, often showing as a mere film of clay. In case the vein splits up into narrow stringers it is said to 'fray out' (Fig. 61). As an example one may mention the branching sub-division of the tin veins of Sau-

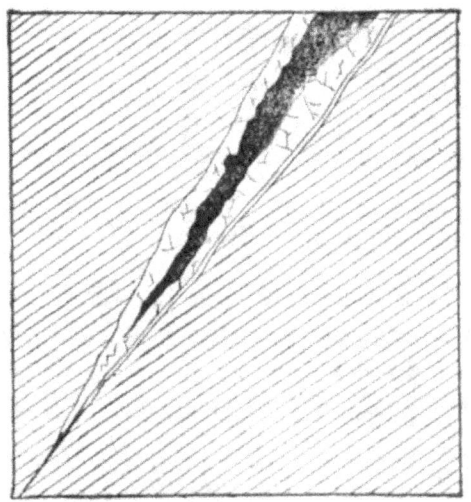

Fig. 60.—A vein wedging out below.   Fig. 61.—A vein dividing below.

berg, near Ehrenfriedersdorf in the Erzgebirge, a short distance before they reach the surface. Very often veins subdivide when they pass from a firm into a more brittle rock. Thus the Freiberg veins very frequently divide on passing out of the gneiss, whose toughness favors the development of a uniform single vein into the quartz porphyry dikes which pass through the gneiss. Figure 62 shows the behavior of the Gottlob Morgengang, at the David Richt shaft near Freiberg, where it passes through such a porphyry dike. The vein retains its full thickness up to the porphyry, but on entering it becomes scattered into numerous more or less parallel stringers. Immediately after leaving the quartz porphyry it resumes its previous condition. It is to be noted, however, that this division into stringers does not take place in all cases where the porphyries in the Freiberg area are traversed by ore lodes. If the vein strikes the porphyry at a blunt angle it may pass through it smoothly and without material impoverishment.

A similarly instructive example described by T. A. Rickard[1] occurs in the silver veins of the Enterprise mine of Colorado. These divide and grow barren in the limestone, but cut cleanly and with a fine body of rich ore through the sandstones that alternate with the limestone. Similarly, in the tourmaline bearing quartzose elvans (quartz porphyry) of Cornwall,

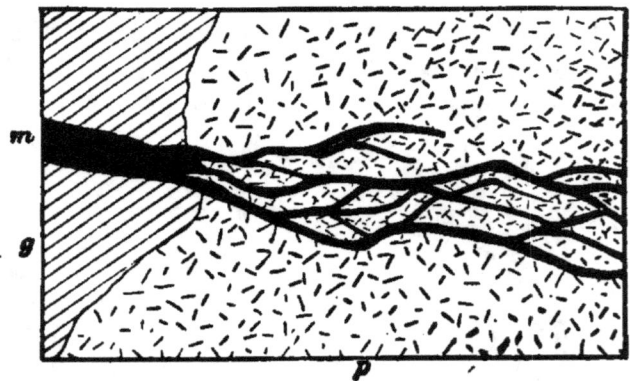

Fig. 62.—Scattering of the Gottlob Morgengang in the David Richt shaft near Freiberg. g, gray gneiss; p, quartz porphyry; m, ore lode.

the veins, according to W. J. Henwood[2], ordinarily divide into countless stringers, which re-unite when the vein passes into a softer rock richer in feldspar. Similar conditions are observed in the veins of the Spitzenberg near Silberberg in Silesia, where the vein passes from the gneiss into the overlying graywackes.

A very gradual fading and 'dying out' of veins does not occur in true vein fissures, but only in veinlike ore-bearing streaks in eruptive rocks, as in the case of ilmenite streaks in the norite of Ekersund in Norway (see p. 34).

Conversely several individual vein-stringers may converge and unite or 'gather' into a larger vein. This will be more fully discussed in the section or vein intersections (p. 132).

A *cross stringer* or *diagonal* is one which connects, at an acute or obtuse angle, two parallel or diverging veins (Fig. 63), as, for example, a stringer between the Burgstadter main vein and the Kranicher vein on the upper Harz. A crescentic stringer, on the contrary, runs out from the main vein at an acute angle, to return to it again at an acute angle some distance away (Fig. 64). The English-speaking miner calls the block between two such branch lodes a *horse*, and the outer vein about it, on a corresponding

[1] *Trans.* Am. Inst. Min. Eng., 1897, Vol. XXVI, p. 197.

[2] *Trans.* Roy. Geol. Soc., Cornwall, 1843, pp. 219-225.

analogy, a *rider.* Typical branching lodes are found, for example, at the Bockwieser main lode on the upper Harz. Side stringers or companions (Fig. 64) are small stringers parallel to the main lode, but not converging toward it, as, for instance, in the Neu Hoffnung Flachen on Himmelfahrt near Freiberg. *Feeders* and *droppers* are small veins or stringers falling into the hanging-wall (*i. e.,* intersecting vertically), or running out and down from the footwall of a vein.

A total cutting off of a lode can be caused only by younger veins or faults or by eruptive masses or dikes.

(d) Length and Vertical Extent of Veins.

In speaking of veins, we distinguish between the *length* along the strike, that is, the longitudinal extension of the vein, and its *depth* or downward extension on the dip. Veins vary greatly in both length and depth. In districts where there are several varieties of mineral veins with different contents, the varieties sometimes differ also in longitudinal extension. Thus in the Freiberg area the barytic lead veins show the greatest length; the length of the Halsbrücker Spates with its associated stringers being 8.4 kilometers (5.0 miles). One of the pyrite-blende lead veins, the Kirschbaum Stehende, and its southward continuation, the Hohn Birke Stehende, shows

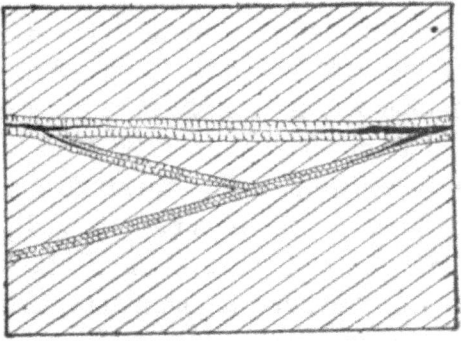

Fig. 63.—Two veins converging eastward, and connected by a diagonal stringer.

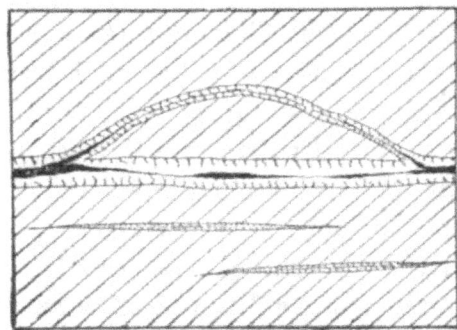

Fig. 64.—A mineral vein with an arched stringer and two side stringers.

a length of 7 kilometers (4.2 miles). The veins of the rich lead formation were in most cases traced only 1 kilometer (0.6 mile), but three veins, Johnannes St., Neuglückstern and the Neue Hohe Birke St., near Bescheert Glück, have a length of over 2 km. (1.2 miles). Finally the longest rich-silver-quartz-lodes were only traced 1.5 kilometers. Much longer veins occur in the Harz mountains. Thus, the Rosenhöfer, Rosenbüscher and Schulthaler lodes form the filling of a compound lode fissure 16.3 km. (9.84

miles) long (von Groddeck), and the Gegenthal-Wittenberger series attains a length of 18 km. (10.8 miles). The mineralized zone or gold belt of California, known as the Mother lode, does not conform to the definition given of either a vein or lode, being rather a system of veins. Its length is estimated at about 67 miles (112 kilometers).

The question "How far do veins extend downward?" is a most important one. Most veins continue downward below the lowest mine workings of the region, or, as the miner says, "to the eternal depths." Very often the older reports on mineral veins stated that this or that lode came to an end at a certain depth. More often, however, the vein ceased to pay and was abandoned. In many cases various difficulties of mining, or finance, prevented the exploration work necessary for its development, and in many instances subsequent workings have led to the rediscovery of apparently lost lodes at considerably greater depth. Thus, in the earlier descriptions of the Bescheert-Glück lodes between Freiberg and Brand they are said to be 'sod runners' confined to the upper strata. This, however, has been disproved by the later workings at that locality. Modern technical appliances have sometimes enabled lodes to be followed down to very great depths without disclosing any notable change in the character of the fissure fillings. Thus, the development of the Adalbert lode at Pribram has reached a depth of 3,640 feet (1,110 meters), or 564 m. (1,850 ft.) below sea-level. The lodes of the upper Harz have been followed to a depth of over 850 m. (2,788 ft.). In Freiberg, the Wernerian doctrine long prevailed that mineral veins are surface phenomena of the earth's crust. It was von Beust who exposed the falsity of this theory. He urgently recommended deep mining on a large scale, so that the Tiefe Rothschönbreger shaft, planned as early as 1838 by von Herder, was put down and the veins worked to a depth of over 650 m. (2,132 ft.). Veins have been followed to great depths in many other districts, for example, the gold veins of the Bendigo gold field in Victoria are mined to a depth of 1.219 m. (4,000 ft.). In general the empiric rule is that veins with a great horizontal extent continue to great depth, while those which extend for only a short distance along the surface do not go far downward.

It must be remembered, of course, that the greatest depths reached in mining operations are exceedingly small when compared with the radius of the earth, and even with the latest mechanical improvements the depths to which mines may be worked will still be relatively small. There is, however, a theoretical interest in the question to what depth ore lodes may possibly extend.

A discussion of his observations on rock metamorphism and of the known conditions of the geothermic gradation, as gathered from thermal springs, led A. Heim[1] to assume that water-filled open fissures may extend to a depth of 3,000 to 4,000 meters without being instantly closed by the tremendous side pressure in the earth's crust. Recently C. R. Van Hise[2] has been engaged on a theoretical investigation as to what depth fissure formation is possible. In his opinion the zone of the plastic condition of the rocks (zone of flowage) begins at 10,000 to 12,000 m. (32,810 to 39,372 ft.), because from there downward the weight of the superincumbent mass is too great to allow even the hardest rocks to retain their form. The fissure formation from there downward is replaced by flowage due to a displacement of the small particles, which is not infrequently combined with solution and recrystallization of mineral substance (granulation and recrystallization).

According to the same author it must not be supposed that at a particular depth, even above that zone, the water in a fissure will be transformed into steam. True, if we assume a geothermic gradation of 30 m. (98 ft.) the critical temperature of water, 364° C., would prevail at a depth of not more than 10,920 m. (35,810 ft.), but this is true only if we disregard the pressure of the column of water. In reality this hydrostatic pressure is at any given depth in a fissure abundantly sufficient to retain the water in liquid form.

Many mineral veins terminate upward and do not show on the surface; that is to say, do not outcrop. This is illustrated by numerous well-described examples, although it must be remembered that errors in this respect are readily committed because the uppermost part of the ore lode may have been disfigured beyond recognition by atmospheric influences. Such a limitation in upward extension is found in the so-called cobalt ridges (Rücken) of Riechelsdorf in Hesse (von Leonhard). In the Freiberg area most of the lodes of the Himmelsfürst mine show no outcrop, not even that of the Silberfund Stehende, which in depth attains large dimensions. At the Segen Gottes shaft and the Moritz shaft, according to E. W. Neubert,[3] the projection of this lode was several times passed through without the vein being encountered. Similar reports are made in regard to the Neu Glück Stehende of the Alte Hoffnung Gottes mine. Finally a very remarkable condition is reported by T. A. Rickard[4] at the Enterprise mine, Colorado. The numerous silver-gold veins cut through gently inclined beds of Carboniferous rocks, but could only be followed upward from the floor of the

---

[1] 'Mechanismus,' etc., Vol. II, 1878, p. 107.

[2] *Trans.* Am. Inst. Min. Eng., February, 1900, pp. 7, 9, 11. Reprinted with papers by Posepny, Emmons, Lindgren, Weed, Vogt, Kemp, Rickard, etc., in 'Genesis of Ore Deposits,' 1902.

[3] *Freiberger Jahrbuch*, 1881, p. 51.

[4] 'The Enterprise Mine,' *Trans.* Am. Inst. Min. Eng., Vol. XXVI, 1896, p. 975.

Group tunnel to the horizon formed by a bed of shale (Fig. 65). Before reaching this horizon the veins became much poorer and were no longer workable; but along the contact or bedding plane there are flat and very rich orebodies whose composition is similar to that of the lodes. According to Rickard's explanation the somewhat plastic shales, offering resistance to the opening of fissures, were not cut by the veins, and the solutions rising through the open fissures were dammed back by the shales and spread out along this horizon. The carbonaceous material of this shale led to the precipitation of the metallic compounds contained by the stagnant solutions, thus giving rise to bedded ore masses.

A vein may also fray out upward before reaching the surface, only its smaller branches outcropping, as was the case in the famous silver-tin veins of the quartz-trachyte Cerro de Potosi in Bolivia (A. W. Stelzner). At Nag-

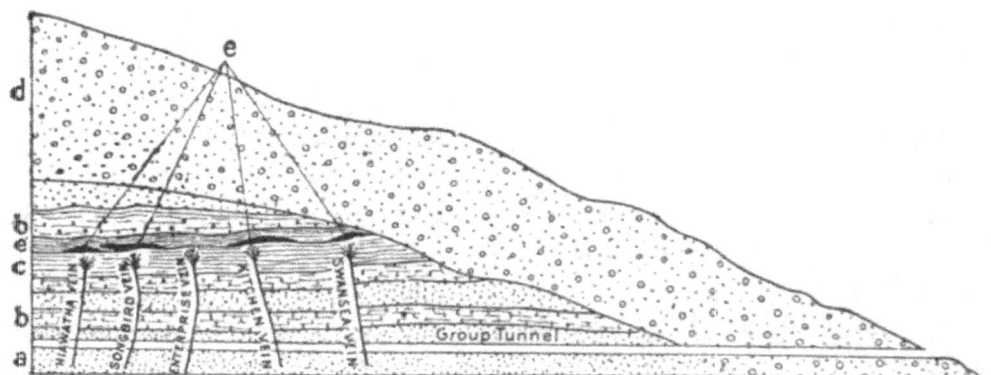

Fig. 65.—The termination of mineral veins at the top of the Enterprise mine. (Rickard.)

a, sandstone; b, limestone; c, shale; d, younger overlying rock; e, ore-beds in the so-called contact.

yag, too, many fissures fail to reach the surface (J. Grimm). The outcrop of a vein is not always distinctly recognizable at the earth's surface. If the predominant gangue is a hard material, resisting weathering, a vein may project as a rocky crest or a low wall, as is so often the case with gold-quartz lodes, which for that reason are called *reefs* and *ledges*. On the contrary, a vein may be marked along the surface by a trench-like depression if the vein filling consists of carbonates and easily decomposed ores. (See also description of gossan formation below.)

The outcrop of a lode, like that of a bed, may be exposed or covered, that is to say, concealed by younger deposits. Pleistocene alluvium is particularly apt to form such mantles over lode outcrops, but igneous intrusions may also do so, as is seen at El Oro, Mexico, where andesitic lavas conceal the outcrops.

(e) Structural Relations of Veins to Country Rock.

The most common type of vein is that which cuts across the stratification of the country rock and is, therefore, called a cross vein (Fig. 66). When the lodes have the same strike and dip as the country rock, they are commonly called *bedded veins* (Fig. 67). The type is very common among gold quartz veins, which usually show numerous lenticular swellings of the vein quartz.

It is often difficult to distinguish these bedded veins from true strata. The principal features by which doubtful cases of such veins can be recognized are, 1st, the occurrence of local cut-offs of the stratification; 2d, the presence of small transverse stringers; and, 3d, the presence in the district of genuine fissure lodes of similar or identical composition. The true vein

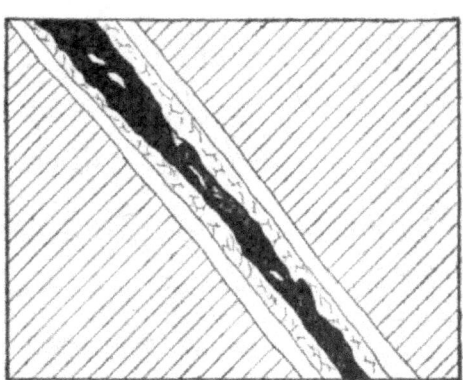

Fig. 66.—A cross vein.

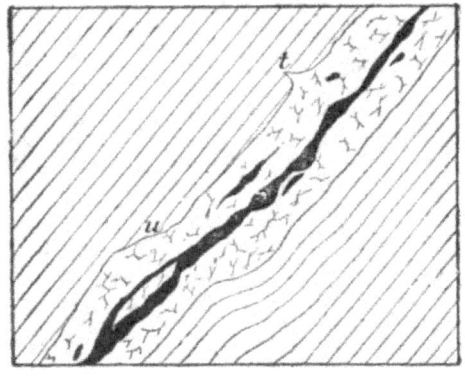

Fig. 67.—A bedded or stratum vein. At u, a local cut-off; at t, a small side stringer.

nature may also be proved by the presence of fragments of country rock in the midst of the vein filling.

A special form of bedded vein is known as a *saddle reef*. The best examples are those of the Bendigo goldfield of Victoria, which have been described by Dunn,[1] Rickard,[2] Pittmann, Samuels,[3] Schmeisser and Vogelsang.[4] As they are a new type they warrant a somewhat detailed description.

[1] Dunn: 'Report on the Bendigo Gold Fields,' Published by the Victoria Mines Department, Melbourne.

[2] Rickard, T. A.: 'The Bendigo Gold Field.' *Trans.* Amer. Inst. Min. Eng., Vol. XX, 1891, p. 499.

[3] Samuels: 'Origin of the Bendigo Saddle Reefs.' *Zeit. f. Prak. Geol.*, 1894, p. 95.

[4] K. Schmeisser and K. Vogelsang: 'Die Goldfelder Australiens.' Berlin, 1897, p. 64.

The Silurian shales and sandstones of Bendigo have been sharply folded into anticlines and synclines and the strata at the flexure points have slipped upon one another in such a way as to form cavities in the arches and troughs. (Fig. 68.)

The reefs are arch-like masses of quartz filling the cavities of anticlines and the bowl or trough-shaped synclines (inverted saddles). The quartz contains both free gold and finely distributed auriferous sulphides, and sometimes encloses angular fragments of country rock. These fragments, and the presence of short transverse stringers running from the quartz bed into the underlying layers, prove the deposits to be true veins (Fig. 69). Sev-

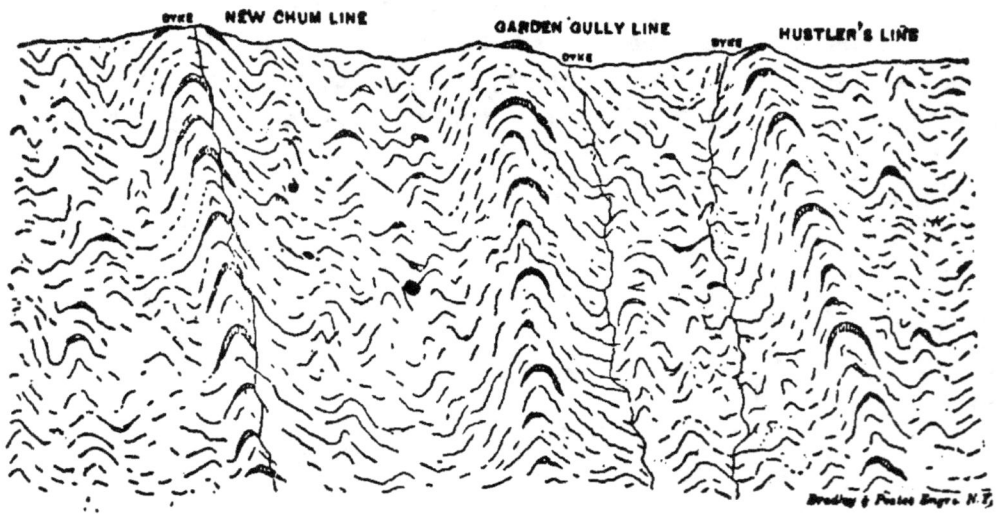

Fig. 68.—Ideal section through the Bendigo goldfield. (T. A. Rickard.)

eral groups of saddle-veins may follow one another at greater or less distances. They are often quite large, the anticline of the New Chum mine being exposed for a distance of 22 km. (14 miles) and to a depth of 975 m. (3,200 ft.) Besides these saddle reefs, ordinary quartz veins of similar mineralogic composition also occur in the district.

Similar anticlinal deposits have been described by E. R. Faribault from Nova Scotia (see under Stratified Gold Ore Deposits).

The famous deposit of Broken Hill in the Barrier range, New South Wales, is, according to E. F. Pittman and J. B. Jaquet,[1] regarded as a great anticlinal deposit, as indicated in the detailed description which is given later.

[1] J. B. Jaquet: 'Geology of the Brokenhill Lode, etc.' Sydney, 1894. R. Beck: 'Beiträge zur Kenntniss von Brokenhill.' *Zeit. f. Prak. Geol.*, 1899, p 65.

Quartz segregations, on a very small scale, often occur in clay slates, phyllites and mica schists, in which the foliations have been forced apart by folding. Sometimes these quartz lenticules contain gold.

Bedded veins are most frequent in platy or shaly or foliated rocks in which the bedding planes are marked by very distinct joints and are not apt to occur in heavy bedded, more brittle rocks.

Bedded veins are simulated when the beds along the walls of a quartz-

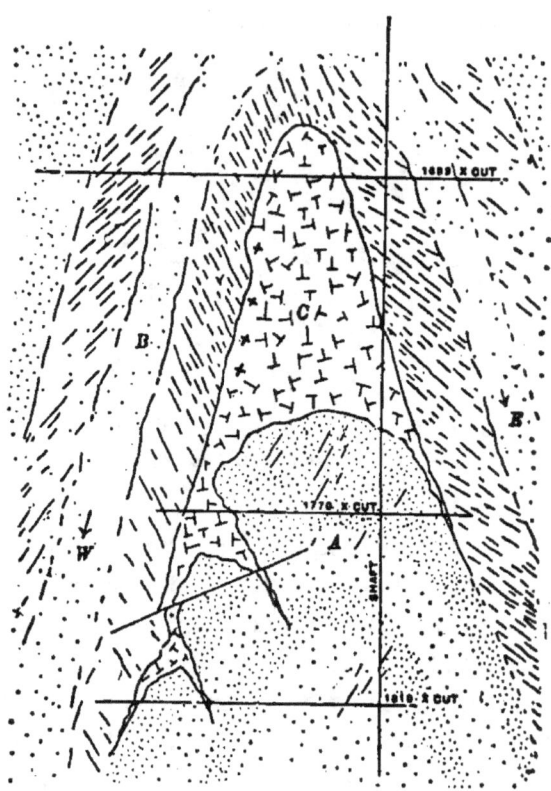

Fig. 69.—Section through the saddle reef of the New Chum Consolidated Mine, Bendigo, Victoria. (T. A. Rickard.)

A, sandstone; *B*, shaly sandstone, with quartz strings; *C*, lode quartz with gold and sulphides.

filled fault are bent so that the layers lying against the vein wall are parallel to it, as figured further on.

Gash veins are commonly understood to be those formed in joints and cross fissures in limestone, which have been widened by the solvent action of circulating water and filled by galena, zinc-blende and other ores. A characteristic feature of this variety of vein is the very short length and depth in strike and dip, and the association with certain calcareous beds, and absence in intercalated slates. Figure 70, after Whitney, illustrates their mode of occurrence; the form is especially common in the Carboniferous limestones of southwestern Missouri. They might, perhaps, be fitly called solution deposits, due to leaching out along veinlets (Auslaugung trümer), that is, fissures enlarged by solution and filled with ore. They belong to the group of irregular cavity fillings, not to true veins.

The *pipe veins,* as certain tubes of ore have been called which traverse

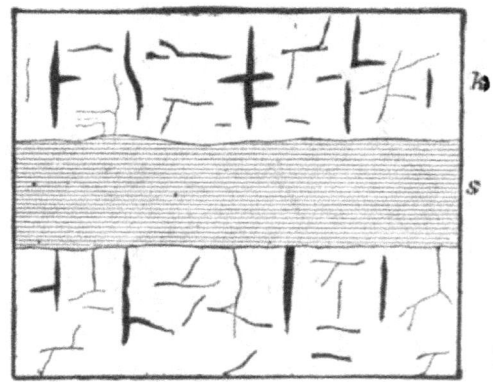

Fig. 70.—Gash veins, after Whitney.
k, limestone with stringers of lead glance; s, slate.

the stratification of limestone obliquely, are of similar class and origin; they present a strong contrast to the cavity fillings called *flats,* which are parallel to the bedding. A more extended application of this term seems to us inappropriate, an opinion already expressed by R. W. Raymond[1].

Columnar deposits are occasionally formed by the stretching and squeezing out of beds between more resistant strata. Thus the ore chimneys of Raposos in Brazil are not true veins, but masses of quartzite impregnated with auriferous pyrite and arsenopyrite.[2] *Chambered vein* is the name given by G. F. Becker to a type of deposit common in the quicksilver deposits of California. Fig. 71 shows the difference between an ordinary and a chambered vein. In this type the vein proper is connected with very irregular, large stock-like orebodies extending from the vein into an insoluble country rock. Their origin is supposed to be due to a condition of varying cohesion of the country rock, the weaker parts of which, when subjected to torsional forces, became broken and crushed. Chambered veins are, however, very different from cavities dissolved in limestones, which are connected with vein fissures in a similar manner.

[1] 'What is a pipe vein?' *Trans.* Am. Inst. Min. Eng., Vol. VI, 1879, p. 393.
[2] G. Berg: 'Goldlagerstätten von Raposos.' *Zeit. f. Prak. Geol.,* 1902, p. 81.

Contact veins occur at the boundary between sedimentary and igneous rocks. Underground water, circulating mainly along joints, and to some extent in pre-existing quartz veins, is checked when it encounters intrusive eruptive masses, because the igneous rocks are less permeable. The flowage thus checked may spread out laterally along the contact plane, because it is the path of least resistance, and in doing so the waters may deposit such substances as they held in solution. In other cases the cooling of eruptive rocks may have formed contraction fissures subsequently filled by ores.

A typical example of such a contact lode is seen in the Haile gold mine in South Carolina, described by H. Credner[1] and illustrated in Fig. 72.

According to J. H. L. Vogt[2] well developed contact lodes (with copper ore in quartz) accompany granite dikes which traverse quartzite slate at the Moberg mine in Thelemarken.

Fine examples occur in Idaho, in which quartz veins overlie or underlie dikes of porphyry or basic rocks.[3] In some cases faults form contact veins. A good example of this variety of contact vein is seen in the Johannes mine, near Schwarzenberg, described by Charpentier.[4] An ordinary hematite vein 2 to 17 meters thick has been opened to a depth of 200 meters. Near the outcrop the vein follows the contact plane between granite and mica schist, but in depth it passes into the granite. Its origin through general

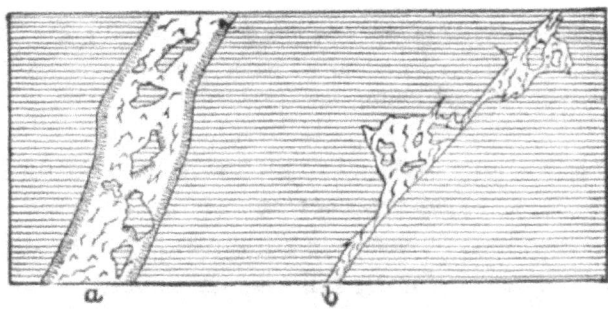

Fig. 71.—Ordinary vein (a); chambered vein (b). (G. F. Becker.)

orogenic causes is shown by its great length; it extends 7 kilometers northward (to Lauter) and passes through alternations of gneiss and mica schist.

The well-known contact lode of the old Haus Baden mine, near Badenweiler, is also a fault fissure. It traverses granite and the Bunter sandstone

[1] *Zeit. f. d. ges. Natur,* 1870, Vol. XXXV, p. 20.

[2] J. H. L. Vogt: 'Zur Classification der Erzvorkommen.' *Zeit. f. Prak. Geol.,* 1895, p. 149.

[3] W. Lindgren: 'The Mining Districts of the Idaho Basin.' Washington, 1898, p. 701.

[4] Charpentier: 'Mineralogische Geographie von Sachsen,' 1778, p. 248.

and contains quartz, barite, fluorspar, galena, copper pyrite and copper glance (G. Leonhard). At this place the Bunter sandstone, the normal stratified rock, is younger than the eruptive rock. The term *contact vein* should properly be restricted to only those which coincide with a real eruptive contact.

(f) Relations of Veins to Each Other.

Several veins with approximately identical strike are called parallel lodes, as, for example, the gold quartz lodes of the Sierra Nevada in California, which run parallel to the longitudinal axis of the range; and the fifteen lead veins of Svenningdal in Norway, which run at right angles to the strike of the strata.

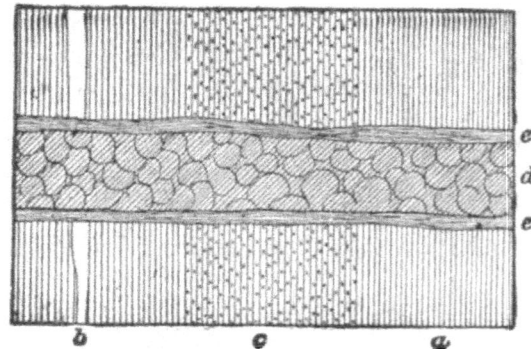

Fig. 72.—Contact veins of Haile gold mine, South Carolina.   (H. Credner.)

a, quartz schist; b, quartz; c, quartz schist with auriferous pyrite; d, diorite vein; e, brown hematite contact lodes.

Radiating veins is a term applied to a cluster of veins diverging approximately from one point. The arrangement of the veins of the upper Harz might serve as an example. (See map sketch Fig. 156.) Typical radiating veins exist in the Cerro de Potosi, Bolivia, tin deposits. Systems of parallel lodes are called *series.*

The orientation of the vein series of Saxony has been studied in detail.[1] The special geologic maps of that kingdom show them as generalized by H. Müller. As may be seen by the map of the Freiberg veins, accompanying this work, two main directions may be distinguished in the strike of the Freiberg lodes: 1. A series with a few deviations between north-south and northeast. Though this series does not coincide exactly with the main northeast strike of the Erzgebirge, it is very close to it, and hence is called the Erzgebirge series. 2. A series striking between northwest and west-

[1] F. C. v. Beust: 'Gangkarte über den inneren Theil des Freiberger Bergrevieres nebst Erläuterungen.' Leipzig, 1842. Also 'Ueber die Erzgangzüge im sächsischen Erzgebirge in ihrer Beziehung zu den dasigen Porphyrzügen.' Freiberg, 1856.

northwest is called the Harz system, because it coincides with the main strike of the Harz. Upon close examination of the special map it appears that the several vein formations distinguished in Saxony (see later) favor one of these directions. The pyrite-blende-lead veins, which predominate at Freiberg itself, as well as the rich silver quartz veins developed at a distance from the town near Bräunsdorf and Gross-Voigtsberg, also have a course parallel to the Erzgebirge; the lead barite veins and many barren lodes run with the axis of the Harz mountains, while the galena and the spathic lead veins have numerous representations in both series, especially in the areas called Bescheert Glück and Himmelsfürst, near Brand.

The Harz (Hercynian) direction is also followed by a majority of the iron ore and cobalt lodes of the upper Erzgebirge, especially in the Schwarzenberg and Schneeberg areas. The formation of all these more or less parallel ranges is evidently connected with the main uplift and folding of the Erzgebirge, which occurred in Permian and late Carboniferous time. In fact, post-Cretaceous mountain-building forces have formed entirely different systems of fissures at other localities, as is so plainly seen in the directions of the two systems of barren fissures so numerous in the so-called Saxon-Switzerland fissures, which, as a whole, form two series similar in orientation to those already noted. The same rule holds true in many other mining areas; thus an exact plotting of the prevailing strike of the veins of a district may lead to important inferences as regards the general structure and *vice versa*. In the Freiberg area, which we have used as an illustration, local orogenic processes, as well as general ones, have evidently been active in forming the network of veins. It should be particularly noted that the gneisses of the Freiberg area form a dome whose summit is traversed by the veins.

When in a certain area the veins strike irregularly in all directions the network of lodes is called a *stockwork*.[1] Thus innumerable tin lodes occur in the granite at Altenberg in the Erzgebirge, and with the intervening country rock which is impregnated with ore, form what is called the "Altenberger Stockwerk." Similar stockworks also occur at several other localities in the Erzgebirge. In Cornwall a tangle of tin veins has penetrated the rock so generally, and at the same time has impregnated it with tin ore so abundantly, that large masses of rock have been mined out as ore, and vast chambers mined out at different levels. Mention may also be made of the Geyer stockwork, whose workings, like those of the Altenberger Zwitterstock, collapsed, forming a great basin.

[1] This German term, adopted in both French and English literature, applies to a mass of granitic rock traversed by a network of small veins, interlacing with one another and traversing the rock in all directions, but the ore also impregnates the interlacing rock. (Louis.)

Many vein aggregates resembling stockworks are also found in the gold ore district of Transylvania; for example, at Offenbanya, where trachyte breccia in the Kreisova stock is penetrated by innumerable gold- and silver-bearing veins.

(g) STRUCTURAL RELATIONS OF TWO VEINS TO EACH OTHER.

Double Veins. When a fissure vein has been reopened and a new fissure formed, usually along one wall, but sometimes through the middle portion of the vein, and this fracture is filled by the deposition of material which may be unlike that of the primary vein, the result is a double vein, which may be composed of different layers or bands. Thus, for example, several of the veins at Himmelsfürst, near Brand, in the Erzgebirge, consist of a band of galena or brown spar ore, and a harder layer, which must be classed as a pyritic blendic lead.vein. A similar condition is observed at the Christian Stehende at Himmelfahrt, near Freiberg. A specially instructive example of double lode is the famous Halsbrücker Spat near Freiberg. It consists of a soft band, composed chiefly of thin layers of barite, containing a slightly argentiferous galena, while a hard band consists of ferruginous quartz with highly argentiferous galena and rich silver ores. These two divisions of the vein diverge northwestward and are worked as separate veins—the soft stringer as the Drei Prinzen Spat, the hard as the Ludwig Spat (H. Müller).

If two approximately parallel veins are, as a whole, separate and only occasionally combine into a double lode, one lode (a) is said to drag against the other (b) (Fig. 73). If after their meeting they remain permanently connected, they are *combined* or are said to have joined.

If two veins cross, either in the strike or dip, the veins intersect. Intersections are further subdivided into:

1. Rectangular intersections (Fig. 74) when the strikes of the two lodes are at right angles.

2. Convergent intersections (Fig. 75) when the veins cross at an acute angle.

3. A dip intersection occurs (Fig. 76) when the veins are parallel in the strike, but intersect in dip.

As the lode cutting across another must necessarily always be the younger, the relations of two lodes to each other is used to determine their relative age.

For this purpose, it is, of course, necessary that the filling of the two veins shall be of different nature, or at any rate that one of them shall show distinct layers or crusts, in order that one may see whether the layers of the one pass through the mass of the other. Fig. 77 shows a simple vein crossing between an older stringer (a) with a younger barite stringer (b).

In very rare cases two intersecting fissures have been filled at the same time. If both veins show a banded structure or crustification, the layers of

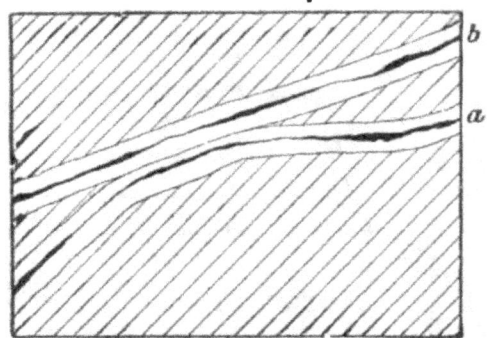

Fig. 73.—One vein (a) drags against another (b).

one vein pass uninterruptedly into those of the other. An excellent example is shown in Fig. 78 recorded by K. A. Külm.[1] It repre-

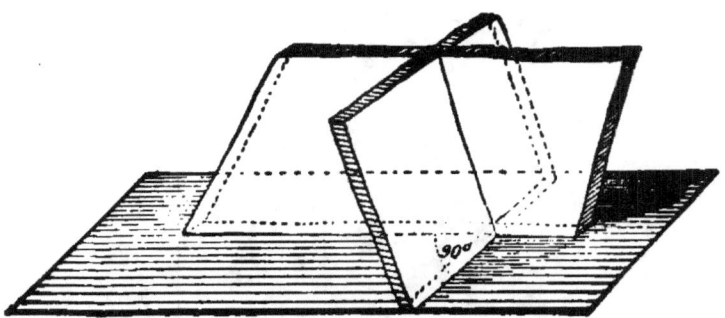

Fig. 74.—A rectangular intersection.

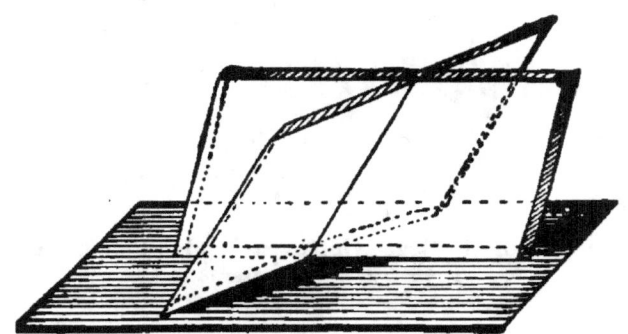

Fig. 75.—A convergent intersection.

sents a crossing of the Karl and the Ludwig veins in the now abandoned Habacht Fundgrube near Freiberg (Bescheert Glück). "The two lodes,

[1] 'Handbuch der Geognosie,' Freiberg, 1836, Vol. II, p. 604.

each of an average thickness of 10 to 15 cm. (4 to 6 in.), join in such a way that the outermost crust of quartz (q) and the second of rose spar (r) pass without interruption from the one into the other, while the third, a mix-

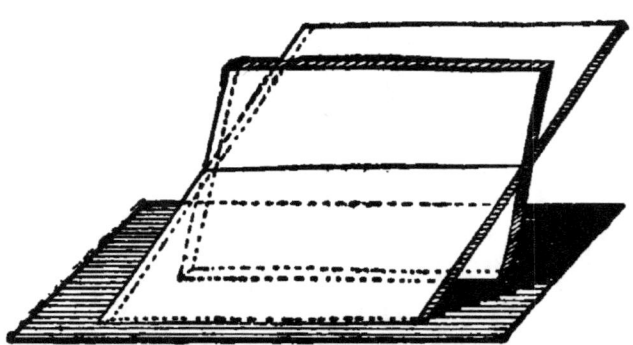

Fig. 76.—An intersection in dip.

ture of rose spar, argentiferous black zinc-blende and argentiferous galena (b), encloses within the space formed by the two lodes a druse-like cavity 18 to 23 cm. (7 to 9 in.) long." This druse is lined with quartz crystals.

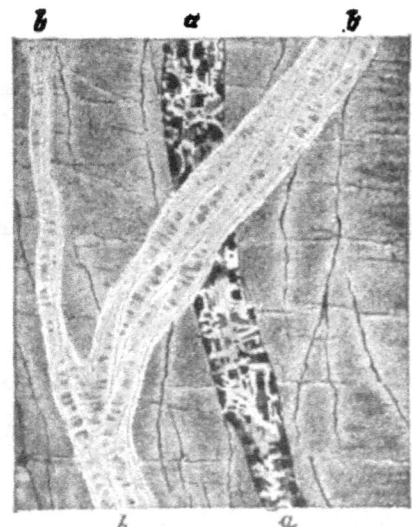

Fig. 77.—Appearance of the Segen Gottes Stehenden vein of the Simon mine at Freiberg. (Weissenbach.)

A similar description is given by K. Dalmer[1] of the crossing of the flat dipping veins with the horizontal tin veins at Zinnwald in the Erzgebirge. In this case, too, there is no intersection, but a coalescence of the vein fillings as well as of the greisen zones on both sides.

[1] 'Erläut. z. S. Altenburg-Zinnwald,' p. 36.

A *deflection* of a vein occurs when one vein drags against another for some distance and then cuts through and along it (compare, on this point, Figs. 79 and 80).

It sometimes happens that the approaching vein breaks up into stringers, only a few of the stringers passing through the other lode and uniting on the other side again to form the lode with its former thickness. This breaking up may take place on one side only, while on the other the lode may at first appear nearly as a single, feeble stringer, as in the example (Fig. 81) from Andreasberg. In this case the deflection is produced not by an ore

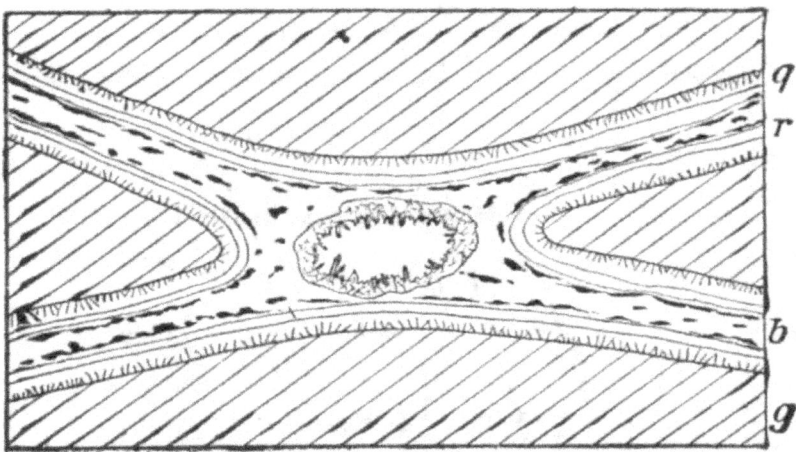

Fig. 78.—Vein crossing with simultaneous filling. (Kühn.)

g, gray gneiss; q, quartz; r, rose spar; b, rose spar with zinc-blende and galena. In the center a quartz druse.

vein, but by a fault (geschiebe) fissure filled with crushed country rock and traversed by many slip planes.

Sometimes a vein splits into stringers as it approaches another, and at that point stops, as shown in the example (Fig. 82) described by von Weissenbach.

The deflection of younger veins against older ones before the older vein is actually reached is due to numerous shear planes or joints approximately parallel to the older fissure and closely spaced, but which may be so fine that they are barely recognizable. The force producing the later vein would be relieved or entirely changed in direction along the shear planes, as well as by the older fissure, since the new fissure would follow the line of least resistance in the rock.

Fault fissures filled by attrition breccia and mashed country rock also cause vein deflections. Thus the so-called Ruscheln (rotten lodes) cause deflection of mineral veins even more frequently than other veins do. By

Ruscheln the Harz miner designates systems of parallel closely spaced, overthrust fault planes filled with completely crushed and ground up coun-

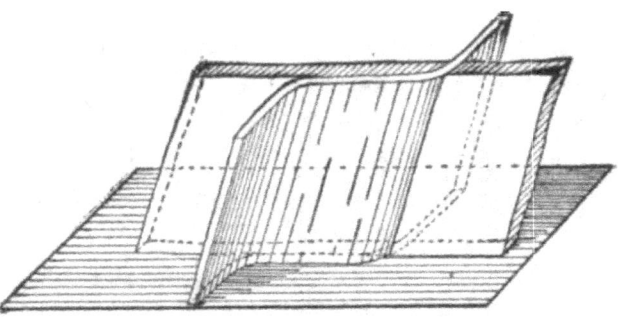

Fig. 79.—Deflection of a vein on passing through another. (Veins crossing without displacement).

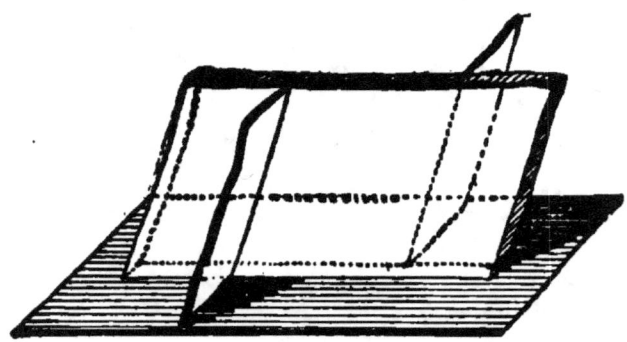

Fig. 80.—Deflection and apparent displacement, with stringer formation.

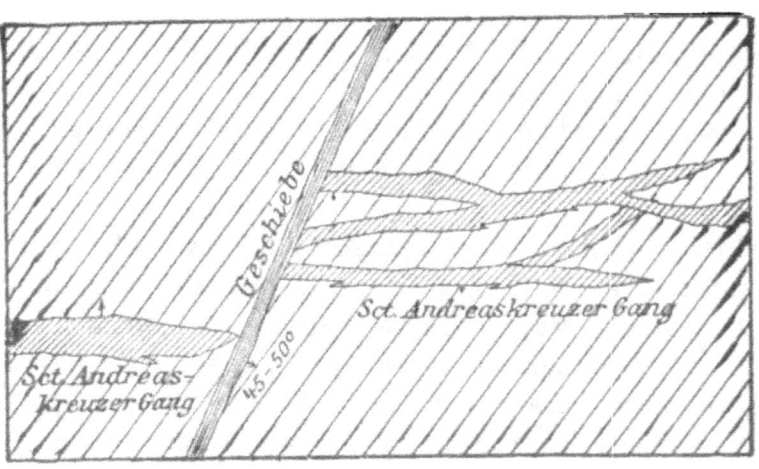

Fig. 81.—Deflection of an Andreasberg vein against an older vein with a scattering into stringers on one side only. (Zimmermann.)

try rock, the so-called vein slate, Gangthonschiefer, and sometimes with solid fragments of graywacke. The construction of these Ruscheln is sometimes disclosed by mining operations. One, for example, the Fauleruschel, near Clausthal, was cut by a drift from the Kaiser Wilhelm, shaft II. Innumerable folds and crinkles were found within it, as well as many well marked slip or gliding fissures.[1] Similar conditions prevail in the 'Lettenkluft' (clay fissure) at Pribram, which is also a reverse fault, against which many of the veins are deflected, also in the 'Bar Flache' of the Himmelfürst mine, near Brand, a zone of barren crushed and altered material which cuts off a great many veins. It will be observed that this feature differs from the cutting off of a vein by a later fault. The mineral vein splits up or ends against the vein of crushed rock.

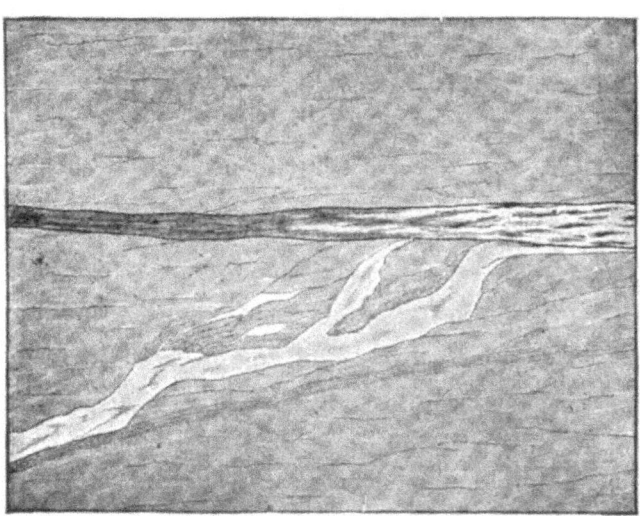

Fig. 82.—Complete splitting up into stringers of the Frisch Gluck Stehende on approaching the Peter Stehende in the Hoffnung Gottes mine. (Ground plan, after von Weissenbach.)

The deflection of dikes of eruptive rock by fissures is of very frequent occurrence. Fig. 83 represents the ground plan of a porphyrite dike cutting quartzitic slate, at Dovigfoss, near Aamot, in Norway. The dike is deflected several times by small quartz stringers. This dike is itself a branch of a larger dike, and it is clear that the course of the walls of this larger dike is also influenced by the quartz stringers.

The discussion of deflection naturally suggests and leads up to the subject of *displacements*, a feature which is most appropriately treated here,

[1] G. Köhler: 'Beiträge z. Kenntniss d. Erdbewegungen und Störungen der Lagerstätten,' *B. u. H. Z.* 1897, p. 217.

since many mineral veins are formed by the filling of fault, that is, displacement fissures.

(h) Displacements.[1]

By faulting we mean a change in the relative position of one block compared with the other, along a plane separating them. Such dislocations fall into two main groups: 1, faults or displacements produced by vertical

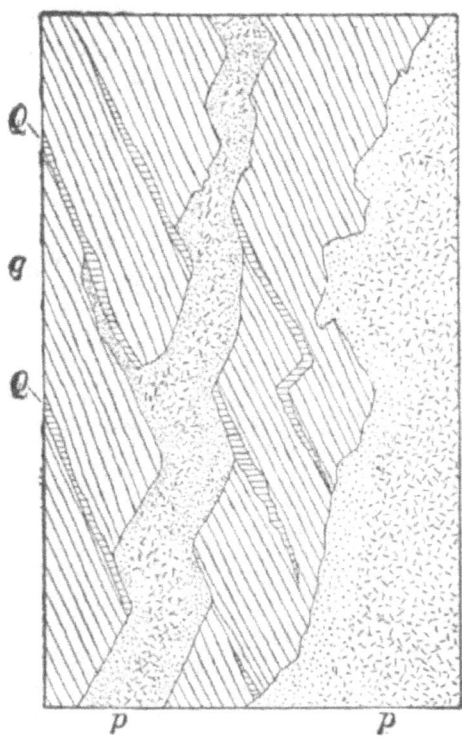

Fig. 83.—Dike of augite porphyrite in quartzose schist at Davigfoss. (Ground plan 1:15.)

q, quartzite slate; Q, quartz stringers; p, augite porphyrite, finely crystalline along the selvage.

movements; 2, faults produced by horizontal movements. The second group is the least frequent and least important. There is a possibility of

---

[1] A very full list of publications about the faulting of veins is given by F. T. Freeland, 'Fault Rules,' *Transactions* American Institute Mining Engineers, Vol. XXI, 1892, p. 491. The more important publications include: J. C. L. Schmidt: 'Theorie der Verschiebung älterer Gänge.' Frankfort, 1810. C. Zimmermann: 'Die Wiederausrichtung verworfener Gänge,' etc. Leipzig, 1828. R. von Carnal: 'Die Sprünge im Steinkohlengebirge.' Karsten's *Archiv n. F.* 1832, IX. H. Höfer: 'Die Ausrichtung von Verwerfungen.' *Oest. Z. f. B. u. H.*, XXIX, 1881, and XXXIV, 1886. G. Köhler: 'Die Störungen der Gänge, Flötze und Lager.' Leipzig, 1886. E. de Margerie and A. Heim: 'Die Dislocationen der Erdrinde.' Zürich, 1888.

combination of the two kinds of movement, and hence cases may arise which may not fit into these two classes.

(a) *Displacement Produced by Vertical Movements.*

This title implies movement along a vertical line. In most cases, however, the movement is not exactly vertical, but oblique. This kind of displacement can be subdivided into two classes: (a) Displacements pure and simple; (b) flexures. The first is caused by a fracture accompanied by a relative shifting of the two parts (Fig. 84); the latter is merely a bending or sharp flexure of a part of a layer or plate of rock (Fig. 85). If the

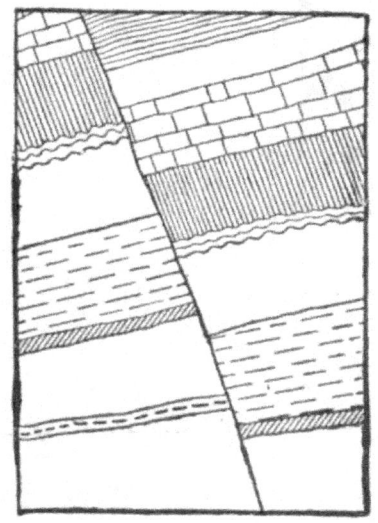

Fig. 84.—Simple fault (section.)

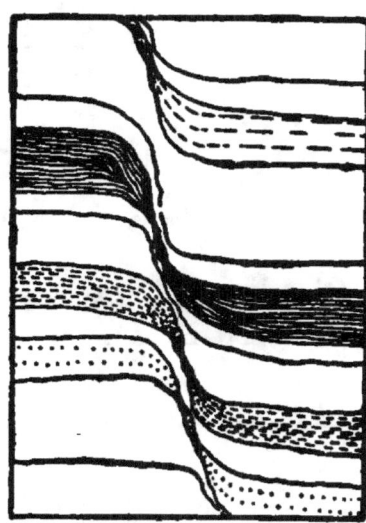

Fig. 85.—Flexure (section).

flexure is so sharp that rupture occurs, the flexure then becomes a true fault, accompanied by a drag of the ends of the strata on both sides (Fig. 86).

I. *General Properties of Simple Faults.*

A few definitions are needed as a preface to a discussion of faults. The distance between the faulted ends measured along the plane is the *displacement* or shift (*f* in Fig. 87). The distance along a horizontal plane between the ends of the faulted vein is called the *heave* (*w* in Fig. 87). The vertical distance between two displaced portions of a vein or bed is called *throw* or vertical downthrow (*s* in Fig. 87). It should also be remembered that the width of the fault fissure must be considered as a factor in discussing the heave of faults.

The horizontal projection of the fault plane is usually a straight line (Fig. 88); sometimes it shows curves, which are merely variations of the

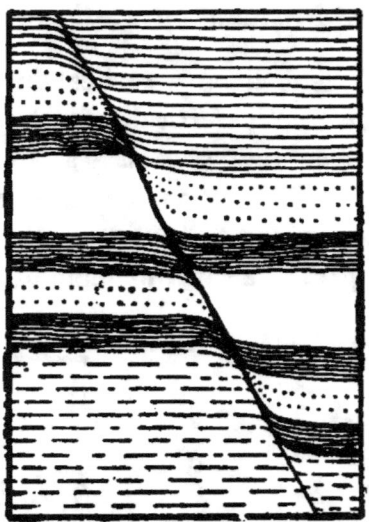

Fig. 86.—Section of a fault becoming a flexure.

general strike. In rare cases closed curves have even been observed (Margerie and Heim).

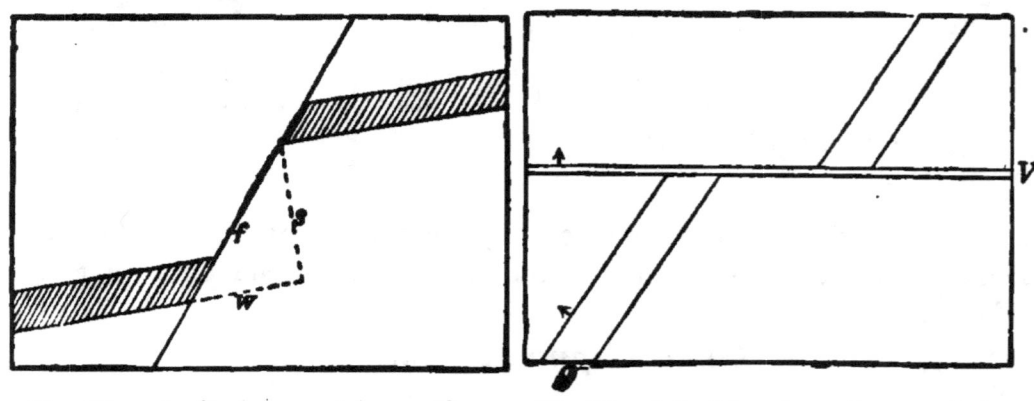

Fig. 87.—A displacement in profile.

f, displacement or shift; s, vertical displacement or throw; w, heave or horizontal displacement.

Fig. 88.—A faulting shown in ground plan.

g, displaced vein dipping northwest; v, fault fissure dipping north.

II. *Filling of Fault Fissures.*

A fault is but rarely a gaping fissure. It is usually filled either by more or less comminuted and ground up fragments of rock from both the walls, or by mineral material subsequently introduced, which may include ore. Very frequently both rock rubbish and mineral deposits are present, especially in faults occupied by mineral veins.

We will first describe in some detail the mechanical fillings which are common in both barren and in ore-bearing fissures. This filling may consist of irregular, mostly angular blocks and fragments, sometimes subangular or even rounded. The material is often held together by a paste or cement of firmly compressed, finely triturated particles, or it may even be cemented together by a mineral deposit. Such *friction breccias,* as this kind of material is called, are sometimes quite thick. In the great Feldbiss fault of the Aachen coal basin, the breccia attains a maximum thickness of 12 m. (40 ft.). The mineral veins of Freiberg very frequently consist of a breccia of gneiss, with scattered stringers of ore. Such a structure is especially frequent in compound veins or lodes. At Freiberg, the spaces between the rounded blocks of gneiss are often filled by pebbles of greatly decomposed gneiss with ore veinlets between. The fragments of rock are often flat, and the slabs lie parallel to the walls. If these fragments exhibit the banding of the gneiss, it is generally seen that the direction of banding does not agree in the different blocks, and differs from that of the solid country rock. Similar conditions were described and figured by G. A. von Weissenbach at the Mächtigen Gang of the Gnade Gottes and Neujahrs Maasen mines at Johanngeorgenstadt, a vein occurring in altered schist (Fig. 89). As he graphically describes it, the schist "appears jammed to the side or somewhat displaced, one part over the other, in the same way that the fibres of wooden props in mines are seen to be jammed by the pressure of the rocks above and squeeze out in a plane at right angles to the axis of the timber."

Sometimes well-rounded fragments of the country rock lie scattered through a finely crushed mass, and by movement within the mass during faulting, and by contact with harder grains or with sharp projections of the fissure walls, they have been scratched and scarred exactly like the striated boulders of the ground moraines of a glacier. Such striated surfaces are especially well developed in the veins of the Himmelfürst mine near Brand. On the Bar Flache, distinctly scratched specimens of country rock were found. In the Daniel Flache, ground-up specimens of this kind occurred in the midst of a clayey mass 0.5 to 2 m. thick, formed by fragments of ore and gangue, and older vein[1] fillings.

Figure 90 represents a fault cobble showing distinct friction striae from the Neuglück Spat drift, No. 15, below adit. The cobbles lie either in the vein clay or among the pebbles of crushed gneiss, both along the walls and

[1] E. W. Neubert in *Freiberger Jahrbuch.* 1881, p. 63.

in the center of the lode. Such friction gravels when compacted and indurated form a conglomerate vein. In such cases the pebbles are sometimes packed close together, as on the Peter Stehende of Christbescheerung mine, near Freiberg. Sometimes they are incrusted over with mineral de-

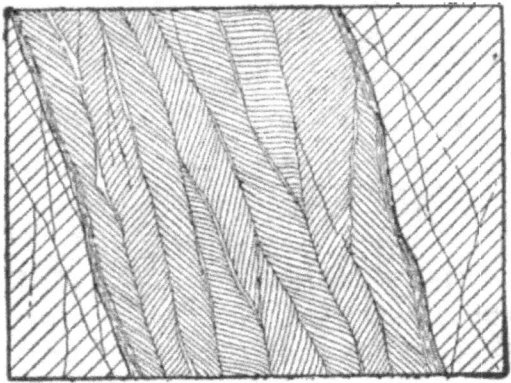

Fig. 89.—Section of a mineral vein near Johanngeorgenstadt. (Von Weissenbach.) Shows flat schist blocks parallel to the lode surface, with planes of schistosity running in various directions.

posits, as on the Eduard Spat of Himmelsfürst, where crystals of stephanite coat pebbles of mica schist. G. A. von Weissenbach[1] described such vein

Fig. 90.—A fault cobble with scratches. From Daniel lode on Himmelsfürst. (A part broken off.) (From nature, two-thirds natural size.)

conglomerates under the term 'Kugelgestein' (ball rock) and very aptly compared the process of their formation by friction movement between the fissure walls to that observed in grinding-mills.

[1] 'Abbildungen merkw. Gang verhältnisse.' 1836, p. 18.

Such conditions are fairly common in the tin veins in quartz porphyry, at Altenberg in the Erzgebirge. The vein filling consists of hard balls of porphyry imbedded in a clayey mass composed of completely altered and ground-up porphyry. Similar formations are known in the Elias vein near Joachimsthal, and from the tin veins of Cornwall (Collins). S. F. Emmons[1] also mentions similar cases, but ascribes the rounding as often due to decomposition resulting from the action of mineral solutions acting upon fragments that had previously been angular. In Fig. 91, we give a picture, after G. A. von Weissenbach, of such a drag or conglomerate vein 10 centimeters (0.33 ft.) thick, which lies along the hanging wall of a vein at Luxbach near Annaberg.

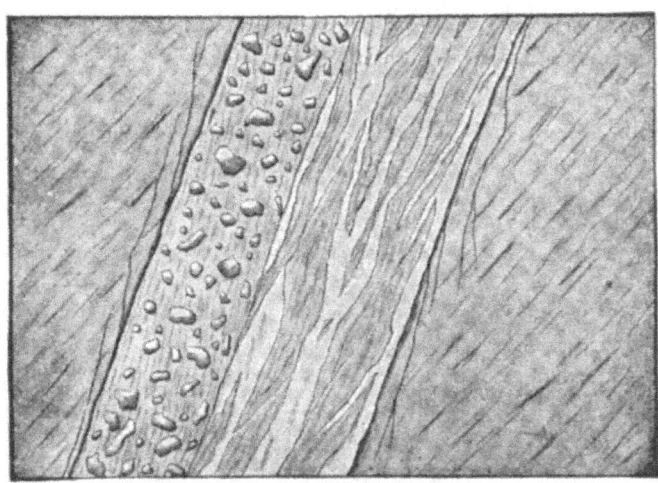

Fig. 91.—A conglomerate vein along the hanging wall of the Neu Unverhofft Glück vein, near Annaberg.

This attrition gravel of veins must not be confounded with those real gravels derived from overlying sedimentary strata, and which, in some cases, have fallen into open fissures, as shown later.

In connection with these vein gravels,[2] mention should also be made of the rare occurrence of rock fragments that have been twisted off cylindrically in fault fissures. The Freiberg mining school museum contains such a piece 15 cm. (0.5 ft.) long, as thick as an arm, and resembling the branch of a tree. It consists of a hard, brittle greisen from Altenberg, which, besides having two rounded faces of cross fractures, is completely covered with

[1] 'Structural Relations of Ore Deposits.' *Trans.* Am. Inst. Min. Eng., 1888, p. 19.

[2] This paragraph in the German original describes distinctions not used in America. The finer particles, consisting of fragments of the size of a hazelnut, and mostly quite soft as a result of advanced decomposition, constitute the secondary layer (ausschram, filings). The so-called Guhrs are still finer, forming a clayey gravel. Lastly we have the vein clays, which are very common, and represent the finest product of rock comminution.

slip striae running obliquely around it. These gliding surfaces (Fig. 92) are even at one point involute, that is to say, two of them follow at a little distance, one over the other, in such way that the slip planes intersect at acute angles. Such peculiar specimens are formed when an upright fragment of rock is dragged between the two walls of the fault fissure in a direction parallel to the longer axis of the fragment, while at the same time it is moved about or revolved about this axis.

Fig. 92.—Drawn-out cylinder of greisen fragment from a fault-cleft at Altenberg.

The *vein clays* are soft, pasty or tough, plastic clays, frequently mixed with larger grains, and often colored black by carbonaceous material. In the blackish vein clay of Verespatak and Maidanpeck, the so-called 'Glamm,' the coloring, according to E. Tietze,[1] is due to copper oxide. Posepny's[2] view that the Hungarian 'Glamm' has been washed into the fissures from the surface seems improbable, and it appears more likely that this material is the result of trituration and decomposition, and was either formed in the lode fissure itself or forced into it from greater depth, as will be explained later on.

[1] E. Tietze: *Jahrb.* d. k. k. Geol. Reichsanst. 1870, p. 321.
[2] F. Posepny: *Idem.* 1867, p. 101, and 1870, p. 273.

Sometimes such a vein clay shows a distinct slaty or schistose structure, due to strong pressure. Such material separates along these cleavage planes into scales which are often as smooth as glass or are covered with parallel friction striae. Such vein clay slates from lodes of the upper Harz, described in detail by A. von Groddeck,[1] are mechanically altered Culm clay slates, which have undergone, especially the variegated varieties, a chemical transformation, as noted later.

The finer pebbles (ausschram) and the clays are often confined to a narrow zone along the wall, in which case they form the clay selvage or gouge. Such a clay selvage is ordinarily welcome to the miner because it facilitates the loosening of the true vein filling from waste material, while the work is more difficult when the lode is "coalesced with the country rock" by crystalline crusts, as the German miner puts it, or "frozen hard to the country," as the American miner picturesquely expresses it.

These friction products are workable when they are impregnated with finely disseminated ores, which are mostly of secondary (*i. e.,* later) origin. In the Erzgebirge such vein clays are colored yellowish, reddish or brownish by metallic compounds, being known as gilben (mountain yellows) or braunen (browns). At Freiberg, in the Thurmhof vein, a yellow vein clay with 0.062% of silver was formerly mined, while in the Friedrich vein the yellow vein clay held scattered particles of wire silver and grains of earthy silver glance, so that the clay contained 1.6 to 1.9% silver. In the Annaberg area, according to H. Müller,[2] these vein-clays often contained a considerable amount of silver, so that they paid to work in the Himmlisch Heer. Black clays containing as much as 5% in silver were also extracted there. [At Butte, Montana, the fault clays often contain ore particles and the filling sometimes consists of balls of pyritic ore, bornite, glance or enargite. quartz and altered granite, so that the entire mass is workable. The black clays often contain as high as 8% copper, but are not mined alone.] Mention must also be made of the fine crystals of native gold in clay of the Farncomb Hill veins of Summit county, Colorado, described by Rickard.

The description of the mechanical material filling fault fissures and the impregnations just mentioned naturally precedes an account of the fault fissures which form mineral veins.

III. *Fault Veins.*

Mineral veins are often true faults. In the Freiberg area it is seldom that this can be shown by a comparison of the rock of the two walls, be-

---

[1] A. von Groddeck: *Z. d. D. g. G.,* 1866, Vol. XVIII, p. 693 and 1869; Vol. XXI, p. 499, and *Jahrb.* d. k. preuss. geol. Landesanst., 1885, pp. 1-52.

[2] 'Die Erzgänge des Annaberger Revieres.' Leipzig, 1894, p. 96.

cause the rock is of very uniform nature, but it is frequently demonstrable when the fault veins cut through older veins.

A good example in which the faulting can be shown by the rocks is seen in the Himmelsfürst mine near Brand, where the Benjamin Stehende cuts through and displaces bands of muscovite gneiss and mica schist intercalated in biotite gneiss; the fault-vein is itself displaced by another mineral vein, the Neuglück Spat (E. W. Neubert, MSS.).

The lodes of the upper Harz are, as is well known, fault veins in which the dislocation is very clearly shown by the different beds of the Devonian and Culm (carboniferous) formations. At Bockswiese and Lautenthal the veins have a Devonian footwall, while the hanging wall is formed of rocks

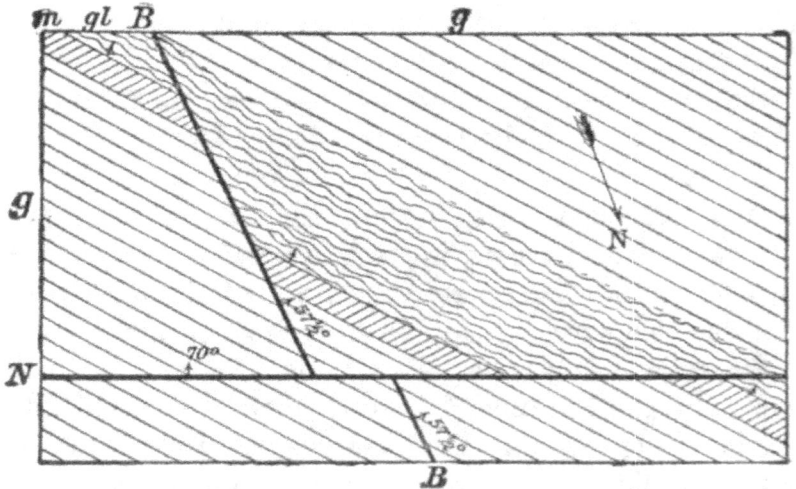

Fig. 93.—The Benjamin fault vein of Himmelsfürst. (Ground plan.)
B, Benjamin Stehender; N, Neuglück Spat; g, biotite gneiss; m, muscovite gneiss; gl, muscovite schist.

belonging to the Culm. The throw, according to A. von Groddeck,[1] is 200 meters. Fig. 94 (after E. Maier[2]) shows a cross-section of the displacements of Paleozoic strata produced by the main Bockwieser (Pisthaler vein and the Grünlindener vein).

Good examples of silver veins filling fault fissures displacing Carboniferous strata occur at the Enterprise mine, Colorado, and have been described by T. A. Rickard.[3]

Even where the uniformity of country rock prevents a direct recognition of displacement along mineral veins, the dislocation may often be inferred

[1] 'Lehre,' etc. 1879, p. 229.

[2] '*Beitrage* zur Geol. des Bockwieser Ganggebietes.' Diss. Ber. d. Naturf. G. zu Freiberg, 1900, Vol. X, p. 2.

[3] T. A. Rickard: 'The Enterprise Mine,' *Trans.* Am. Inst. Min. Eng., 1896, Vol. XXVI, pp. 906-980.

from other observations. Thus the local widenings of veins have been explained since De la Beche's[1] time as the result of dislocation of the vein walls and due to their projections and irregularities. The accompanying Fig. 95 illustrates this shifting of irregular walls, which must necessarily produce prismatic or lenticular cavities. Le Neve Foster[2] uses the same theory to explain the peculiar pipe veins in some Cornwall mines, and Th.

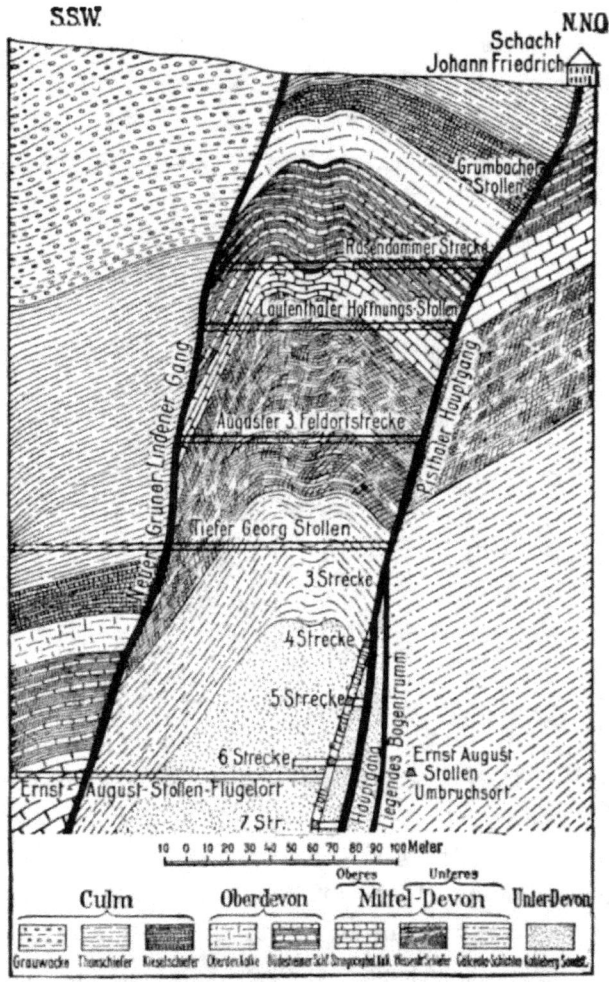

Fig. 94.—Cross-section of displacements of Paleozoic strata. (E. Maier.)

Kjerulf[3] has endeavored to explain the genesis of the Norwegian 'rulers' (erzlineale) in the same way.

IV. *Condition of Fault Walls.*

The sliding motion of one side of a fault over the other, and the conse-

[1] 'Researches in Theoretical Geology,' 1834, Chap. 10, and 'Report on Geology of Cornwall,' 1839, p. 317.

[2] *Trans.* Roy. Geol. Soc. Vol. IX, 1876.

[3] 'Geologie von Norwegen,' Trans. by Gurlt. 1880, p. 294.

quent rubbing of the walls, produces smooth and evenly polished faces. Such polished surfaces are called slickensides and sometimes show truly mirror-like surfaces, if the material itself has a metallic luster, as, for example, the magnificent mirrors of pyrite found in the Confesionario mine (Huelva) or of galena of Inçurtosi (Iglesias, Sardinia). Similar examples may have led the Aztecs to make their remarkable pyrite mirrors. In other cases, numerous parallel scratches are seen on such polished surfaces, either as delicate striae or relatively deep furrows. An example of this kind from the Lade des Bundes lode at Himmelsfürst, near Brand, in which the striations are cut in galena, is shown in Fig. 96. Wherever these features occur they indicate that the fault fissure was not an open one during the move-

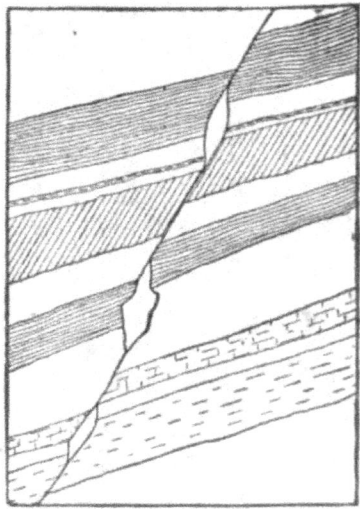

Fig. 95.—An explanation of local vein-widening from a fault.

ment. The direction of the striations on the vein wall is supposed to represent the direction in which motion has taken place, but one must not assume that they are always parallel to the dip of the fault fissure. This is by no means the case, as has been shown by H. Höfer. This view (at one time generally prevalent) arose from the Schmidt-Zimmermann fault rule, according to which the rocks in the hanging-wall of the displacement fissure slid down parallel to their former position. This rule does indeed hold for many cases, but it is by no means universally valid.

Fault striations often form an acute angle with the line of dip; indeed, cases occur where they are almost perpendicular to the dip. Such a condition, for example, has been described by H. Höfer from the ironstone mine, near Ober-Zeiring, near Judenburg, in Styria, and from the Hodritsch lode area near Schemnitz in Hungary. In the fault fissures of the upper Harz region the fault slips or gliding planes show horizontal striae, as has been known since Fr. Ad. Romer's time. Finally, very fine instances of horizontal friction striations several meters long have been observed by the present author in the main great fault in the sandstone of Lusatia.

The direction of fault striae is not always the same over the entire wall of the same fault fissure.

Numerous deviations occur and several systems of lines and furrows may even cross, and in one cluster of many parallel faults the striae may have various directions on different surfaces.

The striations are not always straight. The slipping may have taken place in rapidly varying directions, and in such cases the striae may form curiously winding curves, as in the specimen exhibited before the Naturalists' meeting at Vienna in 1892 by Ed. Suess.

Fig. 96.—Striated surface of galena from Lade des Bundes lode in Himmelsfürst mine. (From nature. One-fifth natural size.)

In this case the intricately tangled scratch lines showing on a polished fault plane surface of black calcareous slate, from Radotin in Moravia, were evidently caused by minute but hard projecting points of the opposite wall of the fault fissure. This is borne out by the complete agreement of the various scratch figures (reproduced in Fig. 97 from a self-printed plate, with the kind permission of the discoverer, J. J. John, and the first describer, Ed. Suess). Suess very appropriately styled the specimen a natural earthquake autograph, for it may readily be imagined that its formation was due to an earthquake tremor.

In other cases it is probable that the curved scratches were produced by hard grains and fragments present in the fissure filling itself, and not by a projection of the opposite walls. Intersecting striations can also be produced by a change of position of the rock fragment, while it was pushed along in the fissure.

The reader is cautioned not to assume that the length of the striation represents the extent of the movement. On the contrary, the length of such a scratch represents in most cases the combined length of the scratched surface and that of the dislocation (movement). It is only in rare cases that such a scratch can be supposed to be caused by a single hard grain, that is

Fig. 97.—Earthquake autograph, gliding surface with intricately tangled striae on limestone from Radotin. (A print from the stone itself.)

to say, geometrically speaking, by a single point over which the surface was drawn.

All these considerations show that these "handwritings on the wall" (Rickard) cannot always be deciphered as easily as might seem at first sight. At any rate, a test made by H. Höfer seems to apply to many cases. If, in passing the hand over an extensive gliding plane, you receive the impression of perfect smoothness, your hand has been moving in the direction in which the opposite rocks moved; if you feel any roughness the hand is moving opposite to the rock movement. The reason is that the gliding planes have the same relief on a small scale as the glacier-polished surfaces of rocks, as is shown in the following diagram (Fig. 98). In this diagram, ground-up material has been indicated as filling the fault fissure. Such material may, however, be practically absent.

When a fault passes transversely through thinly bedded rocks the edges of the strata are often bent downward (or upward) in the direction of the movement. In some veins the walls of the open fissure may have been coated with mineral crusts which remain intact on one wall, while along the other the mineral crusts were broken by later fracturing and the ends of the beds bent around (Fig. 99).

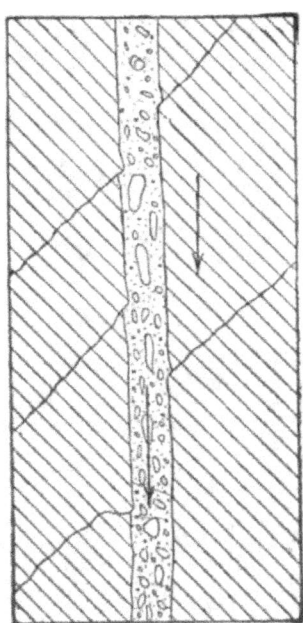

Fig. 98.—Ideal section of a fault fissure to illustrate Höfer's rule. (The displacement and grinding took place in the direction of the arrows.)

V. *Torsional or Turning Movement During Displacement.*

A consideration of slip surfaces or planes, whose striation is not always coincident with the dip, shows that the Schmidt-Zimmermann rule has many exceptions. In these exceptional cases the dislocation sometimes may be inferred from observations of a different kind.

H. Höfer has properly revived an observation communicated long ago by J. F. W. von Charpentier.[1] In the Himmelsfürst mine, near Freiberg, at the junction of a Morgengang with the Schneider vein, 120 feet below the deep level of that locality, the two veins intersect without change of

---

[1] 'Beobachtungen über die Lagerstätten der Erze,' Leipzig, 1799, p. 103.

either strike or dip. On the contrary, 240 feet lower, the course of the Stehende vein changes at its junction with the Morgan vein, and a so-called dragging of both lodes could be noticed. This is also observed at a depth of 360 feet, except that there the union of the two lodes is maintained for a greater distance than in the first instance. Thus the ends of the faulted vein on opposite sides of the fault approach each other in depth, and actually meet 120 feet below the drift. This proves that the displacement is not parallel, but due to a twisting (torsional) movement. Similar examples have been recently found in various places. In the Freiberg area a turn or twist of this kind is shown in the displacement of the Peter Stehende of the Alta Hoffnung Gottes mine. This vein is dislocated by the Flache fissure, dipping 70° east. In the upper part, the two sections thus formed (the northern of which is entered on the maps as Einigkeit Morgen-

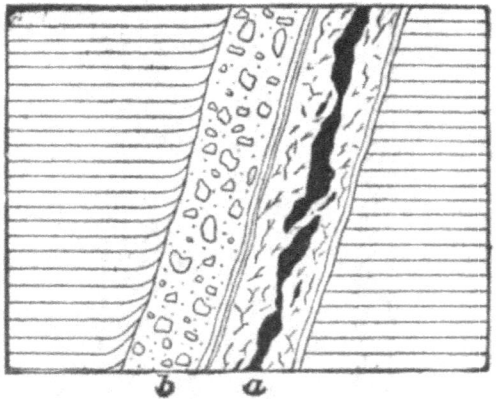

Fig. 99.—Section of a fissure with sharply cut footwall, while the hanging wall shows the strata bent.
a, older ore-bearing stringer; b, younger friction breccia.

gang) are rather far apart. The distance between them along the Flache fissure is 13 meters; in the counter drift, only 11 meters; in the seventh, only 5; in the ninth, 0. Thus if we could see a considerable extent of the striations and slickensides of the fault-wall, they would represent segments of a circle.

Among more recent examples may be mentioned the great displacement called Münstergewand Feldbiss, near Aachen, which, according to Höfer, is connected with a distinct twist. [Also at Butte, Montana, the great Anaconda vein is displaced several hundred feet vertically by a fault showing a torsional movement.]

VI. *The Surface Evidence of Displacements.*

A fault does not usually show as a cliff or wall on the earth's surface. That it does so at first is evident, and scarps are formed by fissures due to the sinking of the earth's surface as a result of coal mining, but such superficial scarps though sharply accentuated at first are very quickly leveled by the rainfall. In the first stage of this leveling process trench-like depressions are often formed as the rainwater washes away the loose material filling the displacement fissure, and a secondary subsidence thereupon occurs near the surface, as shown in Fig. 100 and 101. Such a section is often shown by faults resulting from coal mining, provided the fissures have a favorable dip.

Fault scarps do sometimes form natural features, especially in dry countries where erosion makes but slow progress, as, for example, the displacement scarps of the Colorado plateaus described by Powell. In most cases, however, all traces of the fault scarp are quickly effaced and merely indirect topographic indications are presented; such as the course of brooks that have followed a fissure, or a line of springs along the displacement. If, on the contrary, the fissure is filled with loose, permeable material it may gather the ground-water from afar, become completely saturated in its lower portion, and give rise to dangerous outbreaks of water in mines. Though a displacement may not be indicated topographically, yet in most cases it can be detected by the interrupted or diverted course of the rock outcrops.

VII. *Different Kinds of Simple Displacement.*

If we include displacements of sheets and beds with those of veins, we may distinguish the following cases based on the relation of the down-thrown block to the strike:

1. *Strike* faults, or longitudinal displacements in which the course or strike of the fault coincides with the strike of the strata or of the lode (Fig. 101 and 103). Two cases are possible, (a) normal faults dipping toward the down-thrown block (Fig. 102), (b) reverse or overlap faults in which the opposite is true (Fig. 103). Strike faults of tilted beds are recognized on the earth's surface by repetitions of the outcrops of the different beds, often with omission of several members, as shown in Fig. 104.

2. *Cross* faults striking at or near a right angle with that of the displaced blocks (Fig. 105 and 106).

3. *Acute angled* or *diagonal* faults, in which the course of the fault forms an acute angle with the strike of the displaced block. (Fig. 107 and 108.)

Furthermore, displacements may be classified in accordance with the relative direction of displacement of the thrown block. According to this we have three cases:

1. *Normal or gravity* faults are ordinary dislocations in which the hanging wall block has slipped down along an inclined fault plane; according to

the Schmidt-Zimmermann rule this takes place in the direction of the dip and the two displaced blocks maintain a parallel position. As shown on p. 151, this rule has many exceptions. In plan the faulted beds occupy more space horizontally than they did before dislocation, *i. e.*, an area greater by the amount of the horizontal throw than its original area (extension dis-

Fig. 100.—Recent displacement fissure unchanged by rain-water.

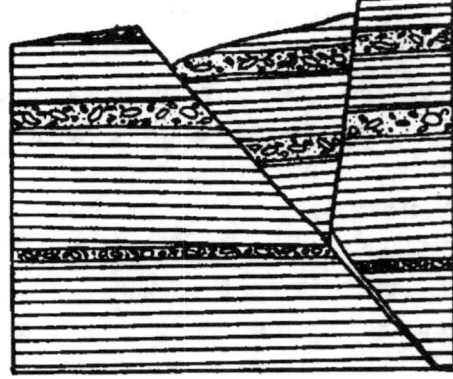

Fig. 101.—The same with a block dropped down.

placement of Margerie and Heim). Most faults belong to this first group. (Fig. 109.)

2. *Vertical* faults. In these the effect is the same whichever side is down-thrown (Fig. 110).

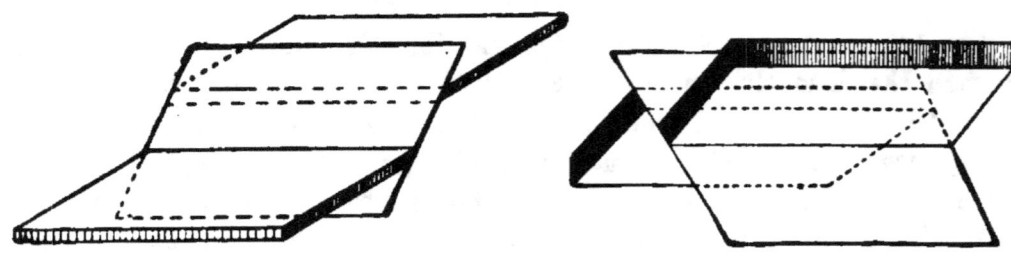

Fig. 102.—Model of a normal strike fault.

Fig. 103.—Model of a reverse or over-lap displacement.

3. *Overthrusts* or *reverse* or *overlap* faults, when the hanging wall block has been shoved upon the block that lies beneath the fault plane. As this is accompanied by a compression of the beds, so that they occupy a small-er space than before, this might, following Margerie and Heim, be called a compression displacement. (Fig. 111.)

Very fine examples of overthrusts are found near Johannesburg in the Transvaal, where the beds of gold-bearing conglomerate are dislocated by

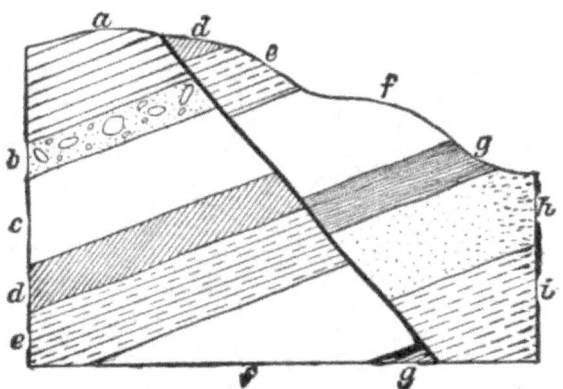

Fig. 104.—Section of a strike fault.

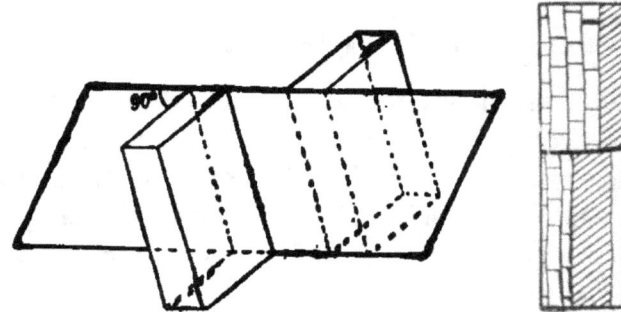

Fig. 105.—Model of a cross fault.

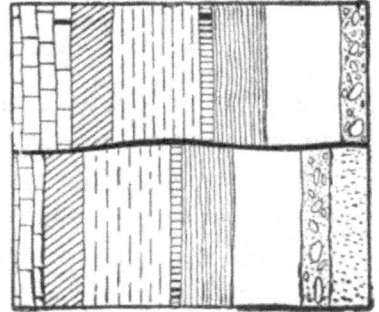

Fig. 106.—Plan of a cross fault

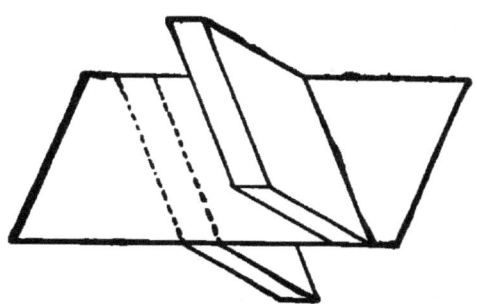

Fig. 107.—Model of diagonal fault.

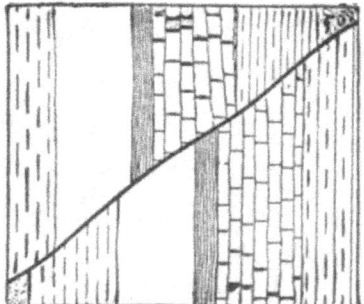

Fig. 108.—Plan of a diagonal fault.

several overlap faults. (Fig. 112.) This example also proves that overthrusts, though common in strongly folded mountain ranges, can and do occur in regions of open and comparatively slight folding.

Overthrusts may also occur along vein fissures, as, for example, in the ore-bearing lode of the Gute Hoffnung mine, near Werlau on the Rhine, and in the Schwebende of the upper Erzebirge in Saxony.

Fig. 109.—Normal displacement. (Section.)　　　Fig. 110.—Vertical displacement. (Section.)

VIII. *Special Types of Displacements.*

Many displacements are the result not of a single fault with down-throw or overthrust, but of a series of parallel fault fissures with successive displacements in the same direction. Groups of such faults, all in the same direction, are termed *step faults* or *distributed faults*. A good example is the great trunk fault, known as the "Rother Ochse," which traverses the

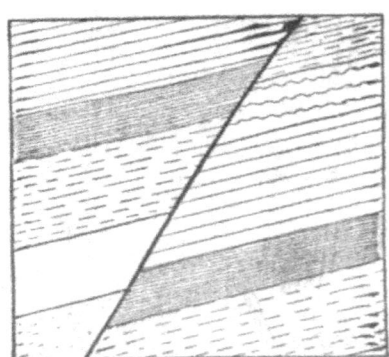

Fig. 111.—Overthrust fault. (Section.)

northeast flank of the Permian coal basin in the Plauen basin near Dresden. At Zschiedge, R. Hausse calculated the total throw of the many individual faults as 350 m. (1,148 ft.). Many partial faults have produced a complete chopping up of the main coal seam, since the brittle coal breaks into slabs and blocks that can readily turn about in the rather soft country rock.

The coal miners of Germany designate the thin blocks or slabs dropped down by step faults, or lying between two parallel faults, as *graben* (moat). Good examples of this character of displacement occur in the Mansfeld copper shales. In the vicinity of the veins such dislocations are naturally quite rare.

A block that has remained stationary between faults is called a *horst* (upthrust), or if narrow and long, it is called a *Rippenhorst.*

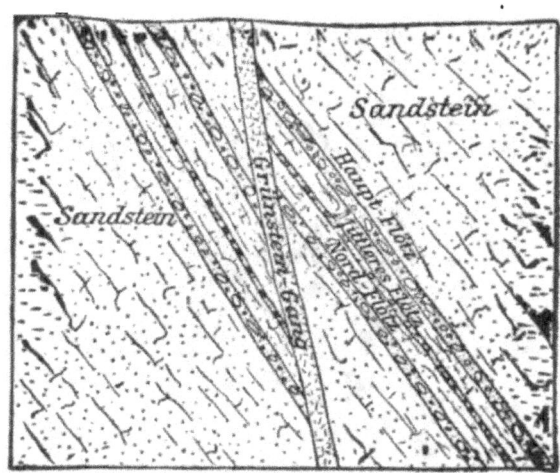

Fig. 112.—Reversed faulting of the gold-bearing conglomerate measures of the May Consolidated mine near Johannesburg. (Schmeisser.) (Section.)

IX. *Simple Displacements.*

Displacements that are perfectly apparent occur when a vein of considerable thickness passes diagonally through another, as illustrated in Fig. 113.

X. *On the Location of Faulted Parts of Veins.*

1. Before assuming that a displacement exists when a vein is lost in a mine working, its possible deflection must be considered, viz.: Whether the lode has not merely changed its direction and followed an older fissure. This supposition would be proven true if stringers occurred in the filling of the supposed fault fissure. These stringers would then have to be followed by an exploratory drift in order to locate the extension of the deflected lode beyond the transverse fissure.

2. If the vein is not deflected, but is actually displaced by a younger fissure, the first thing to do is to look in the fault filling for fragments broken off from the dislocated vein. If such fragments are entirely wanting on one side of the vein, while on the other they occur in a cluster, or fan, or are strung out along the fault forming a fan or string, a drift should be driven along the fault in this second direction (Fig. 114). It should also be noted that in a faulted bed the bending or change of dip next the fault may indicate the direction of the movement. It points in the direction the sheet has been moved. (See Fig. 99.)

3. If fault striations are found on the fault wall their character and direction will often enable one to correctly infer the direction in which the missing part of the lode has been moved. By passing a hand over the slick-

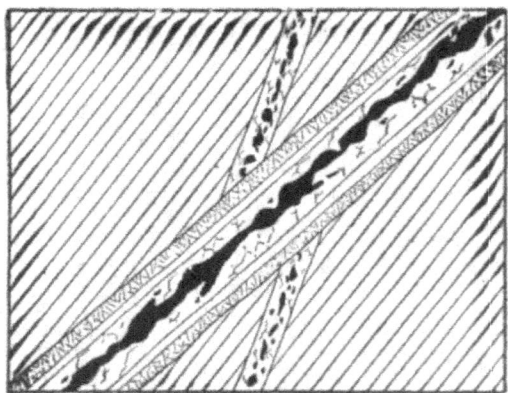

Fig. 113.—Displacement of a vein by a younger vein passing through at an acute angle. (Ground plan.)

ensides one may ascertain whether there exists a difference in the roughness, as set forth on page 150. The opposite wall has been moved in the direction in which the wall feels smoother.

4. If in a lode the hanging wall strata differ from those of the footwall, it may be assumed that the displaced portion of the lode has been moved downward. If the faulted vein has unlike rocks on opposite sides (contact vein) the drift should be driven across the fault to determine the nature of the rock at that point. If the hanging wall country is found across the fault the throw is normal and the displaced block has moved down. This rule is even more important in stratified deposits, and for this purpose a carefully made section of the overlying and underlying beds should first be made.

5. If the faulting is in a district where several other faults occur, which have been accurately located and their characteristics known, the new case

should be treated as analogous to those already known, for ordinarily in the same vein district the faults all show movement in the same direction. This, according to H. Höfer, is especially true for the lateral displacements along many faults, which may be alike over an entire vein area; thus according to H. Höfer, in the ore district of Littai in Carnicia they are left-sided; at the western Harz they are right-sided. At Chañarcillo, according to Moesta, they are left-sided.

6. When special indications of this kind are lacking it is necessary to attempt a graphical solution of the problem by means of a diagram based on geometrical principles. For such cases the Schmidt-Zimmermann rule is used, a rule which in practice is found to be true in the majority of cases. The procedure is as follows:

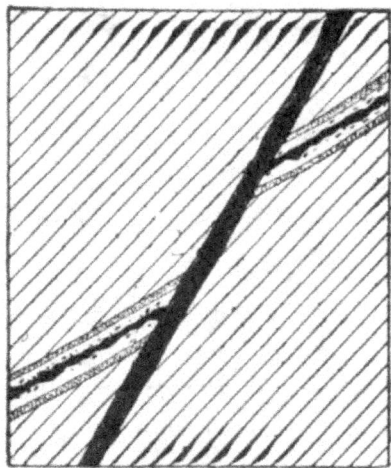

Fig. 114.—Fragments of the cut-off lode in the clayey filling of a displacement fissure as guide in tracing the lode.

First construct the projection of the line of intersection of the vein and the fault fissure. An example will show how this is done in a simple case:

Two intersecting veins, A S and B S, meet at S on the floor of a drift. The vein A S dips at the angle γ_1, the lode B S at the angle γ_2, γ_1 being larger than γ_2 (see Fig. 115). From point S measure any equal distance along each lode, S C and S D, and erect perpendiculars at points C and D. At the point S draw on S C the angle $90° - \gamma_1$. At point S draw on S D the angle $90° - \gamma_2$. Through the intersections E and F of these lines with the perpendiculars at C and D draw parallels to the corresponding vein, namely, through E parallel to A S, and through F parallel to B S. The two parallels intersect at the point G, which connected with S represents the line of intersection of the two lodes.

The line of intersection in the case of normal displacement is situated within the obtuse angle; in the case of left-handed or reverse displacement it is within the acute angle, and is always closer to the vein having the higher dip.

Having constructed the projection of the line of intersection, the following general rule is followed in order to find the missing vein:

The displaced lode is to be sought beyond the fault within the *obtuse* angle between the line of intersection and the fault.

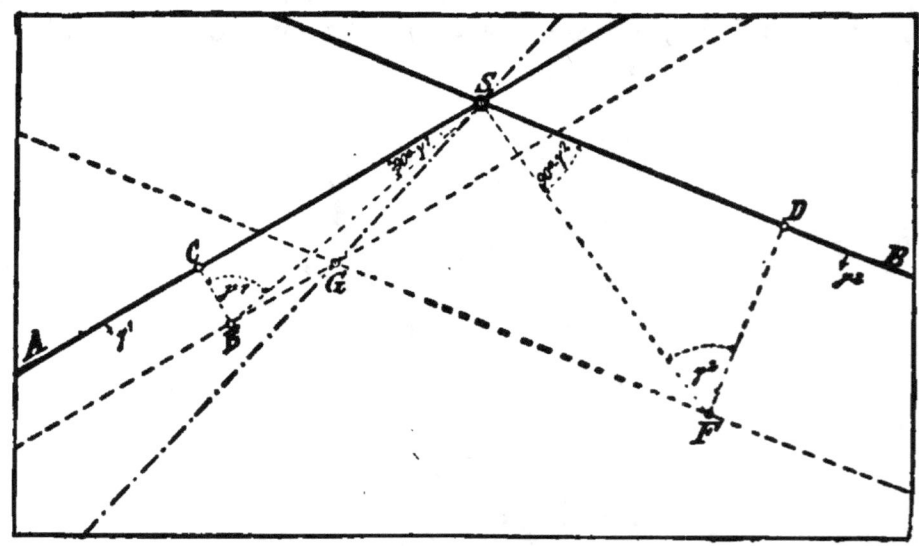

Fig. 115.—Construction of the line of intersection.   (Explanation in the text.)

In Fig. 116, for example, a lode has been cut off at B by the fault VV_1. KK_1 is the line of intersection between the two. The direction of the arrows indicates the direction in which the fault block must be sought following along the fault.

For the better understanding of the rule we reproduce a portion of the plan map of the Himmelsfürst mine, near Brand (Fig. 117), showing the actual recovery of the Silberfund vein which was faulted by the Neuglück vein at I. The dotted line indicates the projection of the intersection, which was drawn from the known intersection on the Halbvierzehnten level through the points II and III, intersections known in the higher drifts and projected on the map. The Neuglück vein was followed along its hanging wall, according to the rule, in the direction of the arrows a-d, and the Silberfund Stehende was actually recovered a short distance away. At point II the situation was more complicated because the vein was first faulted by a parallel fissure of the Neuglück Spat, and then once more by the latter itself. Hence at point II it was necessary to apply the rule twice in succession.

XI. *Displacements Due to Horizontal Movement.*

It has already been shown that in true faulting lateral movement is an occasional but a subordinate feature. We now take up a small group of faults in which a transverse horizontal displacement is the predominant feature. They are termed simple *shifts* (Verschiebungen), and the steep dipping fault planes along which horizontal movement takes place may be called slip planes (Blätter leaves).

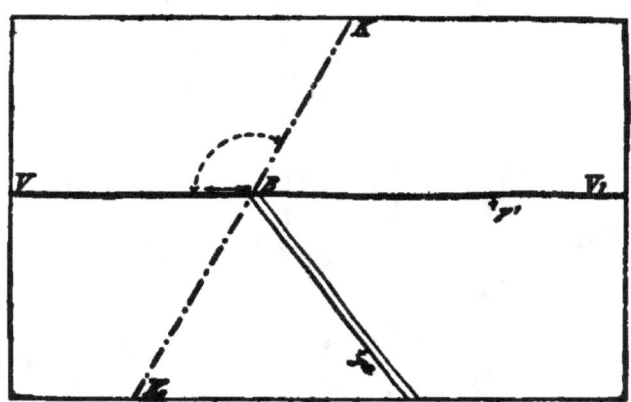

Fig. 116.—Diagram to illustrate the rule for tracing displacements.

G. Kohler has described shifts of this kind from several lodes in the Harz, such as the Samsoner lode at St. Andreasberg (Fig. 118) and the Altenseg-ener main lode. They are still commoner in stratified deposits, as, for example, in the Rammelsberg ore deposit.

In the recovery of faulted and shifted veins, rules 1 to 5, given for ordinary faults, apply in a slightly changed form.

(i) Fissure Formation.

Fissures are planes of disruption; hence dicission spaces (from discindere, to tear apart) must be discriminated from the cavities formed by the leaching action of water on readily soluble rocks, called dissolution spaces by F. Posepny. Fissures sometimes show clear proof of strain and shear (or a tearing apart) in the distorted fossils found in the country rock immediately adjoining. Such phenomena have been described by Harkness from the vicinity of Cork, and their significance discussed by Daubrée.

In some cases the fissures are formed by forces originating in the rup-
tured rock itself; in such cases the fractures are usually confined to this
particular rock mass, occasionally passing outside for a few yards. This type

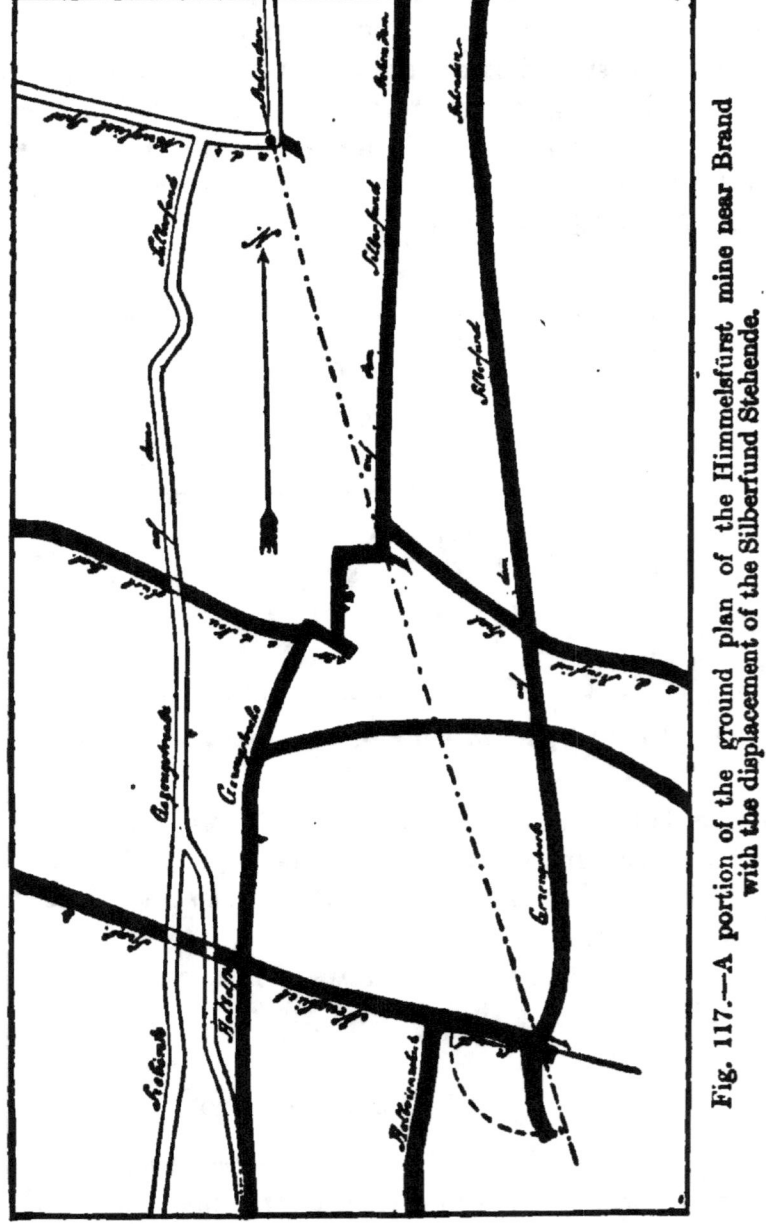

Fig. 117.—A portion of the ground plan of the Himmelsfürst mine near Brand
with the displacement of the Silberfund Stehende.

of fracture, called *entokinetic* fissure (A. W. Stelzner), falls into two sub-
groups.   In the first, the fractures are contraction fissures due to a diminution of volume of the rock mass as a consequence of cooling or drying (cooling fissures and drying fissures).   The second type originates through an

increase of volume in consequence of the absorption of water or some other chemical process as in serpentinization (dilation fissures of A. W. Stelzner). To a much wider extent fissures result from general mountain building forces, and may then be called, following Stelzner, *exokinetic* fissures. Many fissures cannot be readily classed in any of these groups because they are due to complex causes.

A. Daubrée[1] calls all disruption fissures lithoclases, and makes a further distinction between simple joints (diaclases), that is to say, fissures without demonstrable displacement, and paraclases, that is, fissures with displacement.

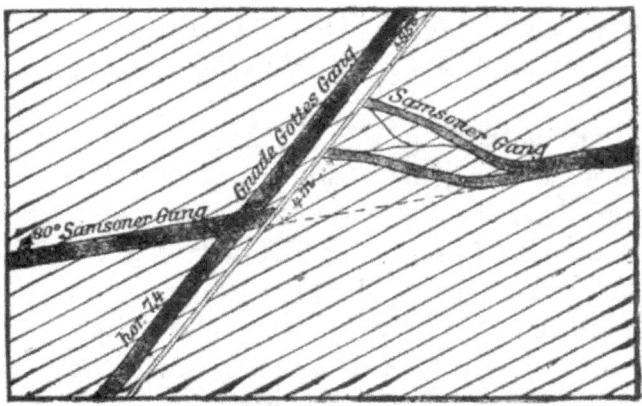

Fig. 118.—Shifting of the Samsoner lode in drift 41 at St. Andreasberg. (G. Köhler.) (Ground plan.)

The classification adopted and the nature of the formation of fissures will be rendered clearer by a number of examples:

a. Entokinetic Fissures.

I. Contraction Fissures.

1. *Cooling Fissures.*

It is well known that eruptive masses after cooling and solidifying from a molten condition are commonly traversed by numerous contraction cracks, which are often quite regular in occurrence, producing the characteristic columnar, laminated or spherical structures, while in other cases the fractures occur without any recognizable order.

Mineral veins resulting from the filling of such cooling (contraction) fissures are represented by the flat tin veins in the Zinnwald granite dome

[1] 'Synthetische Studien zur Exper. Geologie.' German translation by Gurlt. 1880, p. 234.

of the Erzgebirge. That granite masses are traversed by spherical contraction fissures due to cooling, which often conform approximately to the gently rounded surfaces of the hills, is a well-known fact. In quarrying out slabs for sidewalks from the granite quarries of Bautzen these fissures are often of great practical importance,[1] because along them even the fresh, unaltered, unfissured rock may be readily wedged into slabs. At Zinnwald such fissures were apparently filled with the gangues and ores of the tin formation, especially quartz, lithia-mica, tin-stone and wolframite, at a time when the cooling had not yet been completed. As shown in the cross-section, Fig. 119, these veins have the form of a pile of inverted saucers, the superimposed veins being near together. Eleven of them are being worked in the granite itself. A few very low grade ones were also found in the Teplitz quartz porphyry, into which the flat granite dome has penetrated from below.[2]

A case in direct contrast to this example of fissure formation is found near by in the mines of Altenberg, where a granite stock is traversed by innumerable small fissures utterly irregular in course, from which the rock has been enriched with tin ore. These fissures, together with their impregnation zone, form, as a whole, the so-called Zwitterstock (tin ore stock). Most of them are probably simply fissures of cooling.

On the other hand, it is difficult to decide whether the network of lodes in the propylites of Hungary is or is not due to fissures of cooling, though the fact that the veins of that region also occur in the adjoining Eocene sandstone is against the cooling theory. The mineral veins of Nagyag, however, are not fissures of contraction, according to B. von Jnkey.

If an eruptive dike is fractured on cooling by a succession of transverse fissures occurring close together they form a so-called ladder vein (Leitergang). A well-known example of this is found in the Waverly mine at Victoria, Australia, where a dike of decomposed diorite is traversed by numerous stringers of gold-bearing quartz. Some of them are forked, and most of these stringers stop at the selvage, though a few continue a short distance into the adjacent slate[3] (Fig. 120).

In the same way, according to Th. Scheerer, a granite dike cutting through mica schists at the Näsmark mine at Thelemarken, Norway, is itself cut by numerous transverse stringers of quartz containing copper ores, and accompanied, according to J. H. L. Vogt,[4] by zones of greisen.

[1] O. Herrmann: 'Steinbruch's industrie und Steinbruch's geologie.' Berlin, 1899, p. 112.

[2] K. Dalmer: 'Erläuterungen zu Sect. Altenberg-Zinnwald der geol. Special-karte von Sachsen.

[3] As a result of the cooling of the highly heated contact zone.

[4] *Zeit f. Prak. Geol.* 1895, p. 149.

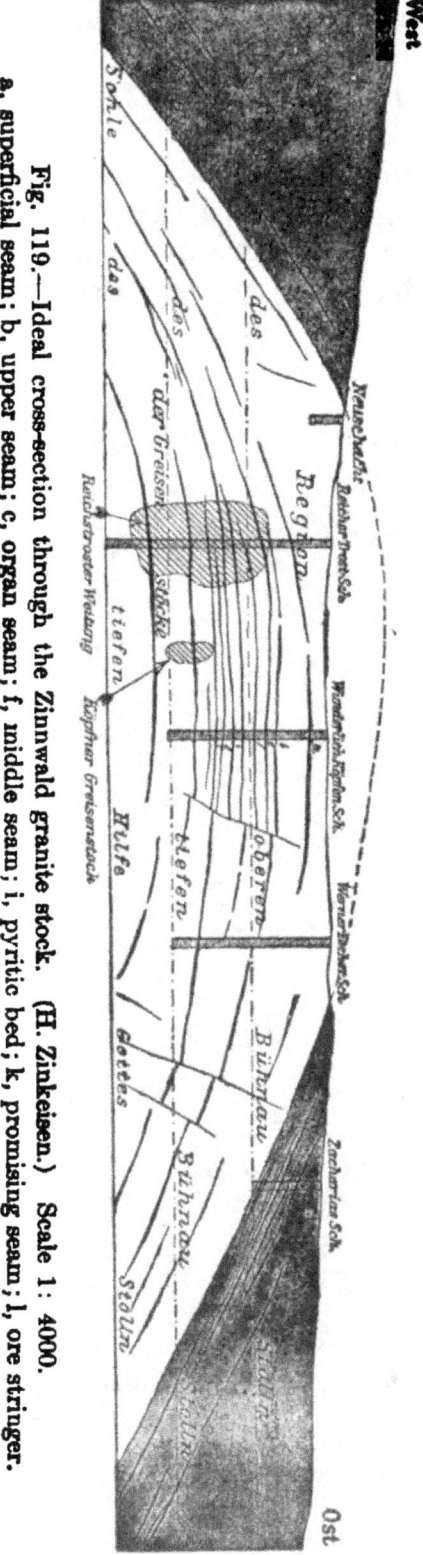

Fig. 119.—Ideal cross-section through the Zinnwald granite stock. (H. Zinkeisen.) Scale 1 : 4000.
a, superficial seam; b, upper seam; c, organ seam; f, middle seam; i, pyritic bed; k, promising seam; l, ore stringer.

Typical ladder lodes are also seen in the gold-bearing quartz stringers of the dikes of fine-grained granite cutting schistose rocks at Berezovsk, near Ekaterinburg in the central Ural.[1] A part of the plan of the working of the mines of that locality (given in Fig. 121) shows the many short side drifts run after the gold stringers, all starting from main levels driven along the body of the dike (Polo). This plan gives a clear idea of the arrangement of these cooling (contraction) fissures, which but rarely extend into the country rock (Fig. 121). Exactly the same phenomenon is found elsewhere in the Ural in the diorite and serpentine dikes of Pyschminsk.[2]

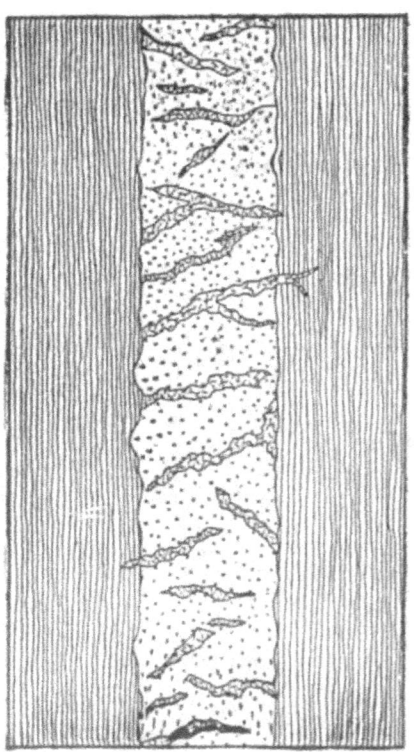

Fig. 120.—Section showing a diorite dike at the Waverly mine, Victoria, with a ladder vein of gold quartz. (Phillips-Louis.)

II. *Desiccation Fissures.*

That fissures may be formed by the drying of sedimentary rocks is well known, but it is difficult to prove in individual cases that veins have this manner of origin. The small fissures in the spherosiderite nodules of the

[1] 'Die Golddistricte von Beresow u. Mias am Ural.' *Archiv f. Prak. Geol.*, II, 1895, p. 499.
[2] F. Posepny: 'Genesis,' etc., p. 102.

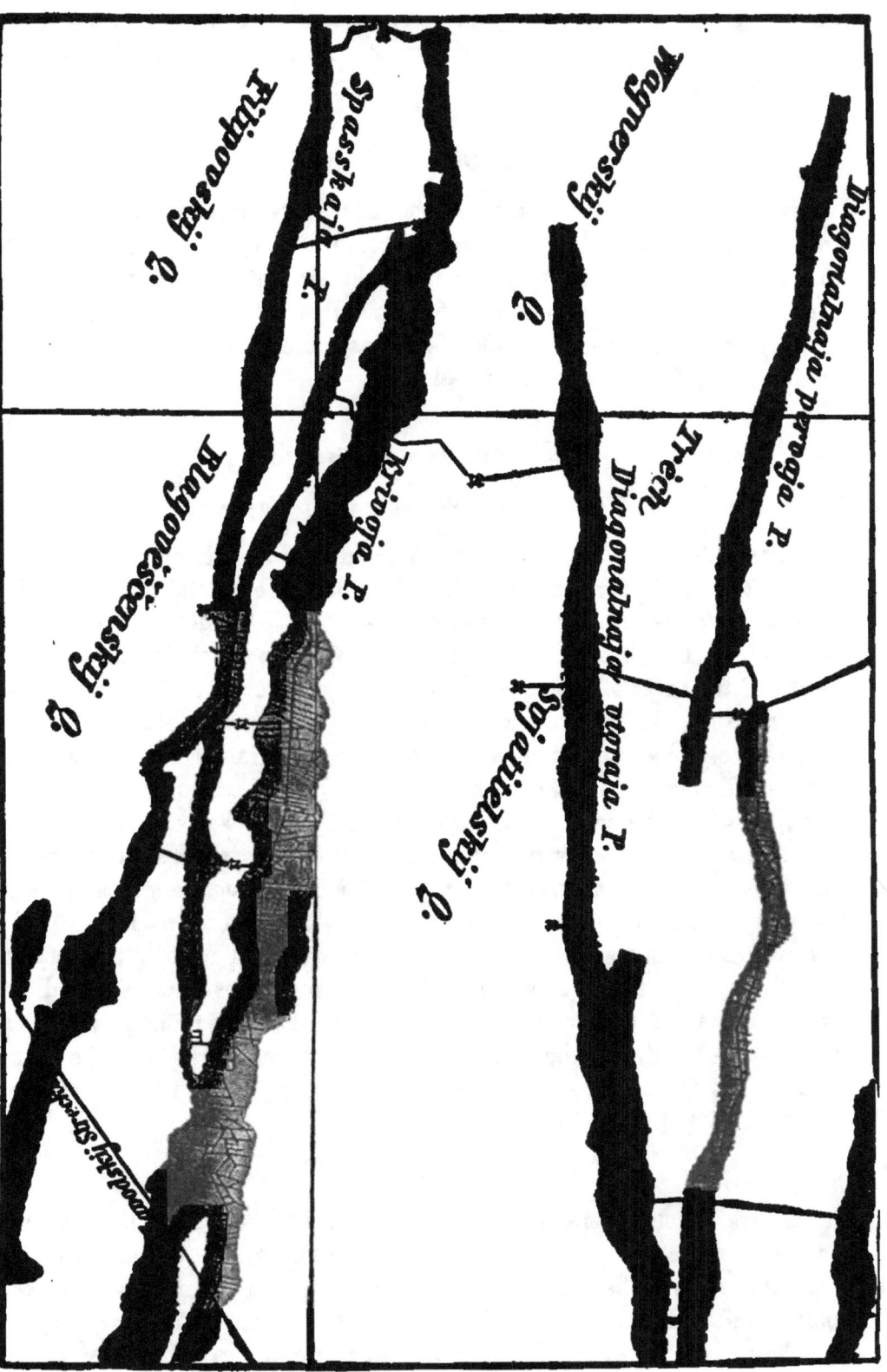

Fig. 121.—Ground plan of a part of mine area of Berezovsk. (F. Posepny.)

Carboniferous formation, for example, those occurring at Zwickau, which often carry zinc-blende and galena, are undoubtedly a result of contraction due to drying.

2. *Dilation Fissures.*

The best examples of these are the numerous fissures and slip planes which are so often found traversing stocks of serpentine. In such cases the cause of the increase in volume was the absorption of water in the serpentinization of the olivine rock. Such fissures have often received deposits of garnierite and other secondary iron and nickel ores, as in New Caledonia and at Frankenstein in Silesia.

β. Exokinetic Fissures.

I. *Fissures of Collapse and of Expansion.*

Collapse fissures (Einsturz spalten) are due to the caving in of the strata overlying a cavity formed by the leaching out of rock salt, gypsum or limestone. Instances of mineral deposits capable of such interpretation are not known. In a broader sense, this class of fissures includes those in great blocks of the earth's crust dropped down by the removal of great masses of material by volcanic eruptions, a feature not infrequently found on the concave side of mountain ranges. Certain ore-bearing lodes found in the eruptive districts of the Carpathian arch may be of this origin.

Expansion (tension) fissures are formed in beds overlying olivine rocks, which swell up on serpentinization, or above anhydrite lenses, that alter into gypsum. This may be the origin of some garnierite stringers, which are said to continue from the serpentine into the country rock.

Expansion fissures are more common in stratified rocks forced up into great arches by lateral pressure. The vein system of a mineralized tract may be explained in this way, as for example, that of Freiberg, which, as a whole, is situated over the summit of a great dome of gneiss. The fissures produced by folding may be summarized in a very general way, as follows:

II. *Fissures Due to Folding.*

Fissures develop, especially in the more brittle strata of a rock series, when the rocks are pressed together into folds, anticlines and synclines. Thus this class of fissures, like the folding itself, is due to the progressive cooling and contraction of the deeper zones of the earth's crust. They may be subdivided into *strike, intersecting* (at acute angles) and *cross* fissures, according as their course conforms to, 1st, the strike of the folded beds; 2d, crosses the beds at an acute angle; or, 3d, crosses the fold at a right angle. It is probable that many mineral veins owe their origin to such rock flexures.

Strike fissures are almost always faults, generally overthrust faults. A. von Groddeck[1] mentions mineral veins in the Rhenish schists as examples of strike faults along folds. Such fissures are certainly not so frequent among veins as those crossing the folds at either an acute angle or a right angle.

The formation of these fissures is the result of an unequal distribution of pressure during the folding process. The intensity of a thrust acting vertically or obliquely against a fold may undergo a uniform diminution at slight intervals; for example, when an unusually resistant rock is

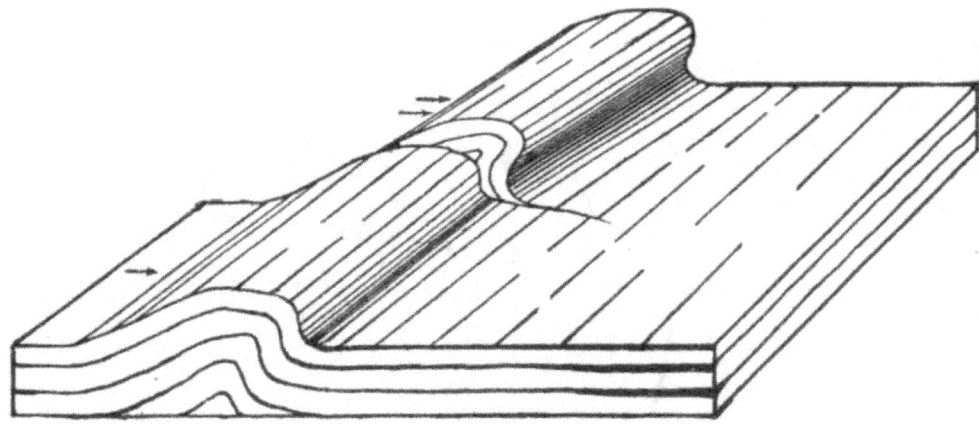

Fig. 122.—Model exhibiting the origin of fissures across folds.

encountered, or when pre-existing folds and fissures effect a deviation. In such cases a fissure forms between the more compressed and less compressed areas, as may be observed by an examination of the sketch-plan shown above (Fig. 122).

Foliation fissures are also produced by folding. If a pile of paper sheets be pressed together from the sides, some sheets will spread over each other; in the same way the layers of schistose rocks when pressed will fold up. The cavities thus formed have the shape of flat lenses, are generally curved irregularly, and may be filled with quartz containing ores, especially auriferous pyrite. Such *foliation fissures* are common among the gold-bearing quartz veins of the crystalline schists of the Appalachian region and Nova Scotia.

[1] 'Lagerstätten der Erze,' 1879, p. 316.

This class includes also the remarkable saddle lodes of Australia, described in detail previously.

III. *Compression Fissures.*

Some rock masses are unable to form folds when subjected to pressure. The hard and brittle eruptive rocks especially are not able to yield by folding, and in consequence are crushed and split. The cracks thus formed are appropriately called fissures of compression.

Our conception of this process is largely based on the beautiful experiments of A. Daubrée.[1]

In these experiments, cubes of brittle rocks are by simple pressure broken

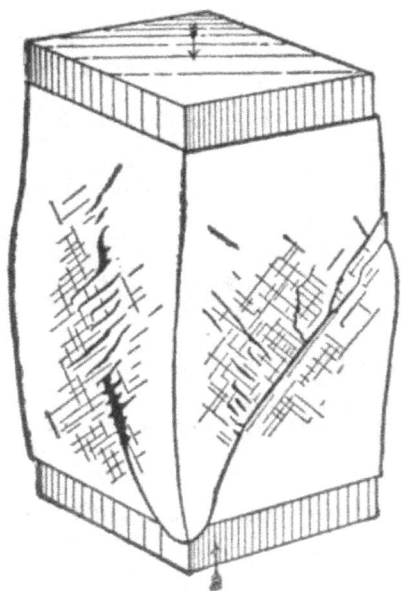

Fig. 123.—Artificial compression fissures in a prism.   (A. Daubrée.)

into irregular prismatic pieces arranged at right angles to the plane of pressure. In this case the direction of the fissure is parallel to the direction of pressure. In nature, however, the masses subjected to pressure are more or less plastic. Daubrée endeavored to approximate this condition in his experiments by using prisms consisting of a mixture of gypsum, wax and resin. The result was the formation of principal fissures, oblique (45°) to the direction of pressure, along which movement had taken place. These cracks were accompanied by a great number of lesser fissures on each side and running either parallel or at right angles to the main fissure

[1] A. Daubrée: 'Études Synthetiques de Geologie Experimentale,' 1879, p. 235.

(see Fig. 123). This cluster of fissures resembles in miniature the network of veins seen at Freiberg or that of the Cornwall tin district. As the mineral districts just named show clear evidences of lateral pressure, it is apparent that the plan of the veins of such an area is very properly comparable with the fissures on the surface of Daubrée's prisms.

Daubrée also experimented with artificial models made of alternate layers of different materials, the whole subjected to a pressure acting parallel to the stratification. In this way he obtained fissures running obliquely to the direction of pressure, accompanied by displacements when there was a chance for lateral displacement. The formation of fissures in this experiment was always preceded by a folding of the layers.

These experiments proved that the forces producing systems of parallel

g. 124.—Formation of regular fissure systems through torsion. (A. Daubrée.)

fissures, whether mere fissures of disruption or of displacement, did not necessarily act in the direction of the fissures. The direction of the force can only be inferred from the arrangement of all the fissure systems of a region. In the case of a system of paired fissures the diagonal seems to usually indicate the true direction. It is assumed, of course, that all the fissures taken into consideration are approximately simultaneous in origin, which, of course, is only in part true of the fissures at Freiberg and in Cornwall.

It also appears that a fissure and a displacement along this fissure may be produced by one and the same force.

Other experiments convinced A. Daubrée that under certain circumstances a torsion of a part of the solid crust of the earth would lead to a development of perfectly regular fissure systems, especially for blocks of horizontal strata, such as plateaux and block mountains formed of level beds. By means of a wrench (Fig. 124) Daubrée twisted a rectangular plate of mirror glass on which a coating of paper had been glued. He obtained two systems of cracks running obliquely to the plane of torsion, one crack being often terminated against another. A consideration of the combined effect of the two forces, lateral thrust and gravity, shows that such torsional stresses and resulting fissures must often occur in nature. The fact that many of the cracks formed in the glass terminate one against the other is noteworthy, as it shows that a fissure dislocating another need not necessarily be younger than the dislocated fissure. It is only by a careful study of the fissure filling that the relative ages of the fissures can be determined.

These experiments of A. Daubrée have been repeated by G. F. Becker,[1] and the importance of their application to the theory of vein formation has been confirmed. The slickensides frequently found on fissure surfaces, even in veins with no apparent faulting, and the spiral bending and twisting of many vein walls, indicate that the fissures originate through torsion. Daubrée also, it may be observed, noted the spiral form of the cracks made in his glass plate.

Fissure systems still more closely resembling those in nature are obtained when provision is made for the asymmetrical and one-sided distribution of pressure and resistance in place of the symmetrical arrangement of Daubrée's experiment.

Lossen[2] ascertained this by accident. Cracks were formed in a pane of glass by the sudden opening out of a window sash caught below. These were mostly diagonal fractures, dipping in part in different directions, together with other curved and transverse cracks. The entire group of fissures bore a striking similarity to parts of the lode system of the upper Harz. Another variety of displacement that is manifestly a consequence of torsion forms the faults with opposing throw, failles a charnière (hinge faults), as they are graphically called by the French and Belgian miners. In this case each part of the faulted block has been elevated at one end of the fissure and depressed at the other end, while in the middle there is a neutral center or point of revolution (Fig. 125).

That the formation of fissures depends not only on the arrangement of the forces, but also to a certain degree on the greater or less brittleness and toughness of the country rock is manifest from the conditions already described (page 119).

[1] G. F. Becker: 'Torsional Theory of Joints,' *Trans.* Amer. Inst. Min. Eng., 1894, Vol. XXIV, p. 130.

[2] K. A. Lossen: 'Ueber ein durch Zufall in einer Fensterscheibe enst.,' etc. *Jahrb.* d. k. preuss Geol. Landesanst, 1886, p. 336.

(k) Time Required for Fissure Formation.

Well authenticated instances are known of the formation of vast fissures by a single shock during a violent earthquake. Thus according to B. Koto[1] a fault fissure was formed during a destructive earthquake in central Japan on October 20, 1891. The fissure showed both vertical and lateral shifting of the walls, and extended in a straight line across mountains and valleys for a distance of 64 k. (38.4 miles). At Midori the throw was 6 m. (19.6 ft.) and the horizontal shifting 1 m. to 4 m. (3.3 ft. to 13.1 ft.). This fissure did not gap. On the other hand, great open fissures were caused by the great earthquake in Calabria in 1783 at Monte Sant' Angelo.

Mining operations have exposed open cavities in true vein fissures in the Freiberg mines, as in the Churprinze mine, where druses were often encountered that were so large that a man might crawl into them.

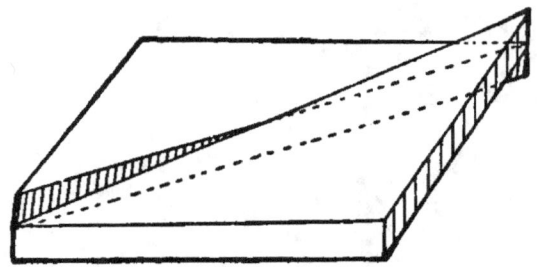

Fig. 125.—Model of a hinge fault.

Very broad veins are, however, not necessarily due to the filling of wide and open fissures. Observations upon veins show that, on the contrary, most large veins are due to an oft-repeated reopening of the fissures, so that only a small fissure space need remain open at a time. This successive fracturing has formed fissures not always exactly parallel to their predecessors, and not infrequently traversing them obliquely, as is proved by the frequent diagonal stringers of later age found in lodes. As an example, we may cite a rich silver quartz vein, the Traugott Spat, near Freiberg. This vein shows many irregular stringers of secondary quartz and ruby silver, cutting an older vein filling, consisting essentially of argentiferous blende, with some pyrite (Fig. 126). These younger quartz stringers are plainly distinguishable from the old vein filling on the selvage wall, whose characteristic comb structure proves it to have been formed before the blende mass.

The younger fissures will sometimes follow the wall or selvage of the old vein for a long distance and then cut directly across the vein. This condition results in the formation of cross stringers, which pass immediately into parallel stringers along the wall. The specimen shown in Fig. 127 is

[1] 'On the Cause of the Great Earthquake in Central Japan, 1891,' Tokio, 1893.

taken from the Komet Stehende on the Himmelsfürst mine, near Freiberg. (Fig. 127.) The cross vein of the specimen consists of white calc-spar, with some pyrite, while the mass of the vein cut by it consists of reddish rhodochrosite with zinc-blende and pyrite.

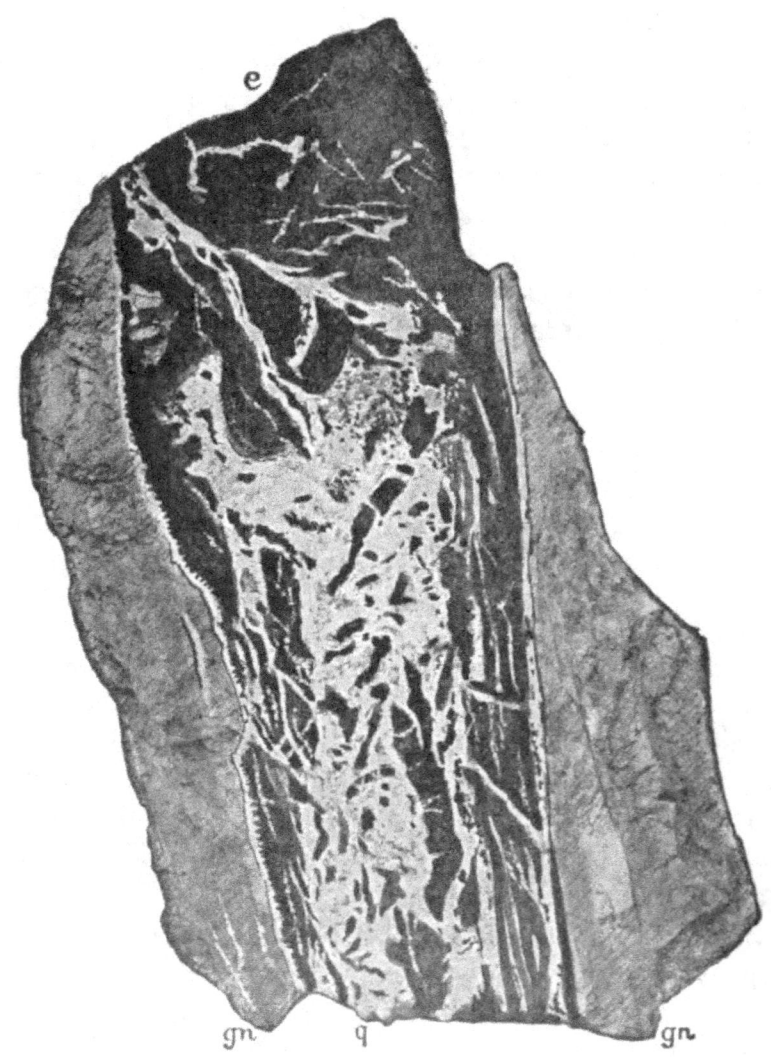

Fig. 126.—View of the vein of the Traugott Spat (rich quartz formation). From nature; an example of a vein reopened many times in succession. (Composite vein.)
gn, decomposed gneiss; q, vein filling, consisting essentially of quartz; e, ore, mainly argentiferous blende, with some copper pyrite, as well as galena.

Clay selvages and false walls are often formed by movements taking place after the vein has been formed, and, as shown by Emmons,[1] they may lead the miner astray or prevent his discovering pay shoots. Drifts are often

[1] S. F. Emmons: 'Origin of Fissure Veins,' *Proc.* Colo. Sci. Soc., 1887, p. 200.

run along such planes of easy breaking, because they are thought to be the true wall of the entire lode, and because, moreover, it greatly facilitates the loosening of rock and ore. In such a case when a cross drift is finally run through the clayey or altered and rotten country rock, inside the false wall, a second layer or stringer may be encountered lying parallel to the one hitherto worked, and which is perhaps much richer. The selvage apparently defining the wall is merely a plane due to reopening of the lode.

(1) The Filling of Vein Fissures.

As already pointed out, it is seldom that vein fissures are filled with ore alone. Most often the vein filling is made up of ore and gangue (*i. e.,* non-

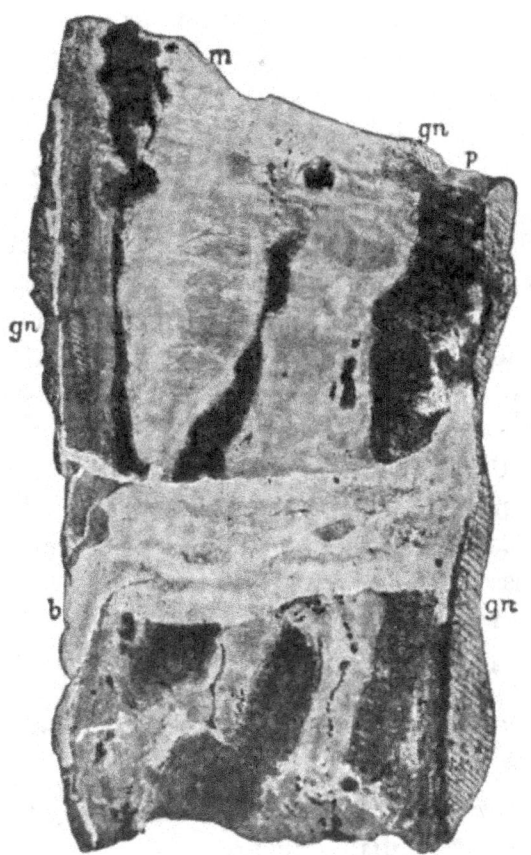

Fig. 127.—View of the Komet Stehende (Carbonspathic lead formation) with a cross stringer.
gn, gneiss; m, manganspar; b, brown spar; p, pyrite, blende and galena.

metallic minerals), together with fragments of the country rock showing various conditions of comminution, trituration and decomposition. In compound lodes this filling is the rule, and in many cases these masses of altered country or veinstuff form the main mass of the vein filling.

The most widely distributed gangue minerals are quartz and the carbonates (calcspar, brownspar, dolomite, magnesite, manganese-spar, mangan-calcspar) also barite and fluorspar. The less common constituents are zeolites, gypsum, orthoclase, lithia-mica and chlorite. It is usually assumed that certain particular species of non-metallic minerals do not occur in veins, and the presence of these minerals in an ore deposit is considered as a proof that the deposit is not a mineral vein. Such assumptions must, however, be used with caution, for it has been repeatedly found that various minerals supposedly absent in veins have subsequently been found in them. This is true, for example, of actinolite, which, according to W. Möricke,[1] occurs together with tremolite, quartz and calcspar as the matrix of the gold-bearing copper veins of La Higuera in Chile, and, according to W. Lindgren,[2] in the auriferous copper veins of Rossland, in British Columbia. Garnet, which was formerly supposed to be foreign to lodes, has now been shown to exist in veinlike orebodies.[3] G. F. Becker mentioned it many years ago as a constituent of certain gold-quartz veins of the Appalachians.[4]

(m) Structure of the Vein Filling.

It is evident that the development of one or more of the minerals of a vein in a peculiar form, as, for example, the coarsely foliated development of calcspar, the sheaves of fibrous pyrolusite, etc., will influence the structure of the lode as a whole, particularly if one of the minerals predominates. As a rule, however, several gangue and ore minerals are present in the vein filling, and the structure of the aggregate is the result of the mixture and intergrowth of all these minerals. The intergrowth and relation of the vein minerals to each other (paragenesis) has been studied very closely megascopically, *i. e.,* so far as recognizable with the naked eye, ever since the time of G. A. von Weissenbach.[5] Very little, however, is as yet known of the microscopic structure of ores, although the result of such studies may throw much light on the question of ore genesis and deposition, and offers a wide field for future investigation. Up to the present time such studies have been confined almost entirely to the ores of gold-quartz lodes. This subject will be referred to later.

While veins show a great variety of structures, the following types of vein structures may be distinguished:

[1] W. Möricke: 'Die Gold, Silber und Kupfererzlagerstätten in Chile,' Freiberg, 1897, p. 27.

[2] *Trans.* Am. Inst. Min. Eng., February, 1900, p. 33.

[3] R. Beck: 'Beiträge zur Kenntniss von Brokenhill,' *Zeit. f. Prak. Geol.,* 1899, pp. 65-71.

[4] 16th *Ann. Rep.* United States Geol. Survey, Vol. III, p. 274.

[5] G. A. von Weissenbach: 'Abbildungen merkwürdiger Gangverhältnisse aus dem sächs. Erzgebirge,' Leipzig, 1836, p. 27.

1. *Massive Filling.*

A massive filling is especially common in 'simple' veins, and the gangue and ore minerals usually show a compact massive development. Many gold quartz veins are of this type, the vein consisting of a uniform and compact mass of quartz containing disseminated grains of gold-bearing pyrite and particles of free gold (disseminated vein structure). Many pyrite-blende-lead veins show this structure, an older fissure being filled with com-

Fig. 128.—View of the Rudolf Stehende (pyritous lead ore formation) with dominant massive structure. (From nature.)
gr, decomposed gray gneiss; q, quartz; b, argentiferous galena; s, pyrite.

pact ore. Fig. 128 represents the Rudolf Stehende on the Himmelsfahrt mine near Freiberg, in which the vein is mainly compact galena.

Examples of a purely massive structure are seen in the small lodes of compact bornite ore encased in gneiss at Hvidesöe (Hvideseid), Norway.

2. *Banded or Crusted Filling.*

This filling in which the different constituents are arranged in more or less sharply defined layers or crusts is a common feature of veins. It is wholly unlike the structure prevailing in mineralized igneous dikes, and is readily distinguished from the stratified structure of sedimentary deposits.

When the crusts are very thin they commonly occur in great number, as, for example, in many of the lead-barite veins in the Erzgebirge. A typical example is shown in Fig. 129 representing a specimen from the Friedrich vein of Memmendorf in the Erzgebirge. The delicate crusts consist respectively of dense barite, fluorspar, zinc-blende and galena.

When the minerals forming a crust are crystalline and the crystals are elongated and arranged at right angles to the layer so that the ends of the

Fig. 129.—The Friedrich Spat (barytic lead ore formation) with definite layer structure. (From nature.)

crystals project into the next following crust or into a free druse space, it is known as 'comb structure,' a variety of crustification confined to mineral lodes. This structure is shown by the primary quartz layers of the Traugott vein in Fig. 126, and it is common in veins of the rich silver quartz class. The ore breccia of the Himmelsfürst mine (Fig. 133) shows this structure, quartz crystals projecting into a dark band of blende.

Occasionally there is a regular and symmetrical development of the crusted structure with a regular succession of mineralogically different crusts, counting from the vein wall to its middle, the succession following a definite order. This symmetric crust-structure is very common in lead-barite veins of the Freiberg district, and is also observable in many tin veins of the Erzgebirge. The symmetric succession may be repeated several times, as shown in the filling of the Drei Prinzen vein of the Churprinz (Fig. 130). In this we may distinguish the following crystalline layers stated in order from the selvages toward the middle:

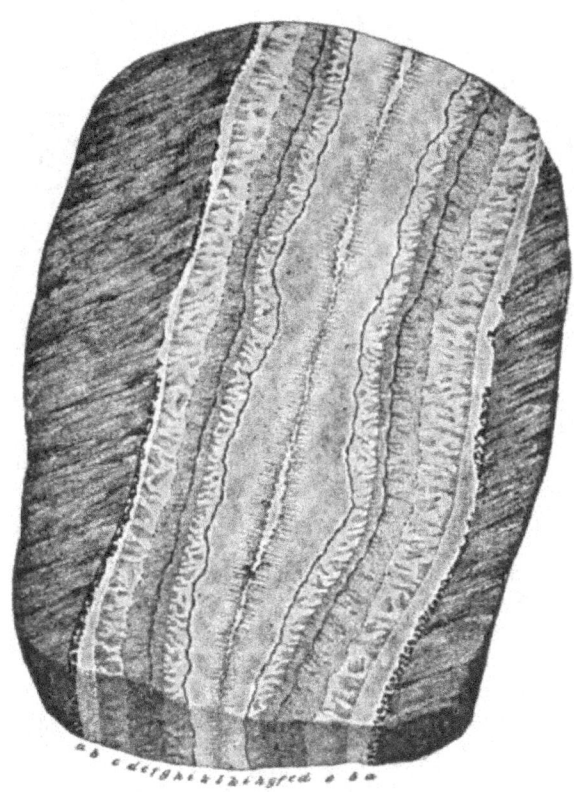

Fig. 130.—The Drei Prinzen Spat (a barytic lead vein) with symmetric crust structure. (G. A. von Weissenbach.)

(a) Brown blende; (b) white quartz; (c) asparagus green fluorspar; (d) an exceedingly delicate band of brown blende; (e) dirty, flesh-red, curved, scaly barite; (f) a narrow band of radiated pyrite; (g) barite like e; (h) fluorspar like c; (i) band of marcasite like f; (k) white calcspar; (l) light wine-yellow calcspar occasionally forming small druses in the middle. The lode traverses gneiss.

The number of layers and of the different minerals and the regularity and symmetry of this specimen are very exceptional. A crust may form at first only on one wall of a lode and afterward simultaneously on both or *vice versa*. Such alteration may even be caused by chemical reaction of pre-existing crusts on the circulating solutions. If a vein is fractured and the old fissure reopened by a crack that does not exactly follow the middle of the vein, the crusts subsequently formed are not symmetrically disposed with those of earlier origin. An example of this is shown in Fig. 131

Fig. 131.—View of the Frischglück Stehende (rich quartz formation). (G. A. von Weissenbach.)

(after G. A. von Weissenbach), representing the filling of the Frischglück vein at Klein-Voigtsberg.

In this illustration we see that, progressing successively from right to left, new openings and new double crusts have been formed, which could not conform to those of the oldest part of the lode. On the other hand, in lodes formed by rising water and at great depths below the surface, it is inconceivable that such unsymmetrical deposition could be dependent on the inclination of the lode unless it is assumed that the water did not fill the fissure, but followed its foot-wall. Such vein fissures must have been completely filled with water, and it is only in the zone above the water level that such partial and unilateral fissure filling is conceivable, as pointed out by F. Posepny.[1]

That the form of the crusts is dependent on the form of the cavity walls is best exhibited by the crusts which rest directly upon the country rock. We may distinguish: (a) flat crustified vein filling; (b) concentric crusted vein filling. The former predominates in vein fissures with approximately rectilinear walls. The latter is found on projections of the fissure walls, and

Fig. 132.—Concentric crusts of the Drei Prinzen Spat (a baritic lead vein).
(From nature.)

is even more beautifully exhibited by the coatings about included fragments of country rock, or of fractured older vein filling of the fissure. Finally, this structure may be developed by the filling of the spherical spaces left in the partly filled fissure, as is very often the case with ore deposits filling cavities in limestones and dolomites.

A fine example of concentric crusted structure along the selvage of a vein is shown in Fig. 132, representing a specimen from the Drei Prinzen Spat of the Churprinz mine near Freiberg.

In this specimen masses of galena are surrounded by exceedingly thin shells or layers of barite and crusts of fluorspar. The kidney-shaped layers

[1] F. Posepny: 'Genesis,' p. 78.

of barite superimposed a hundredfold in some Freiberg lodes were justly compared by C. F. Naumann with the formation of travertine or hot spring sinters.

The gradual encrustation of rock fragments originally lying loose within the vein fissure may sometimes be traced step by step. At first there is only a scanty coating of minerals, so that the fragments are not as yet firmly cemented to each other. Such is the condition of the very imperfect quartz breccias of some tin ore lodes of Zinnwald, cemented by a recent deposit of quartz, scheelite and fluorspar; of the mica schist fragments loosely cemented by barite in the Eduard Spat of the Himmelsfürst mine. More

Fig. 133.—Breccia structure in the Lade des Bundes Flache at Himmelsfürst near Freiberg. (From nature.)

commonly there is complete cementation of brecciated material, but without crustification. Such simple lode breccias are rather common in all large lode areas, as at Freiberg in the upper Harz, and at Pribram. The cement, of course, varies with the character of the lode. Thus, for example, at Pribram dark schist fragments are often found cemented by zinc-blende, brown spar and quartz. In other cases, the breccia consists of fragments of an older lode filling and not of country rock. This is well shown in the ore from the Lade des Bundes Flache (drift No. 8, 1884) shown in Fig. 133.

In this case sharply angular platy fragments of decomposed gneiss and of an older crustified vein filling of quartz, blende and galena are imbedded in a drusy cement of brownspar. The brittle galena is completely comminuted.

If the different ore and gangue minerals form crusts about the fragments, *ring* ores or *cockade* ores are formed. In the Freiberg area such "Spharengesteine" (spherical rocks), as they used to be called, are well known in the Peter Stehende of the Alte Hoffnung Gottes mine and in the Helmrich Spat of the Gesegnete Bergmanns Hoffnung mine. But nowhere are they so magnificently developed as in the upper Harz, especially in the "Ring und Silberschnur," near Zellerfeld. The specimen represented in Fig. 134 was taken from this mine.

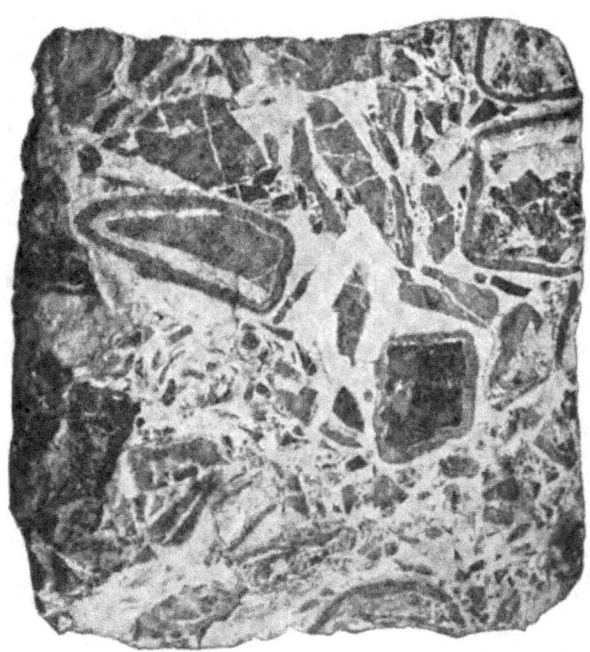

Fig. 134.—Ring ore from the "Ring und Silberschnur" mine in the Harz. (From nature.)

The specimen shows fragments of quartzitic sandstone and dark colored slate surrounded by galena, zinc-blende and quartz. If the succession of the different crusts of the ring ores be examined, it is found that the layers show the same order of succession as those on the walls of the stringers and veins. This is seen at a glance in a comparison of Fig. 135 and 136 (after A. von Groddeck).[1]

Sometimes the first crust formed on the fragments is so sharply defined from the later ones that it appears to constitute the true nucleus or rock

[1] A. von Groddeck: 'Erzlagerstätten,' p. 64.

fragment. Thus, according to Phillips-Louis,[1] the main vein of Huelgeet in Brittany consists of what appear to be quartz pebbles cemented by blende, pyrite, quartz and lead glance. In reality these pebbles are fragments of slate surrounded by concentric shells of quartz resembling fibrous chalcedony. In fact, this radial structure is very common in the individual crusts formed about rock fragments, especially the quartz crusts, forming the so-called cockade ores. In such cases the crust may resemble stellate quartz.

The encrusted balls of such ring ores are sometimes isolated, and there is no contact between the original rock fragments, and hence A. von Weis-

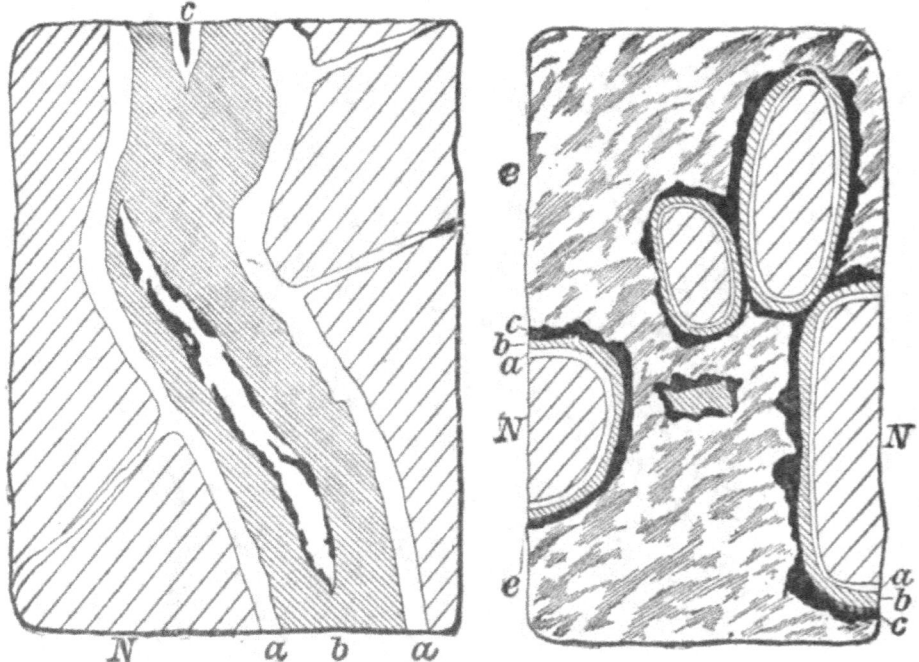

Figs. 135 and 136.—Specimens of vein filling from the Bergmannstrost mine.
(A. von Groddeck.)
N, country rock; a, quartz; b, lead glance; c, zinc-blende; e, calcspar.

senbach[2] and Reich have inferred that fragments originally in contact were forced apart by the crystallizing force of the minerals subsequently segregated, just as fragments of the Alte Mann in the Sauberg, near Ehrenfriedersdrof, according to C. F. Reich, were driven apart by the crusts of freezing seepage water. While the process in itself is physically quite conceivable, it must be remembered that in some cases the only reason why

[1] 'Ore Deposits,' p. 88.
[2] G. A. von Weissenbach: 'Abbildungen merkw. Gangverhältnisse,' 1836, p. 22.
[3] F. Posepny: 'Genesis,' p. 89.

points of contact are not found is that the observers cannot compare a sufficient number of parallel sections.[1]

Ring ores are often of very recent formation, and there are even cases where their formation takes place before our eyes. Posepny mentions the quicksilver deposit of Sulphur Bank, California, where fragments of basalt, sandstone and slate are coated with concentric layers of cinnabar by the hot spring water. If the concentric crusts occur as coatings on the walls of cavities, their form depends on the shape of the little cavity. This is also well shown in the cross-sections of agate amygdaloids. The crusts may in such cases either line all parts of the wall uniformly, or the layers may only be found at the bottom, or finally the two cases may be combined. Very good examples of such cavity fillings are found in the zinc lead deposits of Raibl (see further on), where fibrous blend is found to be made up of thin layers of dense zinc-blende and galena.

Druses occur in lodes as well as in chambers. These druses or vugs are characterized by an inner coat or crust studded with crystals projecting into an open space. They are not always the residual spaces of an imperfectly filled fissure, but may be formed by a secondary leaching out of older crusts or by the solution of fragments of the country rock. Such crystal formations are even possible in stratified deposits; for example, druses formed by the leaching out of plates of anhydrite and the subsequent deposition of calcspar in gypsum deposits. As a rule, however, druses are characteristic of cavity filling.

These vugs are sometimes quite large. In the Freiberg lodes they are sometimes large enough to admit a man's head. They are usually situated in the middle of the lode (see p. 173).

The minerals forming the druse walls show the successive deposition of the various ore and gangue minerals even better than the material filling the body of the vein, the crusts sometimes showing several generations of each ore. One should, however, be cautious in drawing conclusions from such observations, since the minerals found in these vugs are often formed at a much later time than those which constitute the real vein filling. In fact, this deposition in vugs is in some veins known to be taking place at the present day. Undoubtedly, in many instances, the deposition of druse minerals took place under physical and chemical conditions different from those prevailing when the true vein filling was formed, and at a later time, when this section of vein had been raised above the level of ground water, as a result of extensive superficial denudation of the country.

While we have every reason to believe that the open vein fissures were, at some depth below ground-water level, entirely filled with mineral solutions,

[1] F. Posepny: 'Genesis,' p. 88.

this may not always have been the case with the residual open parts of those fissures left unfilled until the stage of druse formation. That this is true is shown by the not infrequent occurrence of later minerals deposited on the upper side of the large crystals of vugs and druses. This is common at Freiberg; thus the large quartz crystals of the Drei Bruder vein at Obergruna are coated on one side only with brown-spar and other car-

Fig. 137.—Group of quartz crystals encrusted on one side with brown spar.
(From nature.)

bonates (see Fig. 137); similarly the quartz crystals of the Selig Trost Flache lode of Himmelfahrt mine near Freiberg are coated with ankerite and pyrite on one side only. These conditions show that descending solutions seeped through the druse spaces, so that only the downward facing surfaces received new secretions, as may be seen with ice crusts along the eaves of houses in the winter.

The new mineral secretions formed in open cavities (*i. e.,* containing only gases or air) produce stalactites at points where the amount of water is sufficient to overcome adhesion and drop downward by gravity. Such stalactitic deposits are most common in ore deposits forming irregular filling of cavities[1]; particularly in zinc-lead deposits in dolomites and limestones. At Raibl stalactitic tubes of galena occur that are as much as ten

[1] J. de la Fontaine: 'Beitrag zu Kentniss Stalak. Vorkommnisse u. deren Genese,' Berne, 1882.

centimeters in length; they consist of a number of delicate, concentric crusts of this mineral, combined sometimes with crusts of zinc-blende and pyrite. The center is often a cavity through which one could blow. These stalactites of galena were, it is true, no longer pendent from the cavity roof, but had fallen and been embodied in a later deposit of dolomite spar. Since the succession of the several crusts is not the same in all, it is probable that they fell at different times.

The fibrous blende of the zinc deposits of Raibl is also in part a stalactite formation, occurring as hemispherical stalactites and reniform and botryoided masses formed on the cavity roof. It consists of very fine concentric scales of blende and galena, as well as of pyrite. This fibrous blende is most beautifully developed in the Schmalgraf mine near Moresnet. Fig. 138 shows a cross-section of fibrous blende from that locality.

At Wiesloch in Baden long stalactites of calamine and galena were formerly found which resembled in form and occurrence the calcspar stalactites of caves. Similar examples occur in the calamine mine of Catavera in Spain. The Freiberg collection contains a specimen from that locality that is a fine, slender, snow-white zincspar stalactite 0.25 meters in length, not counting the missing point.

Stalactite formations are also known to occur in the vugs of veins. They are most common in deposits of manganese and iron, especially psilomelane and limonite. As examples we may mention the stalactite groups of psilomelane in the spathic iron ore lode of the Luise mine at Horhausen (see Fig. 139), and those from the iron ore lodes of the Schwarzenberg region in the Erzgebirge. Stalactites of brown hematite are found in the veins of the Siegen country and in those of Aue in the Erzebirge. Stalactites of pyrite were found at Freiberg in the Johannes vein of Himmelsfürst, and some of pyrite in the Lade des Bundes Flache of the same locality. We ourselves possess a group of marcasite stalactites collected by H. Muller from the workings in the Drei Prinzen vein of the Churprinz mine near Freiberg (above the sixth drift of the Friedrich gallery on the west, about 230 meters below the surface). As lode fissures at such a depth are far below ground-water level, these stalactites could only have been formed within an empty or gas-filled cavity, which renders their explanation very difficult.

The most remarkable occurrence of this kind is, without doubt, the stalactites described by F. Posepny[1] from the Matyas Kiraly mine at Verespatak in Hungary (illustrated in Fig. 140). They are completely studded externally with small quartz crystals, while internal layers of calcspar alternate with layers of rhodonite; in the middle line angular gold wires run

[1] F. Posepny: 'Genesis,' p. 95.

down which sometimes project at the lower gently rounded ends of these stalactites.

Mention must also be made here of the stalactites of crystalline quartz (or chalcedony) 7 centimeters long, from a druse cavity in the cobalt ore lode of Scheneeberg, described by H. Muller.[1]

Stalactites of ore minerals, as well as unilaterally encrusted crystals, must be regarded as records of a descending movement of solutions along vein fissures, as suggested by A. Schmidt[2] for the stalactites of Wiesloch.

Fig. 138.—Cross-section through a stalactite of fibrous blende from Moresnet. (From nature, two-thirds natural size.)

b, lead glance; p, pyrite; z, zinc-blende; outer crust limonite.

Fig. 139.—Psilomelane stalactites from the Luise mine near Horhausen. (From nature.)

Of course, one may adopt the alternative explanation suggested by F. Posepny,[3] namely, that, with sufficient pressure, uprising solutions may also be pressed through the roof of cavities when the floor and side walls are not permeable. This mode of origin, however, is undoubtedly of very rare occurrence. The above mentioned occurrence in the Churprinz may be an instance of it.

[1] H. Müller : 'Ueber eine merkwürdige Druse auf einem Schneeberger Kobalt-gänge.' Z. d. d. g. G., 1850, p. 14.

[2] A. Schmidt: 'Die Zinkerzlagerstätten von Wiesloch in Baden,' 1881, p. 94.

[3] F. Posepny: 'Genesis,' p. 93.

In conclusion we may mention, in discussing lode structures, that in some lodes soft clayey masses of crushed material contain nodular concretions of ores and of gangue minerals. Examples are seen in the nodules of native copper in a vein of the Vesuvio mine near Salinas, east of Mejillones in Chile; of zinc-blende, in the Malhada mine near Albergia Velha in Portugal; siliceous concretions in vein clay of the Gott mit Uns vein of the Himmelsfürst mine near Freiberg. Such formations could develop only in a medium whose particles could readily be driven asunder by the force of crystallization.

(n) Paragenetic Occurrence of Vein-Forming Minerals.

By paragenesis of minerals August Breithaupt[1] understood "the more or less definite manner of their association," at the same time "laying par-

Fig. 140.—Cross-section through a stalactite of Verespatak. (Posepny.)
g, gold; c, calcite; r, rhodonite; q, quartz; d, druse.

ticular stress on the relative age of the materials where a succession may be recognized."

Examples of such associations, given by Breithaupt, may be mentioned here. Copper pyrite will always be found where bismuthinite is present. In like manner pyrrhotite and copper pyrite, bornite, chalcopyrite and pyrite are always found associated. There is an especially intimate connection, as a rule, between fluorspar, topaz, molybdenite, wolframite and tinstone; of the common manganese and iron ores; galena and zincblende; cobalt and bismuth ores. Such associations have always been of practical importance to the miner and prospector in unexplored regions. The finding of certain minerals, either of ore or gangue, warns the prospector

[1] 'Die Paragenesis der Mineralien,' Freiberg, 1849.

to keep a keen lookout for the associated valuable mineral. If in a quartz vein you find pyrite, arsenopyrite and antimonite, the search for gold in the outcrop, as a rule, will not be in vain.

Quite as important as this association of the minerals is the manner of their intergrowth or succession, and the investigation of these features may throw much light on the genesis of the ore deposit in question. The definite succession of the various chemical crusts, the superposition of the druse minerals in a definite order betrays, in many cases, a progressive change in the chemical constitution of the solutions that circulated in the fissure, and possibly a change of the conditions of pressure and heat in the mineral springs supplying the material for the vein. It is true that the application of these hints to particular cases is as yet impossible. The recent important work of J. Vater and others on the influence of the accompanying solutions is merely the beginning of the exact studies which must be made before any definite conclusions can be reached on this subject. Sometimes the succession of minerals appears quite capricious. Thus, as mentioned by A. Breithaupt, in the beautiful druses of the barytic lead veins, the younger calcspar always rests on barite, never on pyrite. "It occasionally arches away over the pyrite as if it were trying to keep out of its way." In like manner, in the Bescheert Glück mine near Freiberg, the later formed ruby silver of the druses occurs only on white silver ore, not on galena.

A definite succession of minerals recurs with special regularity in the tin lodes of Saxony. The quartz and lithia mica form the basal layer, followed by topaz, cassiterite and wolframite, and finally by fluorspar and scheelite, together with uranium-mica (chalcolite).

In many cases the succession of the minerals is inversely as their relative solubility. "In many fissures and cavities the walls are covered with quartz druses followed by crusts of calcspar. The quartz being far less soluble in water, was of necessity deposited first." (Tschermak[1].)

Conclusions regarding the succession of minerals may often be drawn from the nature of pseudomorphism. The significance of this feature is very forcibly emphasized by C. F. Naumann.[2] "These residual monuments of former mineral bodies, now completely or almost completely vanished, give us a deep insight into the various processes of formation and reformation, destruction and annihilation that takes place successively in the course of time within the lode spaces, and which prove not only a very long duration of the lode forming process, but also a frequent change in conditions and a perplexing variation in the efficiency of each cause."

[1] G. Tschermak: 'Lehrbuch der Mineralogie,' 1894, p. 271.
[2] C. F. Naumann: 'Lehrbuch der Geognosie,' Leipzig, 1866, Vol. III, p. 575.

Sometimes almost all the mineral contents of a lode is affected by pseudomorphism. Thus the silver-cobalt lodes of Schneeberg originally had a gangue of calcspar and barite, which is at present almost always transformed into hornstone, common quartz, chalcedony and amethyst. In the silver-gold ore lodes of De Lamar,[1] Idaho, and the gold-silver veins of the Drumhummon mine, Montana, the original calcite and barite of the gangue has been replaced by quartz, and at the former the country rock and rhyolite have been partly silicified. This replacement of calcite by quartz is now actually taking place in the veins forming at Boulder Hot Spring, Montana.[2]

Neither the work of A. Breithaupt nor the earlier work of W. J. Henwood[3] on the mineral succession observed in definite vein types has led to the discovery of any laws of general application, and the same is true of the more recent studies of F. Sandberger[4] and others. At most the conclusions apply to limited areas only. The influence of local conditions, the number of substances dissolved, and consequently the multitude of consequent reactions of the substances among themselves and with the constituents of the country rock, make the phenomena exceedingly complicated.

(o) Vein Formations and Vein Types.

All the distinctive features of a vein, those of paragenesis and mineral succession, geological conditions, prevailing structure, the nature of its country rock, its age, etc., give to it a distinctive character and a particular type will always recur where lodes have been formed under similar conditions. This fact has been used by different authors in various attempts to group the confusing multitude of mineral veins into a number of types, a group to which additions are constantly made. Various attempts have been made to classify veins according to their physiognomy, with the greatest possible scientific precision. G. A. Werner, and after him S. A. W. von Herder, J. C. Freieslehen and A. Breithaupt, distinguished 'formations'; A. von Groddeck and J. H. L. Vogt distinguished 'types.' These attempts, however, have been but partly successful, owing to the difficulty of assignment resulting from the transitional or composite features in the physiognomy of the veins. Moreover, the same vein may change its character in strike or dip. In studying such changes in dip a rigorous discrimination must, of course, be made between those differences that existed

[1] W. Lindgren: 20th *Ann. Rep.* United States Geol. Survey, part III, 1900, p. 164
[2] W. H. Weed: 22d *Ann. Rep.* United States Geol. Survey, 1902.
[3] *Trans.* Roy. Geol. Soc., Cornwall, Vol. V, 1843, p. 214.
[4] 'Untersuchungen uber Erzgänge,' Wiesbaden, 1882, I, p. 96.

from the beginning and those that were subsequently added, that is to say, between the primary and the secondary differences in the physiognomy of the lode according to depth, as will be discussed in detail further on. It may also happen that two characteristic mineral associations may occur in one and the same fissure, but in separate groups, as, for example, in some lodes of the Himmelsfürst mine near Brand, where one stringer belongs to the pyritic-blendic lead type of deposit, while the other belongs to the spathic lead class; or in Tuscany, where the north part of the Boccheggiano vein is a typical quartzose copper ore, while at the south it must be classed as a pyrite-blende lead deposit.

In classifying a vein by its mineral contents the main point to remember is (as emphasized by A. Breithaupt) to base one's judgment not on one specimen or on a few specimens in a collection, but to determine the general character of the vein from the examination of very many samples belonging, if possible, to the most diverse parts of the vein, and, so far as may be done, uniting the data found in many separate lodes of an area into a general picture after inspection of the occurrence *in situ*. In old mining regions, as at Freiberg, in the Harz, in Cornwall, etc., where scientific observations have been made and recorded for a long time, the grouping of the various vein formations has, after many vicissitudes, at last been fixed in a scientifically satisfactory manner, after discarding the sub-classes formerly used.

The fundamental idea and definition of "lode formation" is due to G. A. Werner, who expressed himself on this point as follows:[1] "I give the name 'lode formation' to all lodes of one and the same origin, whether they occur close together in one region or far apart in different countries." Werner distinguished eleven mineral formations or 'niederlagen' for Saxony alone; Von Herder limited the number for the Freiberg area to five; Freiesleben by establishing subdivisions raised the number to fifty; Breithaupt took twenty for his standard, while von Groddeck established a great multitude of converging types which often can be distinguished only by emphasizing the local mode of occurrence, the nature of the country rock, etc.

A complete summary of the older literature on vein formations is found in the valuable treatise by G. A. von Weissenbach[2] 'on lode formations.'

In the following pages a summary is given of the most important mineral veins of the world. In this summary the arrangement followed is a modification of the one adopted by A. W. Stelzner for the collection of ores from the Freiberg deposits, a system which closely follows the old Freiberg classi-

---

[1] G. A. Werner: 'Neue theorie von der Entstehung der Gänge,' 1791, p. 5.

[2] Cotta's 'Gangstudien,' 1847. Vol. I, p. 1 *et seq.*

ficaTion, which has met with acceptance by many students of ore deposits. The classification of gold deposits and some other details are innovations made by the author.

Following this summary of the leading features of each class and its subdivisions, a brief characterization of each group or formation will be given. This will be followed by a brief description of selected examples. For this purpose preference will be given to the deposits of producing mining regions, to those recently opened and to those deposits which even though no longer worked are world-famous and have been carefully and scientifically studied. In the brief descriptions given, it is impossible in some instances to treat certain types of veins separately and independent of their connection with the general geologic conditions of the mineral area, and it is therefore necessary in such cases to give, in a few lines, a geologic sketch of the whole region.

SUMMARY.

The Different Classes of Mineral Veins and Their Varieties.

A. VEINS CONSISTING MAINLY OF OXIDIC ORES.

I. *Veins of Iron and Manganese Ores.*

1. Veins of spathic iron ore (spathic iron ore formation).
2. Veins of red hematite (red hematite formation).
3. Veins of manganese ores (manganese ore formation).

II. *Tin Veins.*

4. Tin veins (tin ore deposits).

B. DEPOSITS CHARACTERIZED BY METALLIC SULPHIDES.

III. *Copper-Bearing Veins.*

5. Veins with copper ores in a gangue characteristic of tin deposits (tourmaline-bearing copper formation).

6. Quartz veins containing copper ores (quartzose copper formation).

7. Veins containing copper ores in a gangue of carbonates and quartz, together with barite and sometimes fluorspar as gangue (spathic copper formation).

8. Veins of native copper with carbonates and zeolite (zeolitic copper ore formation).

IV. *Silver-Lead Veins.*

9. Veins of predominant quartz with argentiferous galena, zinc-blende, pyrite and arsenopyrite (pyritic lead ore formation).

10. Veins of calcite and other carbonates with argentiferous galena, zinc-blende and rich-silver ores (spathic lead formation).

11. Veins of barite and fluorspar, with galena, zinc-blende and rich-silver ores (barytic lead formation).

V. *Veins Carrying Rich Silver Ores.*

12. Veins of quartz with rich-silver ores (silver-quartz formation).

13. Veins of calcspar with rich-silver ores (silver-calcspar formation).

14. Veins with both copper ores and rich-silver ores (rich silver-copper ore formation).

15. Veins with silver cobalt, nickel, bismuth, and uranium ores (rich silver-cobalt ore formation).

VI. *Gold Veins.*

16. Veins with predominant quartz gangue containing gold ores (gold quartz formation).

    (a) Gold quartz veins with predominant pyrite (pyritic gold quartz formation).

    (b) Gold quartz veins with copper ores (cupriferous gold quartz formation).

    (c) Gold quartz veins with stibnite (antimonial gold quartz formation).

    (d) Gold quartz veins with arsenopyrite (arsenical gold quartz formation).

    (e) Gold quartz veins with cobalt ores (cobalt gold quartz formation).

17. Veins of quartz and carbonates with gold and silver ores (silver-gold ore formation).

18. Veins of quartz and fluorspar with gold ores (fluoritic gold formation).

VII. *Veins of Antimonial Ores.*

19. Veins of predominant quartz with antimony ores (antimonial formation).

VIII. *Veins of the Cobalt, Nickel and Bismuth Ores.*

20. Veins of nickel and cobalt ores in a carbonate gangue (carbonspar cobalt ore formation).

21. Veins of quartz with cobalt, nickel and bismuth ores (quartzose cobalt formation).

22. Veins of hydrated nickel-magnesia silicates (hydrosilicate nickel formation).

IX. *Lodes of Mercurial Ores.*

23. Quartz-calcite quicksilver formation.

B. DESCRIPTIONS OF VEINS OF EACH TYPE.

(-) Veins Consisting Mainly of Oxidic Ores.

(d) IRON AND MANGANESE VEINS.

1. *Veins of Spathic Iron Ore.*

The veins consist essentially of siderite with quartz or calcspar, and subordinate pyrite, copper pyrite, etc., and barite. The alteration of the upper part of the veins into brown hematite liberates the manganese originally present as carbonate, and this is frequently segregated in the form of pyrolusite, manganite, psilomelane and wad.

An important example of this type, now of but little economic importance, is the Stahlberg near Müsen, where a lode 12 to 27 meters thick has been worked since 1313.[1] The vein occurs in lower Devonian clay and graywacke slate. It is cut off to the southeast by a fault, while northwest it splits into three main branches, which also subdivide, and have been followed to a distance of 145 meters. It dips east 80°, and consists of almost pure, or but slightly quartzose, compact spathic iron ore, whose high manganese content renders it desirable for the manufacture of steel. There are but slight admixtures of copper pyrite, pyrite, gray copper and galena. The walls are not sharp, and the country rock is penetrated by fine stringers of spathic iron ore.

In 1901 the mines of this type in the Siegen country yielded 854,008 tons of iron ore. The Siegen I district produced 641,170 tons and the Siegen II district 212,838 tons.

[1] A. von Noggerath : 'Die Grube Stahlberg bei Müsen,' *Z. f. d. B. H. u. S. im preuss. St.*, 1863, Vol. XI, p. 63. Th. Hundt, G Gerlach, F. Roth, and W. Schmidt: 'Beschreibung der Bergreviere Siegen I und II, Burbach und Müsen.' Bonn, 1887, p. 137.

Similar conditions prevail in the vein of spathic iron ore of the Luise mine, near Horhausen, described by Hilt,[1] as well as in the lodes worked by the Krupp Company, in the Georg and Harxborg mines near that place. Pyrolusite, göthite, rhodochrosite, copper pyrite, sphalerite, galena, boulangerite and gray copper ore occur alongside of the spathic iron ore. In the Luise mine an intrusion of basalt has transformed the spathic iron ore at that place into magnetite.

A lode which has become famous for its rare minerals is the siderite lode of the Friedrich and Eisengarten mines of the Hamm district.[2] This vein traverses lower Devonian rocks. Besides siderite it also contains nests and stringers of sulphidic ores, mainly galena and chalcopyrite, also zinc-blende, pyrite and bornite. In the siderite ore of the hanging-wall split of the vein, a great nest of ore was struck in 1884, consisting of filiform niccolite, hauchecornite (a nickel bismuth sulphide) and kallilite (bismuth antimony nickel glance).

Analogous veins occur in the green Paleozoic sericitic and chloritic schists of upper Hungary.[3] Now only spathic iron ore is obtained, especially at Kotterbach in the Zips. The two most important lodes striking west-northwest and dipping steeply south are the Drozdziakow lode and the Grobe lode. The former is at points 25 to 30 meters (82-98 ft.) thick, and is usually split up into two stringers from 2 to 6 m. (6.5 to 19.6 ft.) thick. The lower one carries much gray copper ore and copper pyrite in bunches or in parallel layers in the spathic iron ore, whereas otherwise the structure is purely massive. The coarse-grained crusts of the Grobe vein consist of siderite and barite together with some quartz and calcite and much copper pyrite and mercurial tetrahedrite. Sometimes the barite entirely displaces the siderite of the veins and in isolated cases specularite occurs in lenticular patches. In the ferruginous gossan, cinnabar and native quicksilver are found, besides malachite, azurite and native copper. These peculiarities serve to indicate that the veins are closely related to the spathic copper deposits (q. v.).

No veins of this type are now worked in the United States, though a 6 to 8 ft. vein in gneiss at Roxbury, Conn., was formerly mined. It consists of crystalline siderite with a little quartz and a variety of metallic sulphides. The type is a transition to the sideritic silver lead deposits, such

---

[1] Hilt: 'Die Eisensteinlagerstätte der Grube Louise bei Horhausen,' etc., *Z. f. d. B. H., u. S. im preuss. St.*, Vol. XIII, 1865, p. 13.

[2] Wolff: 'Beschreibung des Bergreviers Hamm a. d. Sieg.' Bonn, 1885. C. Leybold: 'Geogn. Beschreibung der Eisenerzgruben Wingershardt, Friedrich, Eisengarten,' etc. *Jahrb. d. k. preuss. geol. Landesanst.*, 1882, pp. 3-47. R. Scheibe: 'Ueber Hauchecornit, ein Nickelwismuthsulfid,' *Jahrb. d. k. preuss. geol. Landesanst.*, 1891, p. 91 *et seq.*

[3] Faller: 'Reisenotezen,' etc. *Jahrb. d. k. k. montan.Lehranst.*, 1867, p. 132.

as those of Cœur d'Alene, where as a gangue the siderite is worthless. Similar ores occur in the Slocan district, B. C., and Wood river, Oregon.

2. Veins of Red Hematite.

These veins consist of dense, earthy and fibrous red hematite, together with quartz, jasper, ferruginous quartz and more rarely carbonspars, barite and manganese ores.

This class of veins is common in the Saxon Erzgebirge in the contact area about the great granite stocks in the vicinity of Schwarzenberg.[1] Some of the veins occur on the contact plane between crystalline schist and granite, others are in granite itself. The Rothenberger vein is the most important. Besides the minerals mentioned above, copper ores occasionally occur. A notable feature is the frequent occurrence of pseudomorphs of quartz, limonite and red hematite after calcspar, more rarely after barite anhydrite and fluorspar, or of mere impressions of crystals of the last mentioned minerals. These show that the original vein filling has undergone great changes. The vein structure is either massive or brecciated. These veins are 10 or even 20 meters (33—66 ft.) thick, and cut through the barren "quartz-brockenfels" veins and stocks of the region. Among the miners of that locality the expressions 'red,' 'black' or 'brown' stringer were formerly current. This fact indicates the frequent change in the development of these lodes in which red hematite, manganese ores, brown hematite or iron ocher prevailed by turns.

In the Harz, especially in the region of Zorge, a somewhat different type occurs, consisting of veins of red and brown hematite in diabase. The ores were certainly formed as a consequence of lateral secretion, and the veins become barren when they pass from the diabase to silicious schist or graywacke.[2] Their origin is disclosed by the fact that red hematite also fills the interspaces between the decomposition boulders of the diabase.

Many lodes of brown hematite undoubtedly originally contained red hematite, while in others it is impossible to tell whether the original ore was red hematite or spathic iron ore, as, for example, in the veins of Bergzabern, in Rhenish Bavaria, which cut Triassic sandstone.[3] Small stringers of brown hematite are a common phenomenon in all sandstone areas

---

[1] A. Breithaupt: 'Paragenesis,' 1849, p. 195. H. V. Oppe: 'Die Zinn- und Eisenerzgänge der Eibenstocker Granitpartie und Umgebung,' Cotta's *Gangstudien*, II, 1854, p. 133.

[2] Von Groddeck: 'Erzlagerstätten,' 1879, p. 153.

[3] B. v. Cotta: 'Erzlagerstätten,' II, p. 170.

of the world, but while deposits of specular hematite are abundant in America, no true veins of the type are now known to be worked.

3. *Veins of Manganese Ores.*

Manganese ores, especially pyrolusite, psilomelane, braunite, manganite, polianite, more rarely hausmannite and wad, accompanied in most cases by oxide iron ores, are associated with quartz, barite and calcspar as gangues.

There are numerous deposits of this kind in Saxony in the vicinity of Schneeberg, Aue and Schwarzenberg, which are found in part in granite, in part in contact metamorphic schists.

In the vicinity of Langenberg, not far from Schwarzenberg, the veins are found to be directly connected near the outcrop, with very peculiar stratiform deposits of an efflorescent iron-manganese ore. These deposits either fill flat basin-shaped depressions, overlaying the prevailing mica schist cut

Fig. 141.—Profile from Gottes Geschick to Schwarzbach.   (H. Müller.)
gl and ge, mica schist; ks, pyrite bed; k, bed of dolomitic limestone; em, efflorescent iron manganese ore and quartz-brockenfels; e, veins of the iron manganese formation; co, veins of the silver-cobalt ore formation.

by the veins, or they form stocks in the midst of the schist. These stocks seem to be the result of lateral impregnation and metasomatic replacement of certain beds of the country rock (Fig. 141). These deposits are intimately connected with great veins of iron and manganese ores which often swell into stocks called 'quarzbrockenfels' because of their brecciated structure. Cobaltiferous manganese deposits of this type occurring near the outcrop of cobalt veins cutting mica schists in the mines near Graul (Fig. 141) enclose rich and sometimes large fragments of siliceous bismuth ochre. These mines during the last decade showed a considerable output of bismuth and cobalt ores; in 1898 they produced 2,726 tons of these ores. A more detailed description of these efflorescent manganese ore deposits is given by H. Muller[1] and R. Beck.[2]

[1] 'Die Erzgänge des Annaberger Revieres,' Leipzig, 1894, p. 104.   (Erläut d. geol. Spezialk.)
[2] 'Erzlager von Schwarzenberg,' part I, Freiberger *Jahrbuch*, f. d. B. u. H., 1902, p. 64.

The manganese veins of Ilfeld in the Harz mountains occur, according to O. Schilling,[1] in a hornblende porphyrite mass which is occasionally, as at Möncheberg, completely penetrated by their stringers. The veins are for the most part only a few centimeters thick and usually grow poor at 12 meters; only in exceptional cases have they been followed down as far as 60 meters. Their filling consists of manganite, pyrolusite, varvicite, braunite, hausmannite, psilomelane and wad, besides barite, calcspar, brownspar and manganspar. Their manner of occurrence shows that their origin is due to lateral secretion.

In the Thuringer Wald there are deposits of this class at Rumpelsberg and the Mittelberg near Elgersberg in veins in porphyry. Other veins which occur in porphyry in Thuringia possess a brecciated structure, and according to H. Credner,[2] the manganese ores are practically free from gangue. At the present time they are still worked at Arlesberg (Morgenstern mine).

In the Oerenstocker district other veins occur partly in porphyry, partly in melaphyre, and at Friedrichsrode a melaphyre conglomerate forms the country rock of such fissure fillings. At Friedrichsrode fissure veins traverse a melaphyre, the ores occurring in remarkably well defined bands and crusts 1 to 3 centimeters thick, parallel to the vein walls.[3] The manganese minerals alternate with layers of calcspar, and barite is usually intergrown with them.

Manganese veins occur in granite near Wittichen in the Black Forest, west-southwest of Santander in northern Spain.[4]

The manganese lodes of the Veitsch in Styria are quite different. The headwater valleys of the Veitsch, especially the Kaskögerl and Friedelkogel, consist, according to M. Vacek,[5] of Silurian limestones traversed by strike fissures filled with manganese ores. According to analyses by C. von John the latter consists solely of rhodochrosite. These deposits were formerly regarded as beds. Mention may here be made of the lodes at Romaneche in France, to be referred to later. (See under 'Primary Differences in Depth.')

Although manganese minerals are common constituents of mineral veins, workable deposits of this type are rare. This is probably because an ore carrying less than 40% is not ordinarily marketable. The world's production comes mainly from concentration due to weathering.

[1] Geol. Spezialkarte, 'Norhausen Blatt,' 1870, p. 9.
[2] H. Credner: 'Geol. Verh. d. Thür. Waldes und d. Harzes,' 1843, p. 130.
[3] Cited by Breithaupt, 'Paragenesis,' p. 195.
[4] *Zeit. f. Prak. Geol.*, 1897, p. 90.
[5] M. Vacek: 'Ueber die geol. Verh. des Flussgebietes der unteren Mürz,' *Verh.* d. k. k. geol. Reichsanst., 1886, p. 459. C. v. John. *Jahrb.* d. k. k., Reichsanst., 1886, p. 344.

(β) Veins of the Tin Ore Class.

4. TIN VEINS.

I. *General Remarks.*

The most important ore-minerals of tin deposits are cassiterite, wolf-ramite (tungsten), native bismuth, arsenopyrite and löllingite,[1] molybdenite and scheelite; more rarely stannite, bismuthinite, specular hematite, iron spar, chalcopyrite and other copper ores, magnetite, stolzite (lead tungstate), as well as the secondary minerals scorodite, pharmacosiderite and bismuth ochre. In the gangue, quartz and lithia-mica are most common.

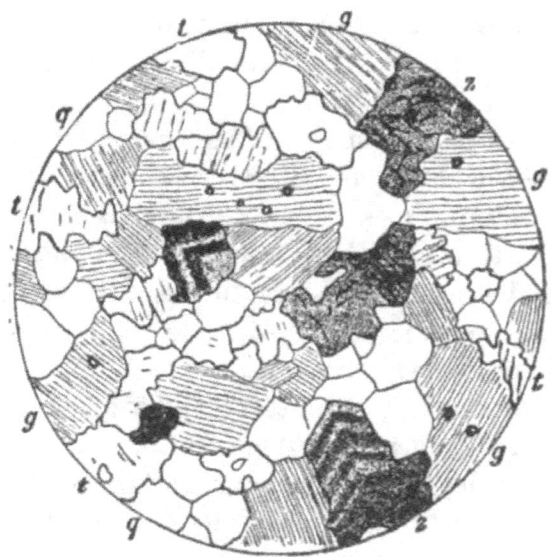

Fig. 142.—Thin section of a greisen from Banka, enlarged fifty times.
q, quartz; g, mica, with dark aureoles around zircons; t, topaz; z, tinstone.
(*Zeit. f. Prak. Geol.*, April, 1898.)

Orthoclase, gilbertite, topaz and its fibrous variety, pycnite, fluorspar, apatite and tourmaline are also frequent. Beryl, herderite PO_4Ca [Be(OH,F)] and triplite PO_4 (Fe,Mn) [(Fe,Mn)] are rarer. Tin veins occur associated with granites that carry lithia-bearing mica and cassiterite among their normal ingredients; sometimes their silicates also contain some tin. A few deposits are found associated with acid eruptive rocks, rhyolites and trachytes. The lodes cut through the eruptive masses, as well as the other country rocks, the rocks showing a peculiar and characteristic alteration adjacent to the veins. By this alteration the feldspars

[1] $Fe As_2$. Probably also leucopyrite, $Fe_3 As_4$.

have been destroyed and in their place quartz, lithia mica, topaz and tin-stone, and often tourmaline also have been deposited. This altered rock is called greisen when derived from granite, the alteration product of the other rocks having no special name, being simply called tin ore. The accompanying figure (142) represents a thin section of a typical greisen of a coarse-crystalline variety.

The structure of the original rock, as, for example, the porphyritic structure of many granites, is often exhibited in the structure of the greisen. The large masses of feldspar formed, in such cases, the main points of attack by the invading tin compound, as proved by the fine pseudomorphs of tin ore after orthoclase which have been obtained from the Botallack mine in Cornwall.

The bands cf greisen and tin ore, though mostly no thicker than a finger or a hand's breadth, and rarely as much as a meter thick, are often the real object of mining operations because of their great abundance and bulk despite their great poverty, while the true veins or fissure filling, being for the most part narrow, rarely pay for working. However, in some cases the true fissure filling is one to three meters wide, consisting in such cases of quartz, having a massive structure or only showing a symmetric banding when considerable lithia-mica is present.

Cassiterite or tinstone, the most important ore of the formation, occurs crystallized in several types: (1) In more or less perfectly developed twinned crystals and crystalline grains, as, for example, in Saxony and Bohemia; (2) in simple columnar crystals called 'needle-tin,' in Cornwall; according to Becke, the hematite-like, so-called wood tin ore of Cornwall and Bolivia belongs to this group; (3) in compact masses suggesting brown hematite, in which form it is thus far known only from Bolivia.

The probable mode of origin of the tin lodes is discussed further on in this work.

Besides the strictly typical examples of this class, there are also transitions toward the tourmalinic copper type, the pyrite-blende-lead deposits and the rich silver ore formations (silver-quartz veins).

Among the numerous occurrences found in nearly all parts of the globe, only the most important are selected as examples.

The European mines are at present competing with difficulty with the mines of the East Indies and Australia. They are in part also actually exhausted. The most important are those of the Erzgebirge of Saxony and Bohemia, and the mines of Cornwall and Brittany.

In the Erzgebirge the districts richest in tin are those of Altenberg, Zinnwald and Graupen; further west, in the vicinity of Ehrenfriedersdorf and Geyer, Eibenstock, Johanngeorgenstadt and Platten. The tin production of the Erzgebirge reached its maximum before the fifteenth century, with about 250 tons a year, and since then has steadily declined.

II. *The Tin Deposits of Altenberg, Zinnwald and Graupen.*[1]

Between Dippoldiswalde and Teplitz the Erzgebirge is traversed oblique-ly by a north-south zone of fracturing along which vast eruptive masses have come up; first, the long extended stock of Teplitz quartz porphyry run-ning north and south, spreading out from the feeding fissure of the erup-tion as a sheet covering both the gneiss, in which the fissure occurs, and also the adjacent Carboniferous and Permian strata. The second eruptive mass is granite porphyry occurring in several broad dikes, one of which forms the eastern boundary of the Teplitz porphyry stock for a great dis-tance. Finally, and latest in time, a number of granite stocks were formed, piercing the two preceding intrusive masses. At Altenberg a small granite stock of this kind pierces the granite porphyry. As the Teplitz porphyry eruptions belong to the period of the Rothliegende, the granite is of post-Permian age. A feldspar (nearest to albite), a dark lithia-bearing po-tassium-iron mica, with cassiterite and topaz as accessories, are its chief con-stituents. Both the granite and its enclosing rock are traversed by in-numerable small stringers of tin ore, each quite insignificant by itself and often hardly perceptible, but accompanied by adjoining zones of ore. Figure 143 represents a piece of Altenberg granite with ore bands of this nature. In some of the fissures the filling consists of quartz and topaz. Many bands that are not thicker than a knife-blade are noticeable only by reason of their whitish color. In the rocks lying above the dome formed by the granite intrusion, these small veins are so closely crowded, and the country rock so strongly impregnated with ore, that is to say, transformed into tin-bearing greisen, that this 'stockwork' was formerly exploited on a large scale by extensive drifts. This Altenberg tin stock, so far as known from ex-posures thus far made, continues only to a depth of about 230 meters (754 ft.) below the summit of the granite dome. At a greater depth there are merely a few narrow impregnation fissures scattered through the otherwise normal granite. as shown in the accompanying section (Fig. 144).

[1] Most important publications: G. A. von Weissenbach: 'Die Zinnerzlagerstätten von Altenberg und Zinnwald,' 1823. Manuscript in Freiberg. H. Müller: 'Bildung der Zinnstockwerke.' Berg u. Hütten. Zeit., 1865, pp. 178-180. E. Reyer: 'Zinnerzführende Tieferuptionen von Altenberg und Zinnwald.' *Jahrb.* d. k. k. geol. Reichsanst, 1878. H. Zinkeisen: 'Geognostisch-mineralogische Beschreibung der Gegend von Zinnwald und Altenberg,' 1888. Ms. in the Kgl. Bergakademie. K. Dalmer: 'Sect. Altenberg-Zinnwald der Specialk. von Sachsen nebst Erläut, 1890 und Ueber den Altenberg-Graupener Zinnerdistrict,' *Zeit. f. Prak. Geol.,* 1894, pp. 313-322. See also publications of R. Beck in *Zeit. f. Prak. Geol.,* 1896, pp. 148-150, and general account by E. Reyer. 'Geologie des Zinnes,' 1881.

The cassiterite content of the ores mined at Altenberg varies between 0.1 to 0.9%. The tin-stone granules are very small, mostly only 0.01 to 0.1 millimeters in diameter, and, as a rule, are not visible to the naked eye. Between 1869 and 1887 the ore treated averaged 0.3% Sn and 0.002% Bi. Only its great quantity makes it possible to work this very low-grade rock. Altenberg was discovered in 1458 and at first, probably during the time when the loose granite débris covering the granite boss was washed, it gave a very rich yield, 5,000 to 6,000 centners of tin per year. The Zwitterstock Company was founded in 1546 and is still in existence.

Fig. 143.—Granite of Altenberg with Zwitter bands. (From nature.)

After various small subsidences the workings completely collapsed in 1620, forming the great basin seen to-day. From that time to the present day mining has been confined to the caved-in area. Real vein mining has been carried on in some of the larger tin ore veins, but only to a very subordinate degree. In 1898 the production of Altenberg had declined to 14 tons of tin, but it has risen again during the last four years.

A different state of affairs exists at Sächsisch Zinnwald and Böhmisch Zinnwald to the south of Altenberg, where not only has tin-bearing greisen been extracted, but a lively mining industry has been prosecuted upon the true fissure filling of the tin lodes of this place. At present the lodes are worked mainly for wolframite and lepidolite, and all the old dumps have

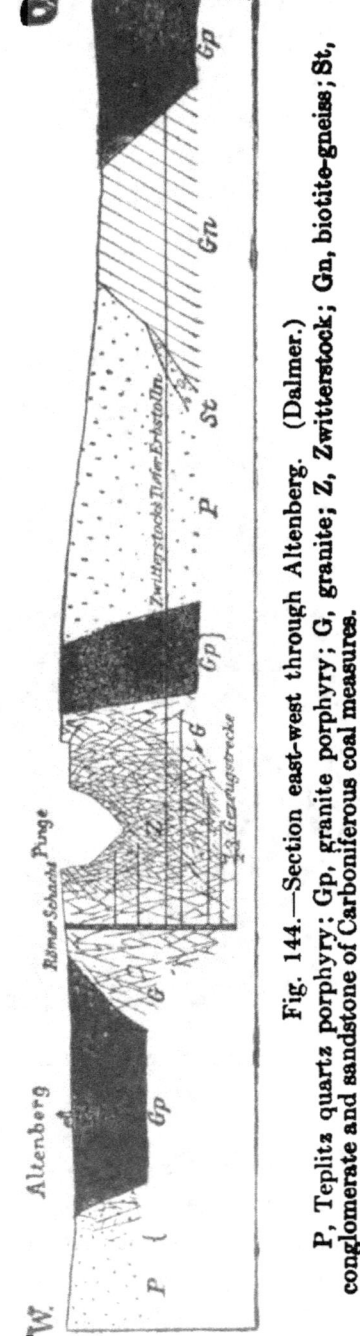

Fig. 144.—Section east-west through Altenberg. (Dalmer.)

P, Teplitz quartz porphyry; Gp, granite porphyry; G, granite; Z, Zwitterstock; Gn, biotite-gneiss; St, conglomerate and sandstone of Carboniferous coal measures.

been turned over several times for these formerly worthless minerals, which are now much in demand.

The mines are exceedingly rich in wolframite, and about 1900 the demand for this mineral gave rise to a rejuvenation of the Zinnwald mine, which was discovered about the middle of the fifteenth century, had its cli-

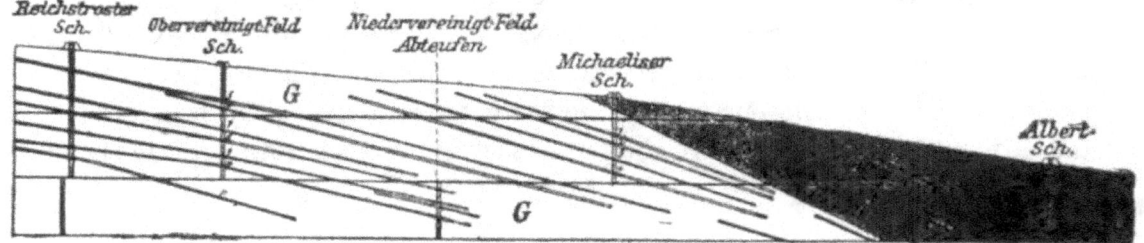

Fig. 145.—Section through the north part of the Zinnwald granite stock.
G, granite mostly altered to greisen. Dark area on right side Teplitz quartz porphyry. 1–4, Michael seam; 5, day seam; 6, upper seam; 7, organ seam; 8, middle seam; 9, upper pyritic seam; 10, lower pyritic seam; 11, lean seam.

max of prosperity about the middle of the sixteenth century, but of late years was merely able to maintain a modest existence. In 1899 Sachsisch Zinnwald produced 50 tons of wolframite.

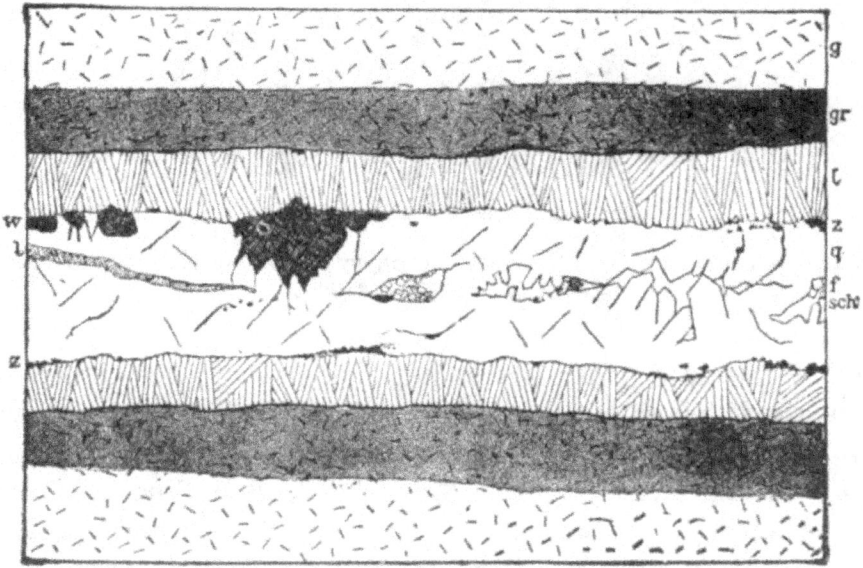

Fig. 146.—Ideal section of a so-called tinstone seam in granite at Zinnwald.
G, granite; gr, greisen; q, quartz; l, lepidolite; z, tinstone; w, wolframite; f, fluorspar; sch, scheelite.

At Zinnwald only the summit of a granite dome intruded in the Teplitz porphyry is seen exposed, the surface of the dome dipping gently in all

directions under the porphyry. Close to the contact the granite sometimes assumes a coarse-grained development, otherwise it is medium-
grained. As shown by the profile (Fig. 119), the upper portion of this
dome contains a number of parallel veins which may be compared
to flat inverted bowls; because of their horizontal or gently dipping position these are called seams or beds (flötze). The veins are not exactly
parallel to the contact plane between the granite and the porphyry, as
the dip is somewhat lower, and the veins therefore occasionally pass into
the adjoining porphyry, where, however, they grow poor (see Fig. 145).

Fig. 147.—Specimen of Teplitz quartz porphyry cut by parallel bands of ore.
(Zwitter.) (From nature.)

Sometimes two such veins converge into a single one. They usually consist of quartz with a selvage zone of lepidolite, on each side, as shown in
Fig. 146. Only rarely is the quartz replaced by orthoclase. Along the
middle of the vein, druse cavities are abundant and are lined with quartz
crystals sometimes 30 cm. long and 17 cm. thick. Some of these projecting
crystals are built up of quartz 'caps' consisting of several layers, between
which minute mica scales have produced planes of parting. Crushed quartz
crystals are often found, whose fragments are again encrusted by quartz,
a phenomena indicating a continued growth for considerable periods. The
cassiterite, also, is not all of one generation, since broken crystals occur
whose fracture surfaces are studded with smaller individuals. Ordinarily
the cassiterite and wolframite are most abundant at the selvage, or between

the mica bands and the quartz, but they also occur in irregular transverse rows of granules and stringers, which rarely attain a thickness of two centimeters. Sometimes the wolframite may occupy almost the entire fissure, as shown in the latest great mining developments. Among the commoner ores, scheelite (tungstate of lime) is the youngest, and is found by preference in the quartz druses, often cementing quartz fragments, and often accompanied by fluorspar. The less abundant minerals of the lodes include topaz (as pycnite), black tourmaline, apatite, uranium mica, zeunerite $(CuU_2As_2O_{12} + 8 H_2O)$, specular hematite, spathic iron ore, stolzite (lead tungstate); finally, copper pyrite, gray copper, arsenopyrite, tin pyrite, galena and zinc-blende. The thickness of the bed-like veins varies between 15 and 70 centimeters, but may also reach 1.5 and 2 meters (4.9 to 6.5 ft.); about twelve of them have become economically important. Some steep-dipping tin veins are also known in the locality. Both horizontal and vertical veins are accompanied by bands of greisen on each side. Extensive drifts, especially in the Reichstroster Weitung, have developed some very large masses of greisen within the Zinnwald granite, which cannot be referred to any recognizable impregnation fissures. Probably they owe their origin to an immense number of minute fissures like those of the Altenberg stockworks. Younger veins of barytic ores also occur. The adjoining quartz porphyry also contains numerous narrow bands of ore traversed by narrow veins of a fine-grained aplite-granite (alaskite), representing later phases of the normal granite of the stock. *The impregnation with tinstone occurred, therefore, before the last phase of the eruptive activity in that region.* The accompanying representation of a platy fragment of quartz porphyry from the Hoffnung Gottes Erbstolln at Zinnwald shows the manner in which the rock is traversed by a great number of parallel bands of tin ore succeeding one another at short intervals (Fig. 147).

The tin veins formerly worked contained 0.2 to 0.8% of tin, and about 1 to 2% of tungsten, while the greisen, which was picked out and treated in the stamp mill, gave 0.2 to 0.5% tin, and was very low in tungsten. In 1898 Sachsisch Zinnwald produced 50.5 tons of tungsten, 116.8 tons of lepidolite and only 1.2 tons of limestone.

In the Graupen[1] region on the south slope of the eastern Erzgebirge tin ore has been produced since the end of the twelfth century, at first from the gravel in the extensive alluvial fan at the mouth of the Graupen valley .

[1] Important publications: Th. Schiller and P. Lewald: 'Das Zinnerzvorkommen zu Graupen,' etc., in H. Hallwich. Geschichte von Graupen. Prague, 1868. R. Beck: 'Erläuterungen zu Section Mückenthürmchen der geol. Spezialkarte von Sachsen.' Leipzig, 1903. A monograph on the Graupen district by the same author is in preparation.

near the present town of Mariaschein. The ores at Steinknochen and at the Mückenberg, near Obergraupen, as well as at the Knotel further down the mountains, were developed next, and for some time mining was also carried on in a number of impregnation fissures at the Preisselberg Klösenberg and at the Zwicken basin. An insignificant mining industry still continues in the Martinistolln.

The tin ores of Graupen are genetically connected with a granite mass and with granite dikes which cut through both the gneiss and in part also the porphyry rocks of that region. A characteristic feature of the Graupen tin lodes is the large amount of orthoclase (micro-perthite) and of fluorspar in the lode filling. Where quartz and lithia-mica predominate, the rock alongside of the veins is altered to a zone of tin ore rich in topaz.

The Schellerhauer granite stock,[1] west of Altenberg, is also associated with tin veins, and further northwest, at Pöbel at the Sadisdorf basin, veins of copper-tin ore were formerly worked, which pass into a granite stock in depth.

In the Seiffen mine, on the other hand, impregnation fissures in gneiss with tinstone and copper ores were worked. No granite was encountered, although it may be presumed to exist at a greater depth.[2]

III. *The Deposits of Geyer and Ehrenfriedersdorf.*

In the western Erzgebirge, near the town of Geyer, a zone of mica schist is broken through by three small granite masses. This rock, which is rich in topaz, is accompanied by tin veins at Greifenstein, Zinnberg and Geyersberg. The most important mines of former days were those of Geyersberg, which have formed a basin on the surface by their collapse. A characteristic feature of the granite of Geyersberg is the 'stockscheider,' a very coarse, crystalline 'giant granite' developed in the porphyry of the stock near its contact with the mica schist. It forms a granite dome or boss, for the most part concealed, but with the apex showing in outcrop as the Weisse Erdenzeche near Aue, where it is decomposed and used as a porcelain clay. This Geyersberg stock, 240 meters wide, whose upper surface dips at 50 to 60 degrees in all directions beneath the schists, is traversed near its top by innumerable tin-bearing fissures varying up to five centimeters thick, striking northeast to east and dipping northwest at 70 to 80°. They are grouped in 19 series, and together with the greisen zones on both sides, which attain 10 centimeters across, are called 'streams,' and are exploited by drifts. These lodes contain quartz, topaz, mica and tinstone; also fluorspar, tour-

[1] F. Schalch: '*Erläut.* Zu Section Dippoldiswalde Frauenstein,' 1887, p. 17.

[2] J. F. W. v. Charpentier: 'Mineralog. Geographie,' 1778, p. 133. H. Müller: 'Die Erzgänge des Freiberger Bergrevieres.' 1901, pp. 130-138.

maline, geyerite and arsenopyrite rich in arsenic; more rarely wolframite, native bismuth, molybdenite and triplite, as well as apatite.

The conditions at the Geyer mines are shown in two figures. One of them (Fig. 148), taken from the work of Charpentier[1], cited below, takes us back to the times of fire-mining. It gives a good view of a series of lodes, showing distinctly how several parallel vein stringers are accompanied on both sides by impregnation zones, which fade away in a cloudy, blurred fashion toward the brighter, normal granite. The second (Fig. 149) is a photograph of a sample from the lode. The broader stringer is about one centimeter thick, and besides the predominant quartz, it carries along the middle a line of dark-looking tinstone and mica. The granite in the vicinity of the two stringers has, with the exception of a small remnant at the left edge, been altered to gray greisen.

Mining began at Geyersberg in 1315. In 1803 the works caved in, forming a great hollow. A quarry in this basin now affords excellent exposures of the vein structure.

Until lately the tin veins northeast of Geyer near Ehrenfriedersdorf were also worked. At the Sauberge these lodes are united into an entire series, the so-called 'rissen' traversing the mica schist, which have long been famous for their fine crystals of tinstone and apatite. They also carry colorless tourmaline, anatase, arsenopyrite, chlorite, fluorite, molybdenite, scheelite, wolframite, gilbertite, herderite and barite. They intersect rich barytic silver veins, and at the meeting the two ores are said to blend; they also traverse older dikes of mica diorite. Granite has not been found here, but may possibly exist at no great depth.

Tin veins were also formerly worked at numerous places in the peripheral parts and within the contact area of the Eibenstock granite massive, for example, at Aue, Sosa, Burkhardtsgrün near Schneeberg, at Auersberg near Eibenstock, at Gottesberg, at Schneckenstein near Auerbach, at Johanngeorgenstadt, and at Platten. At Hengstererben near Platten the Sanct Mauritius mine was opened again in recent time, toward the end of the seventies. The Eibenstock granite carries tinstone microliths among its primary ingredients, as all the granites associated with tin ore deposits (Altenberg, Schellerhau, Banka) probably do. Furthermore, according to Stelzner[2] and others, the silicates composing this granite also contain some tin, namely, the dark lithia-iron mica, 0.32%; the orthoclase, 0.019%; the plagioclase (albite), 0.074% SnO_2. This granite is more-

[1] J. F. W. v. Charpentier: 'Mineralog. Geographie der Chursächs. Lande.' 1778. A. W. Stelzner: 'Die Granite von Geyer und Ehrenfriedersdorf, sowie die Zinnerzlagerstätten.' Freiberg, 1865.

[2] A. W. Stelzner: 'Entstehung der Freiberger Gänge.' *Zeit. f. Prak. Geol.*, 1896, p. 394.

over rich in tourmaline as well as in topaz, a feature typical of the granites associated with tin ores. The adjoining schists metamorphosed into andalu-

Fig. 148.—Tin veinlets of the Geyer stockwork. (Charpentier.)

site-mica rock, etc., have often been impregnated along minute fissures by tourmaline associated with tinstone. Such tinstone-bearing tourmaline schists were formerly worked at Auersberg. The well-known topaz rock of the Schneckenstein, near Auerbach, represents a breccia of such tinstone-bearing tourmaline schist, whose cement consists of quartz and topaz.

The Eibenstock granite area is connected geologically with that of Carlsbad of Bohemia; here, too, at Schlaggenwalde, the contact zone between granite and gneiss contains tin veins closely resembling those of Ehrenfriedersdorf and Geyer. As early as the fourteenth century an active mining industry flourished there, its center being the town of Schoenfeld.[1]

Fig. 149.—Two tin ore veinlets from the Geyer stockwork.

The tin production of Bohemia in the sixteenth century, when it reached its climax, is estimated to have been 500 to 800 tons per year (H. Louis). At present it is hardly worth mentioning.

IV. *The Tin Deposits of Cornwall.*

The tin deposits of the peninsula of Cornwall are of far greater importance than those of the Erzgebirge. In this region the slates, which are

[1]K. Sternberger: 'Die ārarischen Bergbauunternehmungen im böhm. Erzgebirge in Oesterr.' *Z. f. B. u. H.* 1857., p. 62.

mostly of Devonian age, are traversed by five large and several small stocks of tourmaline-bearing granite. Both the slates, locally called *killas,* and the granite are cut by numerous dikes, some as much as 120 m. (393 ft.) thick, of quartz porphyry, also tourmaline-bearing, called *elvans.* These dikes also traverse carboniferous rocks (Culm). The granite intrusions, whose contact generally dips gently below the slates, have caused considerable contact metamorphism, transforming the slates into green, purple and violet hornfels and similar rocks. All these rocks are traversed by lodes of copper and tin ores,[1] which show a great tendency to break up into stringers, and often pass into an exceedingly fine network of veins. These are especially numerous near the granitic masses. Their strike is mostly between east and east northeast; their dip is ordinarily 20 to 50° north. The thickness may rise to 1.5 meters, but is mostly much less. The principal gangue is quartz, with associated orthoclase, tourmaline, chlorite, lithia-mica and some fluorspar. The tin veins contain cassiterite, stannite, copper pyrite, tungsten, blende, arsenopyrite, native bismuth and other rarer minerals; the copper lodes proper also contain gray copper, tennantite, cuprite, native copper, malachite, azurite, pyrite, arsenopyrite and blende. A remarkable feature is the change in the nature of the ore in many lodes when they pass from the slate into the granite, the pure copper becoming tin deposits.[2] This is especially well shown in the longitudinal section (Fig. 150) of the Dolcoath mine, and the cross-section of the same lode (Fig. 151) gives a striking illustration of the stringer formation of those lodes. Some lodes are filled by a breccia enclosing many fragments of the country rock; but the vein filling is generally massive, in some cases showing a symmetric banded structure. The veins are accompanied by zones of impregnation, some of them very wide, which are also worked for tin, and have furnished the main bulk of the ores turned into the furnace.

While the vein fissure itself is often only a few centimeters thick, the 'lode' as worked is several meters in thickness, and is formed of granite altered to greisen.[3] The so-called Carbonas of St. Ives is worked for tin ore to which its high content of tourmaline imparts a dark color. This greisen rock forms very irregular deposits connected by a transverse fissure with one of the main lodes of that locality. These deposits consist mainly of feldspar, quartz, tourmaline and cassiterite, associated sometimes with

[1] W. J. Henwood: 'On the Metalliferous Deposits of Cornwall.' *Trans.* Roy. Geol. Soc. of Cornwall, Vol. V, 1843. H. T. De la Bêche: *Rep.* on geology of Cornwall and Devon, 1839. C. Le Neve Foster has published many papers on this district since 1875 in *Trans.* Royal Geol. Soc., Cornwall.

[2] C. Le Neve Foster: *Mining,* 1883, p. 452.

[3] C. Le Neve Foster: 'On the Great Flat Lode South of Redruth,' etc. *Quart. Journ.* Geol. Soc., 1878, Vol. XXXIV, p. 640.

fluorspar, lithia-mica, copper pyrite and iron pyrite.[1] In the slate, also, the altered rock of the zones of impregnation is rich in cassiterite and is mined, being called *capel*. This dark-colored rock consists mainly of quartz and tourmaline, with short quartz stringers interpolated, and is traversed by small stringers of tinstone and chlorite. These *capels* accompany the cassiterite veins ('leaders').

The most important tin vein in Cornwall is the Dolcoath lode, already mentioned, which for a length of 2.25 miles is worked in the Carn Brea, Tincroft, Cook's Kitchen and Dolcoath mines. The last named mine at present furnishes one-third of the production of Cornwall.

The richness of Cornwall in tin ore was known to the ancients and gave to Great Britain the name of Cassiterides. The greatest annual production of tin ore was 16,759 tons in 1871. It had declined to 12,880 tons in 1894, and is annually diminishing, being but 4,700 tons in 1901. Copper mining, proper, only began in 1700. In 1838 the yearly production was 145,000 tons, but in 1894 it was only 3,370 tons. Copper-tin ore deposits, quite analogous to those of Cornwall, are also found in the adjoining county of Devonshire.

In France, tin deposits of similar nature occur in Brittany; for example, at Pyriac, west of the mouth of the Loire, and at Villeder in the department of Morbihan, but are at present of no economic importance.

The tin veins of Montebras are worked mainly for amblygonite.[2]

Among the occurrences of the Iberian peninsula we may mention those of Santo Tomé, south of Salamanca, and at Cartagena in Spain, as well as those of Ramalhoso near Amarante in Portugal, province of Beira. Some of the Spanish lodes are particularly rich in wolframite. In 1900 Spain produced about 1,958 tons of wolframite.

V. *Tin Districts Outside of Europe.*

Among the tin deposits of other countries, the first and most important are those of Banka and Billiton and those of the Malay peninsula.

The lodes of Banka and Billiton,[3] the so-called "tin islands," occur in granite or older schists. Besides quartz and cassiterite, they nearly all contain magnetite, some also tourmaline, and some tungsten, or else pyrite, and spathic iron ore, so that their mineralogic character presents considerable variety. The remarkable occurrence of true magnetite veins in that

---

[1] W. J. Henwood, *op. cit.*, 21.

[2] J. H. L. Vogt: *Trans.* Am. Inst. Min. Eng., 1901, p. 11.

[3] Th. Posewitz: 'Das Zinnerzvorkommen und die Zinnerzgewinnung in Bangka,' 1886. R. D. M. Verbeek: 'Geologische Beschrijving van Bangka en Billiton.' 1897. R. Beck: *Zeit. f. Prak. Geol.*, 1898, part 4, gives many references.

locality is also noteworthy. The lumps of tinstone weighing more than 1,000 kilograms found in the eastern part of Billiton seem to have been derived from the upper portion of the tin lodes. Alongside of the lodes

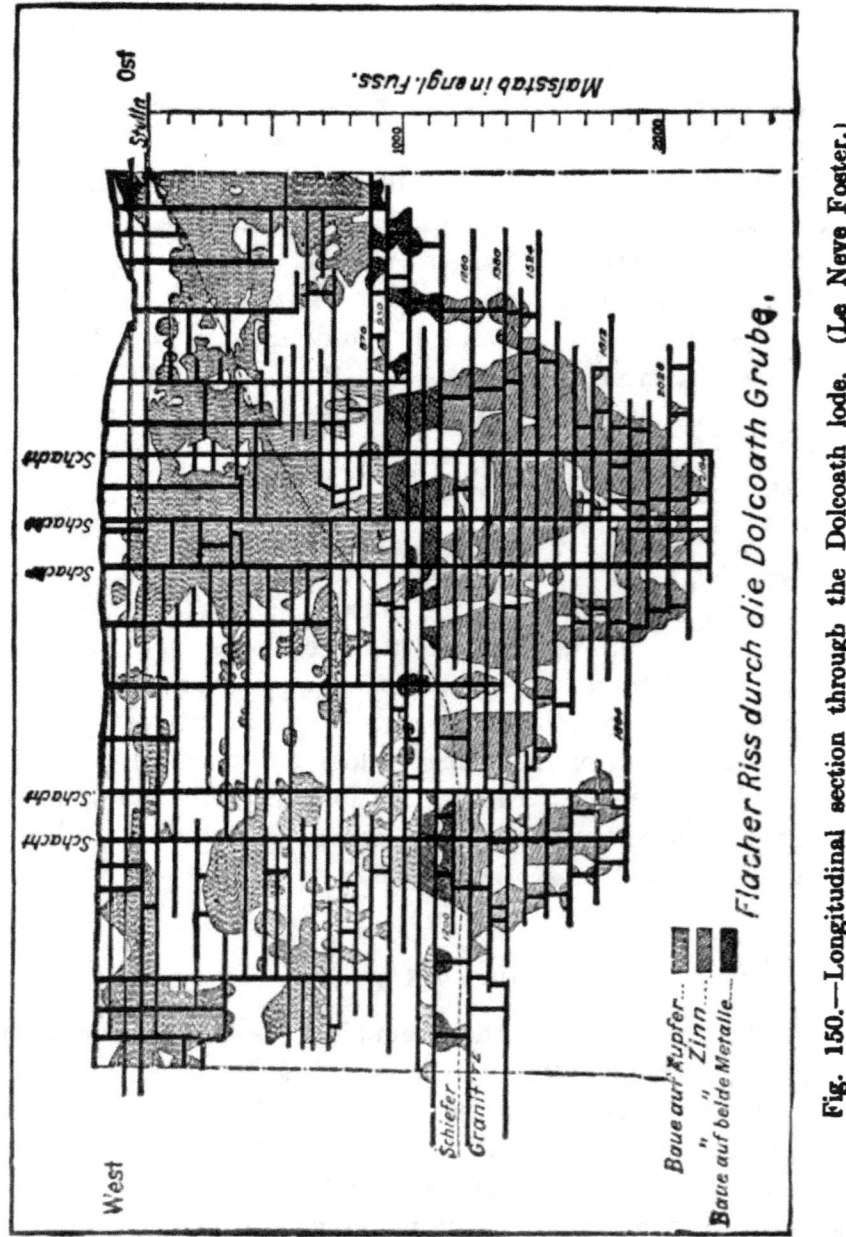

Fig. 150.—Longitudinal section through the Dolcoath lode. (Le Neve Foster.)

the granite has here and there been transformed into greisen, rich in tin-stone and topaz. (For geological occurrence and description see chapter on tin placers.)

On the Malay peninsula special mention must be made of the deposits of Malacca on the west side and of Kuantan[1] on the east side. However, by far the greater part of the supply of Malacca tin is obtained from placer gravels.

Siam and China are also to be mentioned as Asiatic tin producers, and in Japan the Taniyama mine in the Province of Satsuma is located on a tin lode (see 'Copper in Japan').

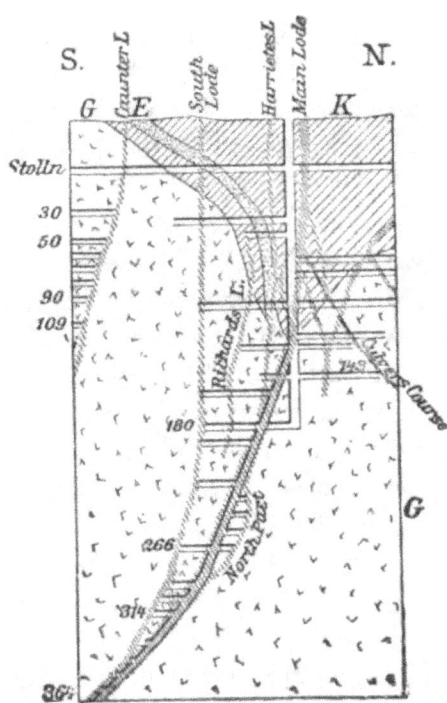

Fig. 151.—Cross-section through the veins of the Dolcoath mine.
G, granite; K, Killas or altered slate; E, Elvan (quartz porphyry) dikes. (Figures give feet.)

In recent decades the Australian tin veins have attained great importance, especially those in the New England district of New South Wales, and at Mt. Bischoff in Tasmania. As the ore from the latter locality comes mainly from residual gravels, we will discuss the Mt. Bischoff lodes in describing the tin placer deposits. Two other tin districts have recently been opened in Tasmania; the first is in the Blue Tier mountains in the northeast part of the island, where, as at Altenberg, the ore is obtained from tin-bearing greisen zones in the granite, holding from 0.375 to 1%

[1] Philips-Louis: 'Ore Deposits.' 1896. p. 601. R. A. F. Penrose: 'Tin Deposits of Malay Pen.,' *Journ. Geol.,* Vol. XI, 1903, pp. 135-154.

of tinstone.[1] The other area lies at Mt. Heemskirk near the west coast,[2] where quartz-tourmaline veins traverse a tourmaline-granite and the adjoining Silurian sandstones and slates. Along these veins the country rock has been hardened and impregnated with tourmaline and tinstone, exactly like that of Auersberg in Saxony.

In the United States, tin deposits are known in South Dakota, South Carolina (near Gaffney), and Texas. The first has already been described. The Gaffney deposit occurs in pegmatite dikes in sedimentary rocks changed to crystalline schists. The tin belt is 35 miles long, extending northward to Lincolnton, N. C. Similar deposits occur on Irish creek, near Roanoke, Virginia. The Texas deposits, as yet of no economic importance, are in the Franklin mountains, near El Paso. The veins are in a mass of intrusive granite that breaks through and uplifts Paleozoic limestones. The fissures are filled by quartz carrying cassiterite and wolframite and have greisen-like walls.[3]

As an African tin ore occurrence, we may mention that described by A. F. Molengraaff from Swazieland in the Transvaal.[4] An old schist is here broken through by granite masses surrounded by contact zones. Within the latter and in the granite itself, the occurrence of pegmatite veins has been noted, which at Embabaan, in the region of the Ryan tin works, have been found to contain corundum and tinstone. The pegmatite veins are 10 to 40 centimeters thick, consisting of quartz, with a lesser amount of feldspar, and carrying tinstone next to the walls in crystals which project toward the center of the lode.

Tin veins were discovered in 1903 in Seward peninsula, the most westerly portion of Alaska. The ores occur in wide dikes of rather fine-grained granite impregnated with fluorite.[5]

The youngest tin veins geologically are probably those of Mexico, which are associated with rhyolites and rhyolite tuffs. According to J. G. Aguilera,[6] besides cassiterite, they contain also specular hematite, topaz, sometimes tungsten, native bismuth and durangite ($Na[AlF]AsO_4$), but are without tourmaline. The most important deposits are near Aguascalientes, Durango, Guanajuato, San Luis Potosi and Zacatecas.

---

[1] W. H. Twelvetree: *Trans.* of the Austr. Assoc. for Adv. of Sc. Hobart, 1902, ref. *Zeit. f. Prak. Geol.*, 1902, p. 276.

[2] G. A. Waller: 'On the Tin Ore Deposits of Mt. Heemskirk.' *Rep.* to Secretary for Mines. A. 2236. Hobart, 1902.

[3] W. H. Weed: 'The El Paso, Texas, Tin Deposits,' *Bull.* 178, U. S. Geol. Survey, 1901. Also *Bull.* 213, pp. 99-102, 1903.

[4] A. F. Molengraaff: *Ann. Rep.* of State Geol. of S. Afr. Republic, 1897. 'The Mineral Industry,' Vol. XII, 1903, p. 332.

[5] A. J. Collier: *Bull.* 225, U. S. Geol. Survey, 1904, p. 221. See also *Bull.* 229.

[6] Mexican Volume *Trans.* Am. Inst. Min. Eng., 1903, p. 325.

VI. *Veins Showing Transitions Between Normal Tin Deposits and Other Classes.*

Examples of transitions from the purely tin-bearing deposits to copper deposits have been repeatedly mentioned, especially those of Cornwall. But there also exist transitions to the class of pyrite-blende-galena deposits. A. W. Stelzner and Schertel[1] were able to prove that the black zinc-blende of the Freiberg veins, in which a tin content had long been known, often encloses microliths of tinstone, and we know from older reports that tin ore was obtained even from Rammelsberg, and near Rosine[2] at Freiberg, in the gossan of the pyritic-zinciferous lead veins, so that the miners spoke of a tin gossan over silver-lead ore-lodes.

There is one example of the transition type between tin deposits and deposits of silver ores, upon which we have ample data, a type which is numerously represented on the high plateau of Bolivia between the 15th and 21st degrees south latitude. According to A. W. Stelzner,[3] this type includes the veins of the Cerro de Potosi (discovered in 1545), Ouro Colquiri, Poopo, Tasna, Milluni, Chorolque [highest mine on the earth, 5,309 m. (17,416 ft.) above the sea]. The gangue consists of quartz, with various carbonate spars and barite. The ore consists of sulphides and sulphates of iron, lead, zinc, copper, silver, tin, bismuth and antimony, but also includes tinstone. Silver and tin ores are so intimately mingled and intergrown that they cannot be separated by hand picking, and it is only after roasting and amalgamation that the tin can be obtained from the residue.

Besides cassiterite, the tin ores include stannite (tin pyrite), together with wolframite and, in places, tourmaline (*chorolque*) and fluorspar (*cloquirri*) also. An interesting fact is the presence of the two germanium-containing ores, argyrodite and franckeite.[4] In the upper workings the lodes were especially rich in tin, and it was only at greater depth that the tin ores were found rich in silver and lead. The cassiterite occurs especially in the form of wood tin and as hematite-like masses. In the Freiberg collection there is a lump of almost pure tin ore from that locality, weighing 93 kilograms. A secondary concentration of the tin content seems to have taken place in the outcrop. The silver occurring in the form of native silver and silver chloride and found in the astonishingly

[1] A. W. Stelzner and A. Schertel: 'Ueber den Zinngehalt der schwarzen Zinkblende von Freiberg.' *J. f. d. B. u. H. im K. Sachsen,* 1886.

[2] J. F. W. v. Charpentier: 'Mineral. Geographie der Chursächs.' 1778, p. 101.

[3] A. W. Stelzner: 'Die Silber-Zinnerzlagerstätten Bolivia.' *Z. d. D. G. G.* 1897, part 1. M. Frochot: 'L'étain en Bolivie.' *Ann. d. Mines* 90, XIX, pp. 186–222.

[4] A. W. Stelzner: 'Ueber Franckeit, ein neues Erz aus Bolivien,' *N. Jahrb. f. Min.,* 1893, Vol. II, p. 114.

rich gossan, continued for centuries to maintain the reputation of Bolivia as the foremost silver-producing country of the globe. These Bolivian lodes are not associated with granite intrusions, but are, for the most part, connected with intrusive masses of dacite and rhyolite.

The tin exported from Bolivia in 1902 aggregated 16,779 tons, the product coming mainly from the departments of Potosi and Orouro. In the latter district the Huanuni mine alone produces every year 3,000 tons of tin ore.

VII. *Tungsten Deposits.*

In connection with the group of tin veins, we may here mention certain veins of tungsten, which, though devoid of tin, yet in their associations show altogether the same features as the tin lodes.

Thus Bodenbender[1] described veins in the southern part of the Sierra de Cordova in Argentina, which occur in granite and gneiss, and consist of quartz with wolframite, molybdenite, pyrite, chalcopyrite, covellite, limonite, apatite, mica, etc.

Tungsten veins are also known at several different points in North America, those occurring at Osceola, in the granite area of the Snake Mountains in Nevada,[2] containing besides wolframite a good deal of hübnerite.

In Portugal, veins of tungsten are exploited in the mines of Covelha (Castello Blanco), etc.

Another end product, or extreme example, of the type of deposit under discussion, which has been called the tin 'formation,' is represented by the pure fluorspar lodes, which, according to J. Valentin, are also worked in Argentina at San Roque in the province of Cordova.[3] The lode materials are fluorspar and quartz, with an occasional addition of red feldspar and a silvery white mica. Similar veins occur in Southern Illinois. According to Bain they are true fissure veins filled with fluorspar, calcite and scanty amounts of lead and zinc sulphides. The veins are 12 to 20 feet wide and cut Carboniferous limestones.

On the contrary, the fluorspar lodes of the Saxon Vogtland are more naturally classified with the copper deposits (which see), although they have some quite characteristic features, which they share in common with the tin veins.

[1] Bodenbender: 'Die Wolfram-Minen der Sierra von Cordoba.' *Zeit. f. Prak. Geol.*, 1894, p. 409.

[2] U. S. Geol. Survey, 'Mineral Resources for 1899-1900,' pp. 300-304.

[3] J. Valentin: 'Ueber das Flusspathvorkommen von La Roque,' etc. *Zeit. f. Prak. Geol.*, 1896, p. 104.

(b) DEPOSITS FORMED ESSENTIALLY OF SULPHIDE ORES.

(γ) Copper Veins.

5. VEINS OF COPPER ORES CARRYING TOURMALINE; INCLUDING ALSO VEINS
CHARACTERISTIC OF THE TIN ORE TYPES.

(Tourmaline-bearing Copper Formations.)

In discussing the characters of certain formations, mention was made
of transition forms between the true tin veins and those of copper deposits, which were observed in many of the Cornwall and Saxon Erzgebirge
veins, in which both tinstone and copper ores occur. Under the present
head, however, we wish to include distinctively cupriferous veins, free from
tin ore, which nevertheless contain, in both the ores and gangues, several
minerals usually characteristic of tin lodes, veins which, like the tin deposits, are connected with granitic eruptions. This type, as J. H. L. Vogt
correctly remarks,[1] represents, therefore, an extreme terminal member of
the class of tin-copper veins already noted several times. For this reason this subdivision of the group of copper deposits has been placed immediately after the tin veins, heading the very important group of sulphide veins although it is not economically important enough to entitle
it to that place when compared with the other types of copper veins, which
are not only far more widely distributed, but are of very great economic importance.

The ores of this class of veins consist of chalcopyrite, bornite and copper glance, with accessory hematite, molybdenite, galena, zinc-blende, tetrahedrite, arsenic, bismuth and uranium ores and native gold, the gangue
consisting of quartz, muscovite, calcspar, dolomite, siderite, fluorspar and
tourmaline, as well as beryl and apatite.

As is the case with the tin lodes, the granite on both sides of these veins
is strongly altered. The feldspar of the rock is destroyed and formed into
quartz, mica and calcite. When altered to a typical greisen, it contains
topaz and lepidolite.

This class of veins is especially represented in the Thelemark region in
southern Norway, where it has been described by Th. Scheerer, T. Dahll,
P. Herter and J. H. L. Vogt.[2] The lodes there outcrop either in granite,
as at Klovereid, or in slate, near a granite massive as at the Aamdal copper works, or they form 'ladder' veins within the granite dikes that cut

[1] 'Zur Classification der Erzvorkommen,' *Zeit. f. Prak. Geol.*, 1895, p. 149.

[2] Th. Scheerer: 'Beiträge zur Kenntniss Norweg. Mineralien,' *Berg u. Hütten Zeit.*,
1845, p. 849 and 'Ueber die Kupfererz-Gangformation Thelemarkens,' *Berg u. Hütten
Zeit.*, 1863, p. 157. Th. Dahll: 'Om Thelemarkens Geologi,' 1860. P. Herter:
'Ueber die Erzführung der Thelemarkischen Schiefer:' Z. d. D. G. G., 1871, p. 377,
J. H. L. Vogt. *op. cit.*

quartzitic slate at the Näsmark mine, or as contact lodes along the borders of a granite dike, as in the Moberg mine. Bedded veins occurring between the planes of stratification of the slates have also been observed at the Aamdal copper works. On the contrary, the Svartdal lodes, in which the copper ores are associated with gold and tourmaline, are in a quartz-mica-diorite, not in granite, and this rock has, like the granite, been changed along the lodes into a greisen-like rock.

The Svartdal lodes show a close resemblance to the gold-bearing copper deposits described by A. von Groddeck, A. M. Stelzner and W. Möricke[1] in Chile, as at Remolinos and Ojancos in the province of Atacama, at Tamaya and La Higuera in the province of Coquimbo, and at Las Condes in the province of Santiago. The Chilean veins contain quartz and tourmaline, forming the matrix of the gold-bearing copper ores; the latter also contains some free gold, and rarely molybdenite and scheelite. They are usually associated with acid or moderately acid eruptive rocks, which in some instances have undergone tourmalinization along the lodes. The copper production of Chile in 1903 amounted to 31,330 tons, which, however, includes the product of other types of deposits.

6. CUPRIFEROUS QUARTZ VEINS.

Chalcopyrite, chalcocite, bornite, enargite and tetrahedrite occur in veins with quartz filling. Pyrite often occurs in the chalcopyrite.

The most instructive European example of this type is probably the Kupferberg district in Silesia, of which we have a detailed description by Websky.[2] The veins outcrop in a hornblende schist and other crystalline schists, which form a zone between the Paleozoic grits of Landshut and a granite massif, and are traversed by dikes of porphyry. The vein filling consists of quartz and hornstone; the ores occur in bunches and pockets, and consist of copper glance, bornite, chalcopyrite, pyrite, pyrrhotite and zincblende, and in druses, of tetrahedrite. In rare cases a flesh-red feldspar is also said to be a constituent, and, finally, in some cases, cobalt and nickel ores. For the most part the deposits are true fissure veins with a good deal of enclosed country rock. Krusch divides the deposits into three classes, viz.: (1) Older veins, with quartz, calcite, gypsum, fluorite, chal-

---

[1] A. v. Groddeck: 'Ueber Turmalin enthaltende Kupfererze von Tamaya.' *Zeit. d. D. G. G.*, 1887. A. W. Stelzner: 'Ueber die Turmalin-führung der Kupfererzgänge in Chile.' *Zeit. f. Prak. Geol.*, 1897, p. 41. W. Möricke: 'Die Gold,- Silber, und Kupfererlagerstätten in Chile.' 1897.

[2] Websky: 'Ueber die geognost. Verh. d. Erzlagerstätten von Kupferberg und Rudelstadt in Schlesien.' *Z. d. Deutsch. geol. Ges.*, Vol. V., 1853, pp. 373–438. P. Krusch: 'Die Classification der Erzlagerstätten von Kupferberg in Schlesien.' *Zeit. f. Prak. Geol.*, 1901, pp. 226–229.

copyrite and some galena, together with chloritized country rock. (2) Younger, pure quartz veins, with disseminated chalcopyrite. (3) **Ore beds** like those of Schwarzenberg, with a floor of actinolite, prase and chlorite, much pyrite and accessory pyrrhotite, chalcopyrite and bornite, with lievrite and magnetite (Einigkeit vein).

A similar deposit described by A. von Groddeck[1] occurs at Rheinbreitenbach on the Rhine, in the Devonian graywacke area.

In the kingdom of Saxony this class of copper veins is found in the Schneeberg mining district, particularly at the old mines of König David and Sanct Michaelis near Oberschlema.[2]

Many of the veins of the Massa Marittima district of Tuscany belong to this quartz-copper vein type, as shown by Lotti[3]; as, for example, the Boccheggiano vein, while others, viz., the Fenice Massetana, Capanne Vecchie, Serrabottini, Sud and Guardione veins also carry blende and galena, and thus form transitions to the pyrite-blende-lead veins.

These deposits also show great variety of form, for besides the true veins just mentioned, irregular stock-like bodies occur, as a result of the replacement of lime rocks along the bedding planes separating dissimilar strata or along faults. Thus, at the plane of separation between the dolomitic Rhaetic limestones and the alternating beds of shale and limestone of the Eocene, there are ore deposits, as, for example, the iron and calamine deposits of Carbonais (Valdaspra), the copper, lead and calamine deposits of Rocchette and Serrabottini-Nord, as well as those of the Speziala, and the copper and iron ore deposits of Cagnano-Pighetti. The lead ore deposits of Montieri and Gerfalco are associated with Liassic limestones.

The veins occupy fault lines, along which the country rock has been much altered. This alteration has produced either silicification with the deposition of copper and iron pyrite and in part a segregation of epidote or else a complete metamorphism into a pyroxene-epidote rock. In some cases (Guardione) this alteration and replacement extends much farther into the hanging-wall than into the footwall. All the veins are of post-Eocene age.

The vein worked at Boccheggiano, an important mining town, northeast of Massa Marittima, is worthy of special mention. The course of the vein is north-northeast, the dip 40 to 50° northeast. The vein has a thickness of 30 m. to 33 m. (98 ft. to 114 ft.) and is mined for 1,200 m. across the Merse valley, being worked on both sides of the stream. It occupies a fault

[1] A. V. Groddeck: 'Erzlagerstatten,' 1879, p. 202.

[2] H. Müller: 'Die Erzdistrict von Schneeberg im Erzgebirge.' 1860. Cotta's *Gangst.*, III, p. 70.

[3] B. Lotti: ' Descrizione Geologico-Mineraria dei Dintorni di Massa Marittima.' Rome, 1893. (Gives other references.)

with micaceous shales of Permian age beneath, and the Eocene limestones and clay shales above. Segregations of epidote occur along cross fissures as far as 50 meters from the vein. The ore consists of quartz, containing pyrite and copper pyrite, with scanty amounts of marcasite, bismuthinite and hematite. The workable ore occurs in three column-shaped shoots pitching northward. The rich ores show, in part, a massive and, in part, a banded structure. Large pyrite cubes occur sparsely interspersed through the ore. In places the vein filling consists of a mineralized slate breccia. The ore contains 3 to 11% of copper, with various other metals, including 0.04% of tin. The gangue is chiefly quartz.

A spring tapped at the lowest part of the shaft, with a temperature of 40.6° C., is at present greatly interfering with the work.

Some of the more important occurrences of other continents should also be noted:

Among these is the copper vein of the Chudak mine in the Altai, 6-7 m. thick, outcropping in the quartz porphyry.[1]

Splendid examples of this type of vein occur in diabase at the Sünik mine, near Kata, south of Schuska, in the Transcaucasia.[2]

Australian Deposits.

Many of the copper veins of Australia are quartz-pyrite veins. The Burra-Burra mines of South Australia, since their discovery in 1845, in the space of 29.5 years, have yielded 234,648 tons of copper ore, with an average content of 22% copper, corresponding to 51,622 tons of copper. Work ceased in 1877, as the two main lodes were worked out. The vein lies on a contact between 'serpentinous' limestone[3] and steeply upturned, thinly bedded limestones. Subsequently the lodes of Wallaroo, in porphyrite and Moonta,[4] also in porphyrite, were discovered, which are also exceedingly large (30 feet) and average 12% copper. The two together produced in 1897 about 5,100 tons of copper.

The Great Cobar copper mines in New South Wales, which recently regained their former importance, work quartz veins in the Silurian slates. The main ledge is in its upper level, 25 to 45 feet wide and composed of nearly solid sulphides. The total copper production of Australasia in 1902 was 29,098 tons.

African Veins.

The most noteworthy African examples are the veins in metamorphic

---

[1] B. von Cotta: ' Kupfergrube Tschudack im Altai.' *Berg u. Hütten Zeit.*, Vol. XXIX, 1870, No. 7, p. 29.

[2] K. Ermisch: 'Die Kupfererze der Sünikgruben', *Zeit. f. Prak. Geol.*, 1902, p. 88.

[3] J. D. Woods: 'The Province of South Australia.' 1894, p. 258.

[4] Phillips-Louis: 'Ore Deposits.' 1896, pp. 691–693.

Devonian slate and granite in Little Namaqualand and Demaraland and on the west coast of South Africa, from which in 1898 the production was 2,438 tons of copper. This class also includes the veins of the Albert silver mine, 50 miles northeast of Pretoria in the Transvaal. These veins outcrop in an area of reddish porphyritic granite traversed by dikes of olivine diabase. The primary vein consists of quartz with bornite, chalcopyrite, silver-bearing tetrahedrite and some hematite (Freiberg collection).

Japan.

The most productive copper veins of Japan, those of Ashio, belong to this group. These lie near Nikko, so famous for its shrines and temples (province of Shimodzuke). According to Yamada (statement in a letter) the Paleozoic (pre-Carboniferous) sandstones, clay slates and hornstones of that locality are traversed by a stock of rhyolite. This rhyolite, which grades at times into dacite, has altered the sedimentary rocks along the contact and encloses many fragments of them. In the rhyolite, but not in the contact rocks, there are numerous northeast or west-northwest to east-west veins dipping steeply north or south. These veins consist mainly of copper pyrite, pyrite, arsenopyrite and quartz. Their walls are usually sharply defined, but sometimes only on one side. Besides the typical copper veins, there are also pyrite-blendic lead veins at Ashio, though these are much rarer. The production of the Ashio mines in 1902 was 6,762 tons of copper. The total Japanese production in 1903 was 31,360 tons of copper.[1]

Butte, Montana.[2]

The prodigious production of the veins of this kind at Butte City, Montana, in recent years, has surpassed that of any other district of the world.

The Butte district is situated in southwestern Montana, in the central part of the Rocky Mountain region. The city, which is built about and over the mines, is the largest settlement of the State, while the neighbor-

[1] Sketch of the Mining Industry of Japan. Published by the Bureau of Mines of Japan for the Louisiana Purchase Exposition, 1903.

[2] W. H. Weed: 'Ore Deposits at Butte, Montana,' *Bull.* 213, U. S. Geol. Survey, 1903, p. 170. S. F. Emmons: 'Notes on the Geology of Butte, Montana.' *Trans.* Amer. Inst. Min. Eng., Vol. XVI, 1887, p. 49. Jas. Douglas: 'The Copper Resources of the United States.' *Trans.* Amer. Inst. Min. Eng., Vol. XIX, 1891, p. 693. R. Vogelsang: 'Ueber den Kupferbergbau in Nordamerika.' *Z. f. d. B. H. u. S. im preuss. St.*, 1891, Vol. XXXIX, p. 248. C. A. Hering: 'Die Kupfererzlagerstätten der Erde.' *Idem*, 1897, pp. 19–22. S. F. Emmons: 'Economic Geology of the Butte District.' Geol. Atlas of the U. S., Butte, Montana, folio, 1897. W. H. Weed: 'The Secondary Enrichment of Ore Deposits.' *Bull.* Geol. Soc. Am., Vol. XI, pp. 179–206, 1899–1900. R. C. Brown: 'The Ore Deposits of Butte City.' *Trans.* Am. Inst. Min. Eng., Vol. XXIV, 1895, p. 556.

ing city of Anaconda, 20 miles distant, is a dependent. The rocks of the ore-bearing area are all igneous, the district forming part of an extensive region of Tertiary igneous activity. The prevailing rock, and the one in which all the veins occur, is a dark basic granite, technically known as quartz-monzonite, which is a part of a great mass of granitic rock. This rock is cut by dikes and irregular intrusions of the Bluebird granite, a white aplite[1] composed of quartz and feldspar, with a little mica. In the copper-bearing area the Modoc porphyry appears in lenticular dikes, traversing both varieties of granite in very irregular fissures. It is a light-colored rock, carrying large and distinct crystals of feldspar and quartz in a dense groundmass, and is technically designated rhyolite-porphyry or quartz-porphyry.

After the intrusion of the Modoc porphyry, extensive fracturing occurred, with vein formation, the veins cutting the porphyry in many instances. After the formation of these earlier veins they were fractured, and silver veins formed, followed by renewed volcanic activity, resulting in the intrusion and eruption of rhyolite, forming dikes cutting across the veins, and in extensive extrusive masses covering the silver veins to the west. The veins of the district, both copper and silver veins, belong to four distinct systems. The oldest lodes have a general east-west course, the Parrot, Anaconda and Syndicate lodes being examples. Another set of veins has a northwest-southeast course, and has displaced the earlier veins. A still later set has an east-northeast course and has displaced both the earlier systems of veins. These veins, in the eastern part of the area, are cut and displaced by a great northeast fault which carries no endogenous ore, the material mined having been broken off from earlier deposits and included in the fault débris.

The silver veins surround the copper lodes on the north, west and southwest. Their course and geologic relations are very similar to those of the copper veins, but their structure and mineralogic character are different. The silver veins contain sulphide of silver, blende, pyrite and a little galena, and commonly contain no copper, save near the border of the copper area, where, though occasional bunches of copper ore occur, it consists of chalcopyrite, and, more rarely still, tetrahedrite, minerals which occur rarely and very sparingly in the copper lodes. The gangue consists of quartz, with rhodonite and rhodochrosite, and shows marked banding and crustification, in strong contrast to the structure of the copper veins.

Several of the copper veins were, as is well known, at first worked as silver veins. The upper portion of the veins consisted of quartz somewhat stained by iron, but not like the great iron gossan caps of other regions.

---

[1] Called 'granulite' by some writers—a name applied by German geologists to a variety of schist, but by French petrographers to aplite.

This extends to a variable distance below the surface, 200 to 400 feet in some instances, where it is replaced by partly oxidized and decomposed copper ores that form the upper limit of the remarkable glance, enargite and bornite orebodies of the district. Carbonates and oxides are rare.

The copper minerals occur in quartz-pyrite veins of remarkable width and extent. The Anaconda ledge is sometimes 100 feet wide, and will average half that width, as will also the Syndicate lode.

Character of the Ores.—The copper ores average 55% silica and 16% iron. About 15% of the tonnage mined is first-class ore, averaging 12% copper; the remaining 85% carries 4.8% copper, and is treated in concentrating mills, the resulting product containing but 15 to 20% of silica, while the copper is increased to 18%.

The ores contain gold to the extent of about 2.25 cents to each pound of copper, with 0.0375 ounce of silver. Native gold has been found upon crystallized glance, and native silver is common in bornite and glance in some mines. It is estimated that the total production of copper ore has been about 31,000,000 tons, averaging 5% copper. The amount of arsenic (and antimony) present is very large, it being estimated that over 32,000 lb. a year pass off in smelter fumes. Tellurium is present in very small quantity in the ores, amounting to 2.375 ounces, or 0.008%, in the crude copper from the converters.

In general, it may be stated that the original mineral-bearing solutions were in all probability hot and ascended through fractures in the granite. The copper deposits are almost entirely replacement deposits. There is a marked association of faulting of the veins with bodies of rich ore, and these faulted areas are wet, so that the miners say: "A dry and tight vein is barren; a wet and crushed one is rich." There is also a distinct genetic relation between ore and country rock, as a result of the deposition of the ore by metasomatic replacement. Thus the Anaconda ledge is low grade where it crosses either the Bluebird granite or the Modoc. porphyry, a feature explainable by the lack of easily replaceable, dark-colored, ferromagnesian minerals in those rocks.[1]

In 1903 Montana produced 272,555,854 lb. of copper. The Anaconda mine alone yielded between 1879-1897 about 470,000 tons of copper.

Cupriferous quartz veins also occur in the Virgilina district along the border between Virginia and North Carolina, but the annual production does not exceed 3,000 tons. The copper occurs as glance and chalcopyrite, more rarely as bornite, in bunches and stringers sparingly distributed through the massive quartz filling of the veins. In the Blue Wing mine the

[1] W. H. Weed: 'Influence of Country Rock,' etc., *Trans.* Amer. Inst. Min. Eng., Vol. XXXI, p. 634, 1901. Also *Amer. Geol.*, Vol. XXX, p. 170, 1902.

vein filling is calcite and the ore bornite. These veins are traceable for several miles by well defined outcrops. They cut soft micaceous schists formed by the metamorphism of pre-Cambrian andesites and volcanic tuffs.[1]

The type of copper quartz veins just described is connected by various gradations with the pyrite blende bearing lead deposits. Several veins in the Freiberg area, proper, belong to such a transitional type. The quartz gangue contains copper pyrite, bornite, copper glance and gray copper, and in the upper portions oxidized ores, Galena, zinc-blende, arseno-pyrite and pyrite occur in lesser amounts. As examples the Gottlob Spat, Franzer Spat and Heinricher Spat at Morgenstern may be mentioned. This was also the character of the now abandoned veins of Junge Höhe Birke and Alte Mordgrube, at least of certain parts of the veins, as, for example, the steep-dipping layers of the Alte Mordgrube.

Veins of intermediate type also occur at Hohenstein[2] in Saxony. They consist of pyrite, arsenopyrite, chalcopyrite and tetrahedrite, in quartz, brown spar and calcite, with some marcasite, galena, zinc-blende and bournonite. These veins are remarkable for their gold content. They are typical copper veins in certain parts of the lode. The two copper minerals named above are carriers of the silver and gold content, the gray copper containing about 0.01% of gold. The Lampertus mine is still in operation.

Similar transitional types are found in other countries; for example, some of the lodes of Sado and Ikuno in the province of Tashima, in Japan, are, according to Yamada, of this nature.

As an appendix to this last group, we may mention the remarkable manganiferous copper lodes, forming a transition of purely manganese veins, found in the vicinity of Muleye in lower California. Their mineralogy has been studied by P. Krusch.[3] They outcrop in Tertiary trachytic tuffs about 110 kilometers north-northwest of the town. In the undecomposed state they carry a manganiferous and cobaltiferous copper glance with gangue of chalcedony with associated gypsum. Farther east there are also found true maganese lodes, with quartz and gypsum as matrix and a psilomelane containing 0.38 to 1.2% copper.

7. Spathic Copper Veins (Gangue of Various Carbonates with Quartz, Barite and Sometimes Fluorspar).

The gangue of these veins includes some quartz, but often consists chiefly, and sometimes entirely, of various carbonates, particularly iron spar, as well

---

[1] W. H. Weed: 'Types of Copper Deposits in Southern United States.' *Trans. Am. Inst. Min. Eng.*, 1900.

[2] H. Müller: 'Ueber die Erzgänge von Hohenstein,' 1879, in Erläut. zu Sect. Hohenstein., p. 28 *et seq.*

[3] P. Krusch : 'Ueber manganhaltige Kupfererzgänge.' *Zeit. f. Prak. Geol.*, 1899, p. 83

as calcspar and dolomite. Barite is very common and is at times accompanied by fluorspar. The ore minerals are chalcopyrite, bornite, glance, tetrahedrite and pyrite. Cobalt and nickel ores and various other ores also occur as accessory minerals.

The lodes of Kamsdorf, near Saalfeld, in Thüringia, have been studied and described by F. Beyschlag.[1] The veins fill fault fissures in the *zechstein* (Permian limestone) formation, and they also continue downward as barren siderite-barite veins, into the strongly folded and tilted Culm slates underlying the *zechstein* formation. In these lodes, barite is the chief vein mineral, carbonates appearing but scantily, and quartz only showing where the fissure is within the Culm. Among the ores an argentiferous gray copper and a non-silver-bearing copper pyrite, together with secondary copper ores, are most common; cobalt and nickel ores are accessory constituents, especially smaltite (cobalt pyrite) and niccolite ($NiAs$). The ores often enclose fragments of the country rock. The structure is both massive and brecciated, and it is only where the lodes descend into the Culm that a symmetric banding of the lode is seen. The veins are richest between the dislocated parts of the bed of Kupferschiefer faulted by the veins. In the upper strata as well as in the Culm the lodes are barren of ore. Near the veins certain beds of the lower and middle *zechstein*, especially limestone and dolomite strata, are replaced for varying distances by spathic iron ore. The veins are now worked for these masses of spathic iron ore, or the secondary brown hematite derived from them. The orebodies at times contain impregnations of copper ore. The diagrammatic section (Fig. 153) is a composite made from several of Beyschlag's profiles. The total product of the Kamsdorf mines for 1898 was 24,760 tons of spathic iron ore and 17,929 tons of brown hematite.

Near Ceilsdorf and Oelsnitz, in the Vogtland of Saxony, similar veins cut Devonian rocks, not the *zechstein*. Analogous veins were worked at the beginning of the 19th century, at the mines of Deichselberg. The mines of Sanct Burkhard and Heilige Dreifaltigkeit (Holy Trinity) were formerly important. Some of the lodes, it is true, are worked mainly for spathic iron ore; in fact, transitions to the non-copper-bearing spathic iron ore lodes may be very frequently observed in this class of copper veins.

Some of the copper veins of Vogtland vary in character, containing a large amount of fluorspar, being, in fact, worked only for this mineral to-day. The Auf der Kunst vein, which is in places over 25 meters thick, is a typical example of this kind. It lies between Schönbrunn and Planzschwitz in the Saxon Vogtland and belongs to the Heilige Dreifaltigkeit

[1] F. Beyschlag: 'Die Erzlagerstätten von Kamsdorf in Thüringen.' *Jahrb.* d. k. preuss. geol. Landesanst., 1888, p. 329.

mine just noted. According to E. Weise,[1] this vein occupies a fault fissure dipping steeply east-northeast and striking north 25° west, cutting Devonian slates, diabase breccias and diabases. The footwall streak carries brown hematite (altered spathic iron ore) and copper ores. Above this lies a thick streak of quartz, carrying fluorspar, brown hematite and subordinate copper ores. The middle band, 2 to 8 meters thick, is formed of white and

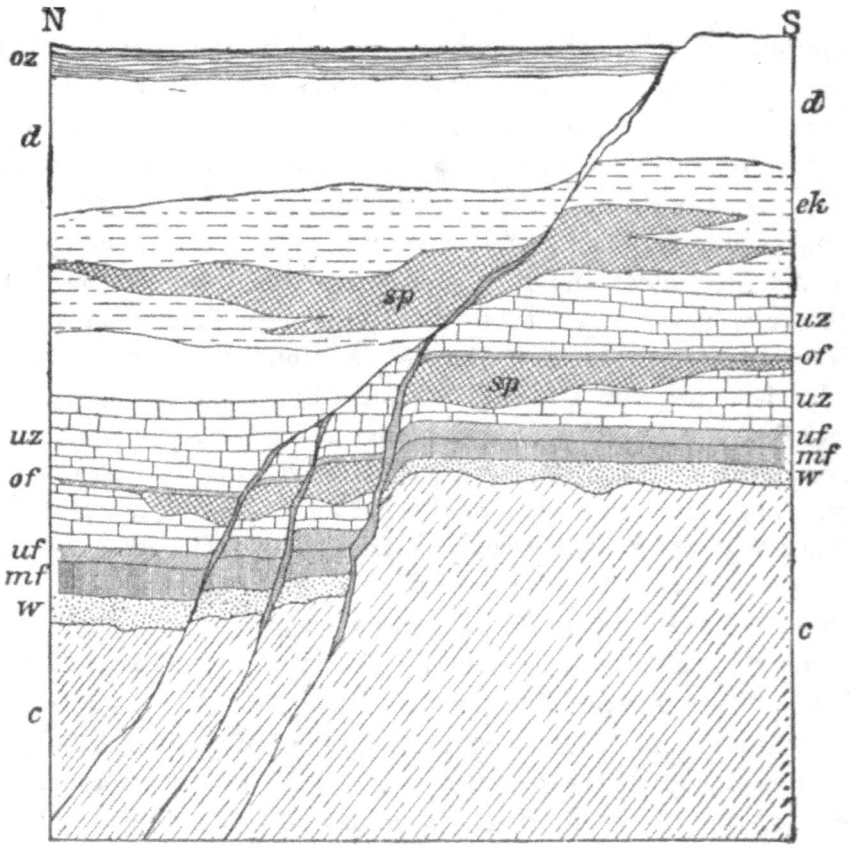

Fig. 152a.—Ideal cross-section through a mineral vein at Kamsdorf. (F. Beyschlag.)

oz, upper zechstein slates; d, dolomite; ek, ferriferous limestone; sp, brown hematite and spathic iron ore; uz, lower zechstein limestone; of, upper bed; uf, lower slate bed (copper slate bed); mf, mother bed; w, zechstein conglomerate; c, Culm schist.

green fluorspar. In the uppermost layer lenticular bodies of copper ore with barite and violet fluorspar occasionally occur imbedded in the barite.

A similar vein is worked at the Brüder Einigkeit mine at Bösenbrunn, near Oelsnitz.

[1] E. Weise: 'Erläut zu Section Plauen-Oelsnitz' der geol. Spezialk. v. Sachsen, 1887, p. 74.

It may be well to point out that this somewhat aberrant development of the spathic copper deposits agrees in many respects with that of the tourmaline-copper veins, and thus approaches the tin deposits in character, a relationship which does not obtain in the case of the typical examples of Kamsdorf. As a matter of fact, tin ore was obtained between 1511 and 1533 from the Vogtland region, where these fluorspar-bearing lodes occur, and even from the Sanct Burkhard and Heilige Dreifaltigkeit, now referred to as copper mines.

The numerous barytic copper veins of the Rhine valley are of purely scientific interest. They occur in the Black Forest, the Odenwald and at Spessart. These deposits are, according to K. von Kraatz-Koschlau,[1] partly of. pre-Triassic age, in part of Tertiary age. They strike northwest, rarely east-west. Their filling consists of barite of two generations, fluorspar, hornstone, chalcedony, quartz, copper pyrite and pyrite, as well as oxidized copper ores.

Among the copper deposits in the Alps this class is represented by a vein, 2 to 3 meters thick, outcropping in the Silurian clay slate of Mitterberg near Werfen in the Salzburg district. The vein filling consists of quartz, siderite, ankerite, copper pyrite and iron pyrite. F. M. Stapff[2] points out the fact that in this lode both copper pyrite and spathic iron tend to form lenticular masses in the quartz, and these ore shoots are at times inclined at an oblique angle to the walls. Sometimes the ores contain intergrown leaves of hematite.

A quartzose variation of this type of vein appears among the copper veins recently described and graphically represented by C. Derler,[3] veins which cut the Silurian clay slates in the vicinity of Kitzbühel in the Tyrol. They were formerly worked at Röhrerbühel and Sinwell, and even to-day some mines are still worked at Schattberg, in the Kupferplatte and on the Kelchalpe. The fissures, which attain a width up to 4 m. (13.1 ft.), are filled with quartz, some ankerite, fragments of slate and copper pyrite, and at times contain also pyrite, gray copper ore, niccolite, chloanthite ($NiAs_2$), zincblende and galena. The chalcopyrite occurs in scattered grains, small streaks, and as occasional large masses of compact ore within the vein filling.

A typical development of this form of lode is found at Altgebirg and Herrengrund, north of Neuschl, in northern Hungary. Quartz, barite, siderite, brown spar, copper pyrite and gray copper form the fissure filling of the

---

[1] K. v. Kraatz-Koschlau: *Abh. d. Hess. Landesanst,* 1897. Vol. III, part 2, p. 55.

[2] F. M. Stapff: 'Geognost. Notizen über einige alpine Kupfererzlagerstätten.' *q. u. Hutten. Zeit.* 1835 p. 6.

[3] 'Bilder von den Kupferkies-Lagerstätten bei Kitzbühel.' Vienna, 1890. Published by J. Graf. Falkenhayn.

Pfeifer vein, which traverses the mica schist, while the Kugler vein runs parallel to the stratification of the Triassic greywacke schist.[1] Similar lodes at Kotterbach, Szlovinka and Göllnitz in the Gömör-Zips schist area contain, besides the minerals mentioned, also some cinnabar.[2]

In the Swiss Alps the lodes of Mürtschenalp, in the canton of Glarus,[3] described by G. Troger and E. Stohr, outcrop in the Sernf conglomerate rock. They are probably of Permian age, and contain besides the copper ores (mainly bornite) also some pyrite, gray copper, hematite and molybdenite, the matrix being finely crystalline dolomite.

This type of vein also occurs in the Japanese copper districts. In the northern and central part of the main island there are five examples, viz.: Osaru-zawa, in the province of Rikutschu, Arakawa and Ani in the province of Ugo, Kusakura in the province of Etachigo, and Ogoya in the province of Kaga. According to a letter from Mr. Yamada, the strata of the Tertiary formation are there everywhere traversed by rhyolites and partly by propylitized andesites. The veins occur in the eruptive rocks. They consist of copper and iron pyrite, with lesser amounts of specularite and lead, zinc and silver ores in a gangue of quartz, calcspar, rhodochrosite and barite. In the Kusakura mine the specularite content increases considerably in depth. Similar veins occur in Paleozoic beds at Yoshioka in the province of Bitshu, and at Sassagatani and Dogamaru, in the province of Iwami, district of Chugoku. They are said to be connected with the granite and diorite intrusions of that locality.

8. Zeolitic Copper Veins with Native Copper.

This type of copper vein is quite uncommon, but, with the mineralogically similar amygdaloid beds and copper-bearing conglomerates, forms the enormously productive deposits of the Lake Superior region. The veins, formerly the exclusive source of copper, are no longer worked. Native copper and some native silver occur, accompanied by calcspar, laumontite, prehnite, apophyllite, natrolite, datolite, desmine, quartz, fluorspar, epidote and chlorite.

A brief description of the entire copper ore district of Lake Superior is given because these veins form a genetically connected whole with the other copper deposits of the locality.

[1] B. v. Cotta: *Erzlagerstatten.* II, p. 304.

[2] G. Faller: 'Reisenotizen' in *Berg- u. H.-Jahrb.* d. k. k. Bergak, 1867, Vol. XVII, p. 132.

[3] G. Tröger: ' Ueber den Kupfer- und Silberbergbau an der Mürtschenalp.' *Berg u. H. Z.*, 1860, p. 305. E. Stöhr: 'Die Kupfererze an der Mürtschenalp.' Zürich, 1865.

The copper region of Lake Superior[1] lies on a long peninsula projecting into Lake Superior and terminating in Keweenaw Point, the copper belt being 130 miles long and 6 miles wide. The eastern side of the peninsula is formed of Cambrian (Potsdam) sandstone, the west side of vast sheets of melaphyre (Irving's diabase) and melaphyre-amygdaloid, with intercalated beds of conglomerate. The age of the melaphyres, which in part bear all the characters of old lava streams, has been much disputed. According to M. E. Wadsworth the copper-bearing strata

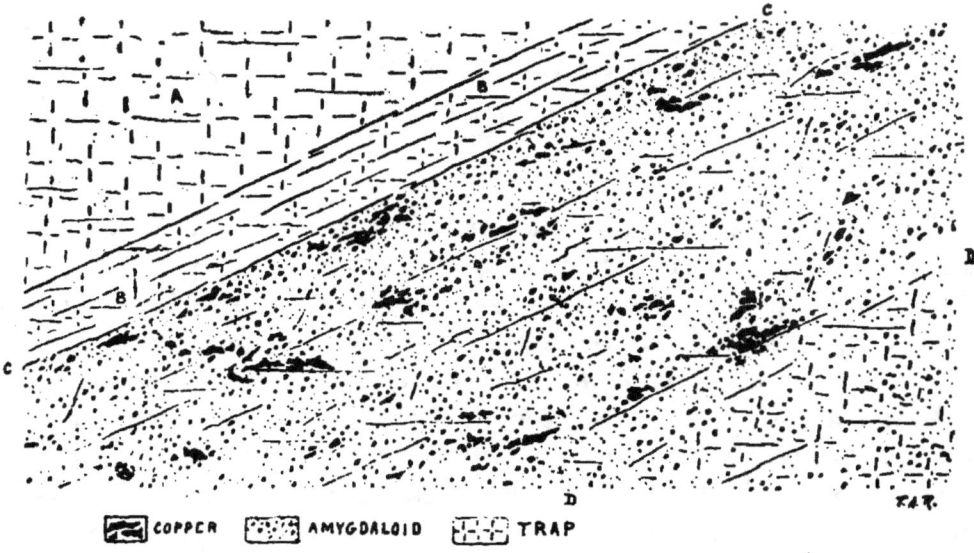

COPPER   AMYGDALOID   TRAP

Fig. 152a.—The Quincy (amygdaloid) lode.   (T. A. Rickard.)

rest on the Cambrian sandstones; according to R. D. Irving and T. C. Chamberlin this overplacement is merely apparent, in consequence of an over-thrust along a plane of dislocation, which in certain portions can be identified with certainty. The amygdaloidal structure of the eruptive sheets has been especially developed in the parts near the surface, while

[1] The following list comprises the most important publications upon this district: Herm. Credner: 'Beschreibung einiger charakteristischer Vorkommen des gediegenen Kupfers auf Keweenaw Point,' etc. *N. Jahrb. f. Min.*, 1869, pp. 1–14. R. Pumpelly: 'The Paragenesis and Derivation of Copper and Its Associates on Lake Superior.' *Amer. Journ. Sc.*, 1871 (3), pp. 188 *et seq.* and 'Geol. Surv. of Michigan,' Vol. I, pt. II. 'Copper Bearing Rocks,' 1873. 'The Metasomatic Development of the Copper-Bearing Rocks of Lake Superior,' *Proc* Amer. Acad. of Arts and Sc., Vol. XIII, 1877–78, p. 253. E. Wadsworth: 'Notes on the Geology of the Iron and Copper Districts of Lake Superior.' Cambridge, 1881. R. D. Irving: 'The Copper-Bearing Rocks of Lake Superior.' Washington. *Monograph*, U. S. Geol. Surv., 1883. T. C. Chamberlin: 'The Copper-Bearing Series of Lake Superior,' Cambridge, 1883. T. A. Rickard: 'Copper Mines of Lake Superior,' *The Engineering and Mining Journal*, 1904, Vol. LXXVIII, No. 15 *et seq.*

the deeper parts show a diabasic structure. On Keweenaw peninsula, the largest and most famous mine is the Calumet & Hecla, one of whose vertical shafts is 4,920 feet deep; the Portage, Osceola, Quincy, Central mines, where the largest masses of native copper were found, and the Copper Falls and Cliff mines are well-known producers. The copper area extends along the south shore of the lake southwestward, where the Minesota mine, in the vicinity of Ontonagon, is especially noteworthy. In this direction the copper-bearing series may be followed into Wisconsin, but no workable deposits are known. On Isle Royale, near the Canadian shore, deposits of similar character were worked.

Following H. Credner, three different kinds of deposits may be distinguished within the copper-bearing rock group:

Disseminated Grains and Amygdules in the Melaphyre Lava Sheets.

Such copper-bearing lavas, called amygdaloids and 'ash beds,' are worked in the Quincy, Franklin, Osceola, Atlantic, Huron and Copper Falls mines. At times the vesicular cavities of the amygdaloids are exclusively filled with native copper, sometimes accompanied by native silver, calcspar, quartz, chlorite, laumontite, prehnite, analcite, epidote, datolite, hematite, etc. The two native metals often occur intergrown into one lump, and hence cannot have been congealed from the molten condition as a magmatic segregation, for in that case they would have formed an alloy. On the contrary, like the constituents of the accompanying zeolites, they seem to have been leached out of the melaphyre itself, in which they may have existed at one time in fine particles of sulphides, subsequently concentrated by secondary process in the existing cavities.

In the Conglomerates.

The conglomerates, according to Wadsworth, are characteristic littoral formations. Their pebbles consist of quartz porphyry and melaphyre, and are often quite decomposed and disintegrated, and this material has been replaced and impregnated with copper. The interspaces between the pebbles are largely filled by native copper, with the same associated minerals as in the amygdaloid. Such conglomerates are worked at the Calumet & Hecla mine.

Filling of True Fissures.

The thickness of these veins is usually 1 to 3 m. (3.3 to 10 ft.), but sometimes rises to 10 m. (33 ft.) The more decomposed and porous the country rock, the richer the ore; the fissures accordingly carry their best orebodies at the points where they traverse the amygdaloids. The great thicknesses given for some veins include the adjacent country rock, which, when strongly impregnated with copper, has been counted as included in the vein proper; occasionally vein stringers group themselves into stock-like aggregates. The fissure filling consists of quartz, calcspar, prehnite, laumontite, apophyllite, natrolite, stilbite, as well as epidote, chlorite and fluorspar, native copper and some native silver. Fragments of the decomposed country rock are always found in the fissures. The sketch by Credner (Fig. 153) gives a typical cross-section of such a vein. The native copper forms jagged and stout branched lumps, sometimes of astonishingly stout dimensions and

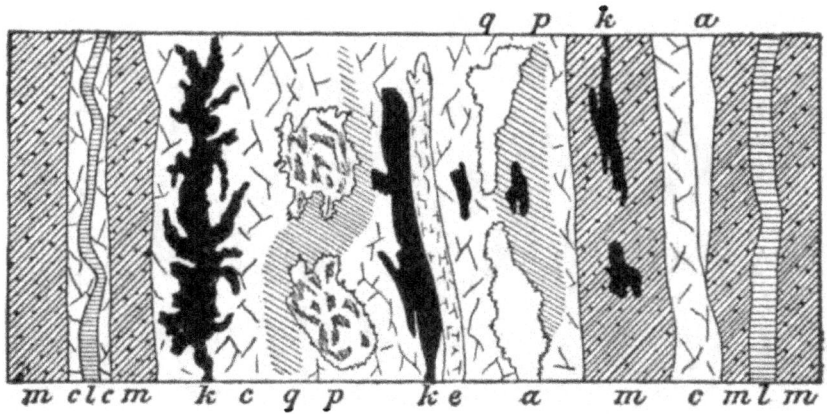

Fig. 153.—Cross-section through Cliff lode. (H. Credner.)

m, amygdaloid; c, calcite; l, laumontite; k, copper; q and p, quarts and prehnite with copper; e, epidote; a, apophyllite.

weighing several thousand kilograms. Mining is now confined to the conglomerate and amygdaloid beds.

Genetically these deposits are best explicable by the assumption of a lateral secretion of the copper ores, which were originally finely distributed in the melaphyres. The only enigma is, why the re-secretion and concentration took the form of native copper. According to Pumpelly, and, later, Van Hise, the copper is reduced from solution by magnetite and iron oxide.

The copper mines of Lake Superior for a long time took the lead in the copper production of the world. The Calumet & Hecla alone yielded 555,000 tons of copper in 1867-1897, all the other works on Lake Superior from 1845-1897 about 440,000 tons. The mines have been carried to astonishing depths; Red Jacket shaft of Calumet & Hecla has attained 1,460 m. (4,920 ft.). At the present day the "Lake Mines" occupy the second place in the world's production, being surpassed only by the mines in Montana.

In 1903 the total production of the works on Lake Superior was 192,400,577 pounds of copper, of which Calumet & Hecla alone produced 76,620,145 pounds, or about 6 per cent. of the world's entire supply of copper.

Copper-bearing amygdaloid melaphyres also occur in the district of Dschida in the Trans-Baikal, carrying native copper associated with opal, chalcedony, calcite, epidote and prehnite.

Native copper accompanied by native silver occurs[1] in diabase porphyries in the Mercedes mine, near Algodones, Chile. The eruptive rock shows propylitic alteration, and holds small amygdules of calcite, an undetermined green decomposition product, native copper and cuprite. Möricke cites

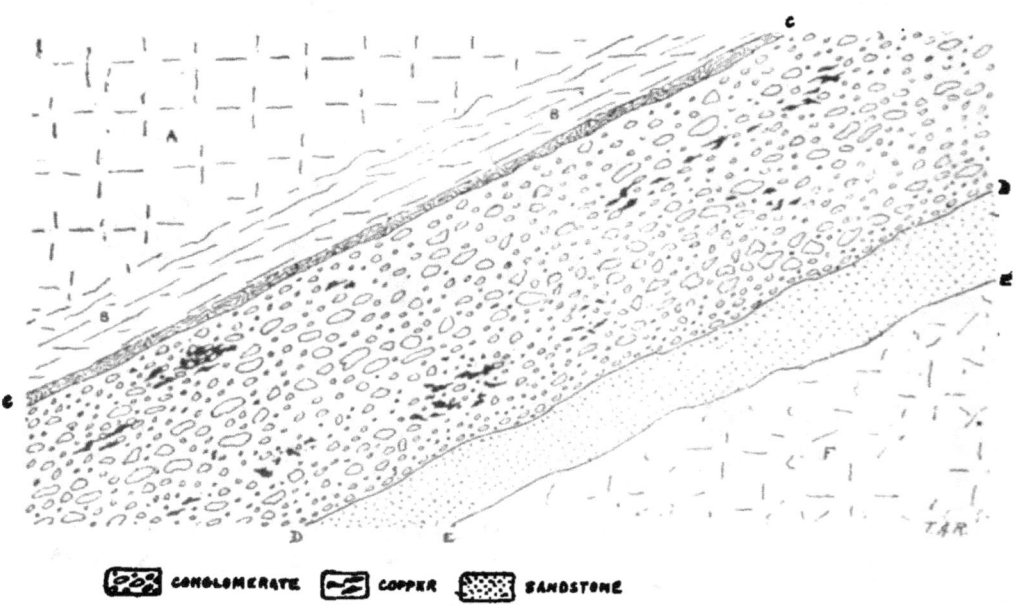

CONGLOMERATE   COPPER   SANDSTONE

Fig. 153a.—The Franklin, Jr. (conglomerate) lode. (T. A. Rickard.)

the Queensland deposits described by Daintree as analogous in character. The latter contain native copper and copper sulphides, with malachite, calcite and prehnite in amygdaloidal dolorites. Native copper is also found with zeolites in basalt in the Faroe Islands.

An interesting occurrence of native copper in feldspathic segregations of the gabbro of Pari, in Tuscany, is described by Lotti.[2]

The deposits at Zwickau, Saxony, seem to be genetically analogous, though not of economic importance, native copper occurring in quartz porphyry and porphyry tuff. These rocks form intercalations in the middle Rothliegende close to an underlying sheet of melaphyre. The copper is

[1] W. Möricke: 'Die Gold, Silver und Kupferlagerstätten in Chile.' Freiberg, Baden, 1898, p. 33.

[2] B. Lotti: *Zeit. f. Prak. Geol.*, 1899, p. 354.

found in thin sheets (rarely as much as 3 m. thick, up to 0.5 m. long and 0.15 m. broad) which usually fill short fissures, that rapidly wedge out on both sides; more rarely it is found in larger fissures which may be connected with fault planes. On both sides of the fissures, in zones 0.5 to 2 cm. wide, the dirty violet colored rock is leached and altered to yellowish white. This is supposedly due to reducing solutions. Spherical reduction spots are also found in the midst of the unfissured porphyry mass, and in the center of these spots native copper was also found at times in grains. Besides the native copper, A. Weisbach[1] discovered domeykite (mohawkite) (Cu_3As) in the quartz porphyry. Especially rich masses of copper ore were found in sinking several coal mine shafts, showing a rather wide horizontal distribution and a close genetic relation to the above mentioned intrusive sheets, intercalated in the 'Rothliegende.'[2]

(δ) Silver-Lead Veins.

The veins of this type are characterized by silver-bearing galena and sphalerite, with high-grade silver ores in subordinate amount. Accordingly as the gangue is predominantly quartz, or carbonates, or barite, S. A. W. von Herder,[3] in treating the veins of the Freiberg district alone, established three subdivisions, which he called formations and which are in the main quite sharply defined, and are as follows:

9. The pyritic lead quartz formation.

10. The rich (high-grade) lead formation, also called brown spar formation or Brander formation, with carbonates.

11. The barytic lead formation, also called heavy spar formation or Halsbrück formation, with barite as the principal matrix.

These Freiberg vein-types have been found to exist with but little variation in the mineral districts of all parts of the globe, so that they are really of universal application. In using the terms originally chosen for the veins of Saxony, it is found that the term high-grade lead formation or brown spar formation proves to be too restricted, and it is changed to "carbon spar lead formation" (i. e., carbonate gangue). The following detailed descriptions fully indicate the character of the various divisions.

9. The Pyritic Lead Deposits.

In the pyritic lead veins, quartz, galena, sphalerite, pyrite, arsenopyrite and chalcopyrite are most abundant. As accessory constituents of the

---

[1] A. Weisbach: 'Arsen Kupfer von Zwickan.' *Neues Jahrbuch f. Min.*, 1873, p. 64. H. Mietzsch: 'Erlaut z. S. Zwickau,' 1877, p. 36.

[2] A. v. Gutbier: Brief comment in *Neues Jahrbuch f. Min.*, 1843, pp. 460–461.

[3] S. A. W. von Herder: 'Der tiefe Meissner Erbstolln,' 1838, p. 17.

gangue, we have hornstone, jasper, ferruginous quartz, calcspar, siderite, more rarely brown spar and chlorite. As rarities, and occurring, as a rule, only in the vicinity of intersecting veins of a different nature, fluorspar, barite, nacrite, tetrahedrite and its argentiferous variety, silver glance, native silver, black silver ore, arsenical sulphide of silver and other high-grade silver ores are found. Witherite and smithsonite, too, are among the very rare occurrences in this formation.

The most important of these minerals to the miner is galena. In the Freiberg area it contains 0.1 to 0.2% of silver; elsewhere the percentage of silver is somewhat higher, but rarely more than 0.5%. Pyrite and arseno-pyrite as a rule are very poor in silver, containing for the most part only 0.005 to 0.02% of it, rarely as high as .05%. The Freiberg sphalerite, always blackish, contains up to 0.03% silver and usually some tin, as noted by A. W. Stelzner and A. Schertel[1], owing to mechanically intergrown tin-stone microliths.

The vein filling has a prevailingly massive structure, in which the ore minerals occur either commingled or in separate compact masses. More rarely the ingredients are arranged in bands.

One of the most important occurrences of this formation is found at Freiberg.

This mining field is a part of a gently undulating plateau, dissected by the moderately deep valleys of the Mulde and its tributaries. This plateau forms the gentle basal slope of the Erzgebirge, which rises to the northwest, while to the southeast a rather steep scarp forms the descent into northern Bohemia. The region of Freiberg, 400 to 450 m. (1,312 to 1,476 ft.) above tide, lies on the northwest side of the range, which trends southwest-north-east. This part of the range has a relatively much greater altitude than that lying to the southeast of the chain, which along the scarp just men- - tioned has been downthrown along a fault zone, commonly called the Bohemian thermal fissure. Biotite gneisses, with some subordinate inter-calations of mica schist, form the country rock of the veins. These rocks at Freiberg are only approximately parallel in strike to the trend of the moun-tains, as they form a great dome or anticlinal uplift, the center of which coincides approximately with that of the city. The summit of this dome of Freiberg gneiss is cut by the ore-bearing fissures.

A great stock of granite breaks up through this dome, not in the center, but to the east of it, at Bobritsch-Naundorf. The eruptive dikes occurring in the field consist of mica diorite, a fine-grained mica syenite and quartz porphyry. These rocks, especially the last named, are all cut by the min-

---

[1] A. W. Stelzner and A. Schertel: 'Ueber den Zinngehalt und über die chemische Zusammensetzung der Schwarzen Zinkblende von Freiberg,' Freiberg, 1886.

eral veins. The porphyry dike of Muldener Hütton plays an especially prominent rôle in the mines. It may be traced with slight interruptions to a distance of about 20 km. (12 miles) from Frauenstein as far as the Nonnenwald, with strike north-northwest and thickness not exceeding 10 m. (33 ft.). The quartz porphyries cut through the high-grade silver quartz veins, but are themselves cut by the pyritic, barytic and rarely also by the high-grade lead lodes. The unfavorable influence of the quartz porphyry on the development of the vein fissures of the pyritic lead veins has already been referred to.

The mineral veins of the Freiberg district belong for the most part to this class of pyritic lead deposits; to a lesser extent to the high-grade silver and to the barytic lead types. In the northern, western and southeastern part of the district high-grade silver quartz veins are also found. As shown by their relation to the above mentioned eruptive dikes, these types of lodes are not of the same age. The high-grade silver quartz veins are the oldest; the barytic lead formation is the youngest. They also show some difference in strike. In general the Freiberg veins strike in two principal directions: (1) North-south to northeast, mostly north-northeast; these are the lodes of the pyritic and of part of the high-grade lead formation, as well as most of the representatives of the high-grade silver quartz formation; (2) northwest to west-northwest; to this group belong the barytic and another part of the high-grade lead ore lodes, as well as many barren lodes. The lodes of the first group are for the most part called 'stehende' (strike south-southwest to north-northeast), in part also 'flache' (strike north-northwest to south-southeast) and 'morgengänge' (strike west-southwest to east-northeast); those of the second group are called 'Spatgänge' (west-northwest to east-southeast), in part also 'flache' (north-northwest to south-southeast. (See page 121.)

The lead veins belonging to the category just mentioned are at the present time worked in the Himmelfahrt mine, owned by the State and situated close to the city, also in the Himmelfürst mine, also owned by the State, situated at Erbisdorf, near Brand, and, until recently, other veins were also worked in the Mord mine division of the Vereinigt Feld mine, near Brand, and in Junge Hohe Birke, between Freiberg and Brand, and still earlier in a very large number of other smaller mines. The veins for the most part have a thickness of only 0.1 to 0.8 m. (0.3 to 2.6 ft.), rarely over 2 m. (6.5 ft.). On the map of the Freiberg veins, accompanying the present work, the names of the most important representatives of this formation are given. Some of them have been found workable to considerable horizontal and vertical distances; for example, Kirschbaum or Hohe Birke Stehende to a distance of about 7 km. (4.2 miles) and a depth of 650 m.

(2,134 ft.) and the Selig Trost Stehende to a distance of 2.2 km. (1.3 miles) and a depth of more than 500 m. (1,640 ft.).

The pyritic lead veins were probably the first to be discovered in the Freiberg field, their gossan being very rich in secondary silver ores and especially in native silver. This discovery must have taken place about 1162 to 1170[1] and was followed by an immigration of miners from Goslar and other points of lower Saxony, who built the "Sächsstadt," the oldest part of Freiberg, which under that name was founded by Margrave Otto den Reichen, of Meissen, being subsequently fortified and further developed. In 1218 Freiberg is named for the first time in a document. The first mines mentioned in the documents, the Gottesgabe, Schöne Maria and Heiliger Gregorius, were probably located on the "Hauptstollngang" (main adit lode). After many vicissitudes, the silver mining industry of Freiberg attained a great period of prosperity between 1795 and 1870, during which time the production reached 45,000,000 to 60,000,000 oz. of silver a year. Somewhat earlier, on Easter, 1766, the Royal Mining Academy of Freiberg had been opened, and in 1776 the Royal Mining School. Despite great improvement in methods, the mining industry of Freiberg has, by reason of the depreciation of silver, ceased to be remunerative, and although the ore production and the output of silver and lead are still very considerable, and the works employ a large number of people, the Government has decided to gradually close down the mines. In 1901 the ore production in the Freiberg mining district was as follows: 11,563 tons silver ores, 6,195 tons arsenious, iron and copper pyrite, 28 tons of zinc-blende, 409 tons of barite. The output of silver was 17,573 kg., that of lead 2,090 tons. For every square meter of lode surface worked, there were produced 0.3 kg. silver and 37.73 kg. lead.

From the beginning of mining up to the year 1896, according to H. Müller, the total production was 5,242,957 kg. silver, worth about $227,-000,000.[2]

Pyritic lead veins also occur elsewhere in Saxony, especially at Schneeberg and Annaberg. At Schneeberg, particularly at Kuttengrund

[1] H. Ermisch: 'Das Sächsische Bergrecht,' Leipzig, 1887, p. xvi.

[2] The following is a list of some of the more important publications upon the Freiberg veins: J. F. W. von Charpentier: 'Mineralog. Geographie der chursächsischen Lande,' 1778. H. Müller: 'Die Erzlagerstätten nördlich und nordwestlich von Freiberg,' Cotta's 'Gangstudien,' 1847, Vol. I, pp. 101-305. W. Vogelsang: 'Die Erzlagerstätten südlich und südöstlich bei Freiberg.' Cotta's 'Gangstudien,' 1848, Vol. II, pp. 19-133. H. Müller and B. R. Förster: 'Gangstudien aus dem Freiberger Revier.' With two plates. Freiberg, 1869. H. Müller: 'Die Freiberger Erzlagerstätten in Freiberg's 'Berg und Hüttenwesen,' 1883, pp. 28-84. 'Gangkarte der Freiberger Bergrevier.' H. Müller: 'Ueber die Erzgänge des Freiberger Bergrevieres.' A monograph accompanying the special geologic map of Saxony, 1901. (A comprehensive work, in course of publication.)

near Lössnitz, a peculiar variety of this type of deposit occurs, in which galena is subordinate to arsenical pyrite. The lodes produced 118 tons of arsenopyrite in 1898. At Annaberg this class exists in an aberrant form, since the deposits contain a predominance of copper ores, and cassiterite appears. Similar transitions to the copper formations were mentioned previously.

In this place one should also mention the many parallel veins which, according to E. Haber[1], traverse the Devonian shales of Ramsbeck, in the mining district of Brilon in Westphalia, the veins in the clay slate southeast of Trier in the Bischofsheim mine and those of the Aggerthal north of Siegburg, in the Rhenish Schiefergebirge.

Other veins which are not quite typical, since they differ from the examples just noted by the constant presence of siderite in the vein filling, are found in the mining district of Diez and Wiesbaden, on the lower Lahn, in Nassau.[2] Transitions to the class of carbonate-bearing lead deposits are also found there. Wenckenbach[3] divided the lead lodes of Nassau into seven reefs, of which the veins at Ems and Holzappel are the most important.

The vein system of Ems comprises all the veins occurring between Dernbach and Braubach. The most important are the Mercur vein at Ems and the Friedrichssegen at Oberlahnstein. It would be more accurate to designate these veins as a vein-group, since, in spite of their close association, they differ in strike and dip. They all occur cutting Lower Devonian clay slate, impure sandstone and quartzite belonging to the Upper Coblenz formation, whose beds strike northwest. The strike of the lodes lies between north-northwest and east-west. The predominant gangue mineral is a firm massive quartz, often cementing fragments of slate, while calcspar and brown spar occur sparingly. In the upper part of the veins siderite has altered to brown hematite, but in depth siderite is always present. The ores consist of argentiferous galena, with sphalerite, together with copper and iron pyrite. In the oxidized zone of the vein, white, green and brown lead ores occur frequently, together with occasional plumbo-resinite, native silver and copper, cuprite, chalcocite, azurite, malachite, lead sulphate and cuprous anglesite; at greater depths, gray copper, cobalt pyrite, nickel-arsenic-glance and bournonite, with other antimonial lead ores, are present. The structure of the veins is partly massive, partly banded. The thick-

[1] E. Haber: 'Der Blei-und Zinkbergbau bei Ramsbeck,' *Z. f. d. B. H. S. u. S. im preuss. St.,* 1894, Vol. XLII, p. 77.

[2] Oberbergamtliche Beschreibung 'der Bergreviere Wiesbaden und Diez,' Bonn, 1893, pp. 91-116.

[3] Wenckenbach: 'Beschr. der im Herzogthum Nassau, etc., aufsetzenden Erzgänge.' Wiesbaden, 1861.

ness may be as much as 10 m. (33 ft.), and, in an exceptional case in the Friedrichssegen mine, as high as 20 m. (65 ft.)

As early as 1158, Emperor Frederick I. (Barbarossa) conferred on Archbishop Hillin of Treves the right to mine silver near Ems. Subsequently the Ems mining industry is said to have been flourishing in the 14th and 15th centuries. New enterprises began in 1743. Since 1791 the mines of Ems have been uninterruptedly in the possession of a company now called Emser Blei- und Silberwerk zu Ems, and have been continuously worked to the present time. The largest production of lead ores in the mining district of Diez, to which the Ems lodes belong, was reached in 1878-1880, with 22,539 to 24,611 tons per year.

The great Holzappel lode may be traced, according to Bauer,[1] from Holzappel on the Lahn, across the Obernhof, to the west of Singhofen, across Dahlheim, Ehrenthal, Werlau and Norath as far as Peterswalde, a distance of 50 km. (30 miles). In recent years the ore shoots in this lode at the Gute Hoffnung mine, near Werlau, described by L. Souheur,[2] have attracted much attention. Here the clay slates, sandy slates and sericite schists belonging to the lower Coblenz stage of the lower Devonian are traversed by a main lode with a parallel hanging-wall vein 10 m. (33 ft.) from it, both veins striking east-northeast, with a dip of from 50 to 90° southeast. The thickness is from 0.3 to 4 m., averaging 1 m. The fissure filling consists of quartz, sphalerite, galena and chalcopyrite, with fragments of country rock, and subordinate amounts of siderite and pyrite. The lode cuts the strata at an acute angle. The adjoining schists, as well as a parallel diabase dike in the overlying strata, are partly sericitized and are then called 'white rock' (weisses Gebirge). The lode is dislocated on its dip by flat dipping overthrust faults (so-called Deckelklüfte). Mention may also be made here of a similar vein at the Altglück mine, near Uckrath, not far from Sieg.

The structural relations of similar pyritic lead veins to eruptive dikes traversing Silurian shale is especially noteworthy in the Katzbach region of lower Silesia, and at Eisenberg, near Altenberg. According to Rosenberg-Lipinsky,[3] the veins accompany steep dipping dikes of olivine kersantite (olivine-mica-diorite) in such way that the kersantite either forms the middle of the lodes or is confined to one side. Dike-like intrusions of quartz porphyry also occur, and run at times parallel and close

[1] Bauer: 'Die Silber, Blei und Kupfererzgänge von Holzappel an der Lahn, Wellmich und Werlau am Rhein.' *Karsten's Archiv* f. Bergb., etc., Vol. XV, 1841.

[2] L. Souheur: 'Die Lagerstätte der Zink, Blei und Kupfererzgrube Gute Hoffnung bei Werlau am Rhein,' *Jahrb.* d. k. preuss. geol. Landesanst, 1892, p. 96.

[3] V. Rosenberg-Lipinsky: 'Beiträge zur Kenntniss des Altenberger Erzbergbaues,' *Jahrb.* d. k. preuss. geol. Landesanst, 1894, pp. 161-182. With list of previous works on the district.

to the ore-bearing fissures. The fissure filling of the mineralized veins consists of clay, fragments of shale and olivine kersantite, arsenious-copper pyrite (arsen-kupfer-schwefelkies), galena, tetrahedrite and rarely of zinc-blende, antimonite, boulangerite, epiboulangerite and bournonite, as well as quartz. Of the ores just named, arsenopyrite is by far the most abundant, constituting about 40 to 50% of the ore. The copper pyrite is gold-bearing. This gold content and the richness in arsenopyrite, as well as the presence of gray copper, establishes a close relationship between these lodes and those of Hohenstein in Saxony, whose more highly copper-bearing variety was described previously. The Altenberg mining industry, which is said to have been flourishing in the thirteenth century, was forgotten for a long time and again resumed lately.

As a typical example in Austria, the old and famous lodes of Kuttenberg[1] in Bohemia may be noted. The region consists of gneisses, in part overlain by Cretaceous strata. The gneiss is traversed by a great number of veins which may be grouped into 18 series, with uniform north-south strike and steep dip east or west. The filling is of quartz, hornstone, fragments of the country rock, calcite, ankerite, pyrite, zinc-blende, argentiferous galena and arsenopyrite, with rare native silver, zincite, boulangerite, proustite, siderite, cronstedtite and lillite. The district is 6.5 km. (3.9 m.) long and 4.5 km. (2.7 m.) broad. The mining industry dates back to the beginning of the thirteenth century, flourished in the fourteenth century, and its revival was recently attempted.

The lodes in the pre-Cambrian clay slate at Mies,[2] Bohemia, which are distinguished by uncommonly large druse cavities, form a transition to the barytic lead deposits by reason of the presence of a little barite.

Among Hungarian deposits, the Schemnitz lodes belong to or are closely related to the pyritic lead veins. As they are, however, rich in gold and silver, they will be included in that class of veins instead of the one treated here.

In Spain veins of pyritic lead ores are of considerable importance, especially at Linares in the province of Jaen, in the eastern part of the Sierra Morena; at L'Horcajo in the province of Ciudad Real, and at Castuera in the District of Badajoz.

The mining field of Linares is, according to Caron,[3] about 12 km. (7.2 m.) long and 9 km. (5.4 m.) broad, lying north of the town on an arid plateau and rocky hilly country south of the Sierra Morena proper. The

[1] F. Katzer: 'Der Kuttenberger Erzdistrict,' *Oesterr. Z. f. B. u. H.*, 1896, p. 247 *et seq.*

[2] F. Posepny: 'Der Bergbaudistrict von Mies,' Vienna, 1874.

[3] Caron: 'Bericht über eine Instruktionsreise nach Spanien im Jahre 1878.' *Z. f. d. B H. u. S. im preuss. St.*, 1880, Vol. XXVIII, p. 119.

whole region consists of granite, which on the south and west is overlain by clays, sandstones and conglomerates of uncertain age while at the north it adjoins the Silurian shales, forming the high mountains. The veins are in the granite, only few of them continuing into the shales. Their strike is in the main northeast and their dip is quite steep. In general they are relatively thick, reaching eight meters in La Cruz lode. The veins are often quite long, exceeding 1 km., while the Alamillos lode is 6 km. (3.6 m.), and La Cruz and Arrayanes veins 4 km. (2.4 m.) long. The veins are exceedingly rich, as gangue and country rock form but a small amount of the fissure filling. The total thickness of a lode is often filled with the galena, not very rich in silver it is true, but mingled with very little blende and a small amount of pyrite. The gangue is mostly quartz, with small amounts of calcspar, dolomite, barite and siderite. Copper ores occurred in the outcrop. The average silver content of the ores is stated to be 180 gm. (5.3 oz.) silver per ton. The Arrayanes mine, belonging to the State, and La Fortuna, belonging to an English Company, are the principal mines. The ancient Phoenicians, Carthaginians and Romans had mines in this region. The old shafts situated north of Linares, on the Cerro de Val de Infierno, are called to this day "Pozos de Anibal" (Hannibal's wells). The most brilliant period of recent mining in that district was about 1889, when the production of lead in the province of Jaen rose to 188,325 tons.

Deposits of this class also occur in France. A great vein which occurs in the Silurian shale of Pontpéan, in the vicinity of Rennes, and the lodes of Huelgoat and Poullaouen (Finistère) in granite and in the adjoining Paleozoic shales are the most noteworthy. The two last named localities were formerly the most important lead mines of France, but they are now abandoned.[1]

The main vein of La Touche mine, in the neighborhood of Rennes,[2] is of considerable geologic interest and deserves a brief description. The lode occurs in granite and has been traced for about two km. (1.2 m.); some of its orebodies have been opened up to a depth of more than 100 m. Its strike is north-south, its dip 70° east. It is a very characteristic double vein. In all parts of its known extent a gray breccia-like quartz, barren of ore and 5-10 meters thick, forms the lower and larger streak of the vein. In the hanging-wall of this, but only about certain ore-shoots, the true ore-bearing branch vein is developed. The ore consists of argentiferous galena, argentiferous zinc-blende and pyrite with quartz, chalcedony and horn-

[1] Fuchs and De Launay: 'Gîtes Mineraux,' II, p. 498-509.

[2] According to a brief communication of W. Frhr. von Fircks, and specimens in the Freiberg museum.

stone as gangue minerals. The structure is either banded or brecciated. A peculiarly mottled ore, occurring in the vein, consists of sharp angled splinters of the quartzose lower stringer, which are coated with several layers of chalcedony and cemented by zinc-blende and hornstone. Parts of the country rock are not infrequently found enclosed. The granite in the hanging-wall is decomposed to a distance of 1.5 meters and impregnated with finely divided ore; in the immediate vicinity of the lode itself, the granite has frequently been altered into a clay. Away from the ore-shoots, the upper streak of the vein narrows from a thickness of 1-2 meters down to a thin hornstone layer.

Examples of this class, very rich in silver, occur, according to Vogt, in Svenningdalen district of northern Norway.[1]

In the United States examples of this class occur at Bingham,[2] Utah, the veins consisting of quartz and galena, together with pyrite, chalcopyrite and blende, with a gossan which was rich in silver. This class also includes the veins of the Clear Creek district in Colorado.

10. Veins of Spathic Lead Ore.

In this class the gangue consists essentially of the carbonates, calcspar, brownspar, rhodochrosite, siderite and quartz. The ore minerals are argentiferous galena, argentiferous sphalerite, and, less often, pyrite, marcasite, tetrahedrite, both with and without silver, together with high-grade silver ores, especially ruby silver and argentite. The structure is usually imperfect and but seldom well banded. The ore minerals frequently occur sprinkled through the gangue.

In the Freiberg district proper, this class of veins is especially developed in the Himmelsfürst and Bescheert Glück mines near Brand (see page 245), where they exist as high-grade lead ores or the brownspar formation. The high-grade 'noble' veins of that locality are, according to H. Müller,[3] distinguished by their composition, consisting of brownspar and rhodochrosite, galena, sphalerite, argentiferous and ordinary tetrahedrite, antimony-silver blende, silver glance and native silver. The subordinate minerals are quartz, hornstone, calcspar, siderite, pyrite, arsenopyrite and copper pyrite, stephanite and polybasite; nickeliferous pyrrhotite, native arsenic, pitchblende (Uranpecherz), etc., are rarely observed.

[1] J. H. L. Vogt: 'Söndre Helgeland,' Norges Geol. Undersög., No. 29, Christiania, 1900, and *Zeit. f. Prak. Geol.*, 1902, pp. 1-8.

[2] J. M. Boutwell: 'Ore Deposits of Bingham, Utah,' *Bull.* 213, U. S. Geol. Survey, 1902, pp. 105-122, also 'Economic Geology of the Bingham District, Utah,' by Boutwell, Keith and Emmons, Professional paper No. 38, U. S. Geol. Survey, 1905.

[3] 'Litteraturangaben,' p. 249.

The galena contains between 0.4 and 0.6%, and in some cases as high as 2% silver. A very characteristic feature is the so-called glazed blende, a dark zinc-blende with microscopic inclusions of silver glance (Glaserz) and possibly other high-grade silver ores, having a silver content of as

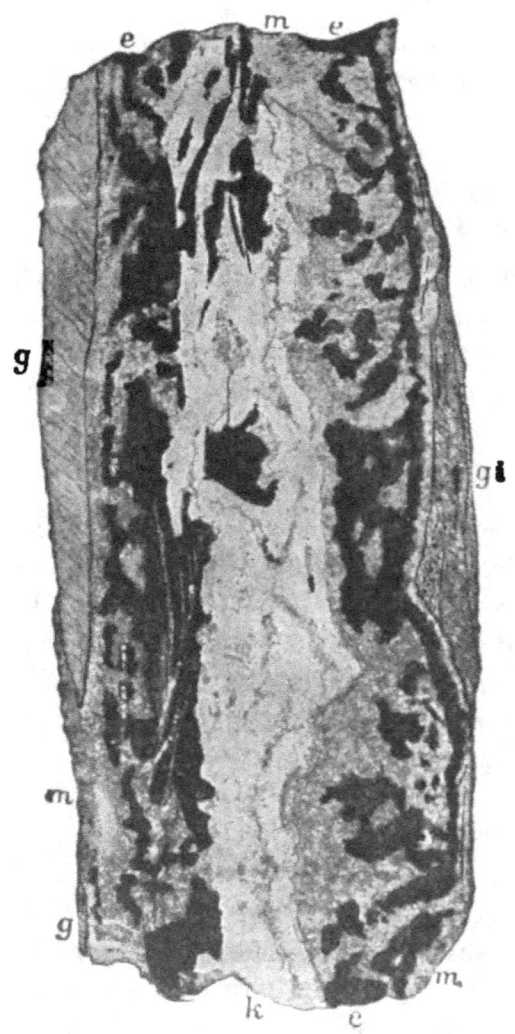

Fig. 154.—Section across Verbogen Flachen in the Himmelsfürst mine (carbon spar lead formation). (From nature.)

g, gray decomposed gneiss; gi, gneiss impregnated with pyrite and copper pyrite; k, calcspar and brownspar; m, rhodochrosite; e, argentiferous blende and some gray copper.

high as 1.5%. Even the pyrite in this formation often contains as high as 0.2% silver. When a banded structure is developed in these lodes, zinc-blende and galena, together with rhodochrosite and brownspar, are usually found concentrated immediately along the selvages, while in the middle calcspar, and at times quartz predominates. Fig. 154 shows the manner of development of these veins, whose thickness usually amounts to 0.08-0.75 meters, rarely over 1.5 meters.

The reproduction of a thin section, Fig. 155, gives an idea of the manner of microscopic intergrowth between the gangue and ore minerals.

Most of the lodes of this kind, including the most important examples in the Freiberg district, have a low strike (11.4 to 12.4 hours) (see page 130), with flat dip of 40 to 60° west. According to H. Müller, another important parallel reef follows the main direction north-northeast to northeast and

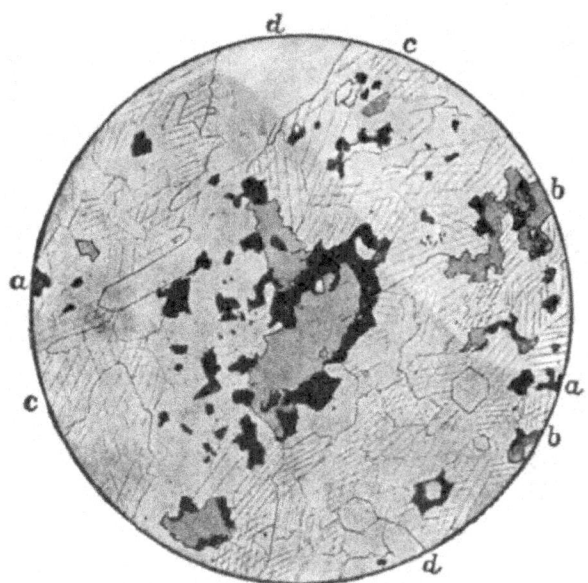

Fig. 155.—Thin section of ore of the Silberfund St. auf Himmelsfürst.
(Magnified 50 diameters.)
Shows the inner core of zinc-blende (b), with argentite and gray copper (a), and calcite and brownspar (c), as well as quartz (d).

dips steeply 70 to 90° southeast or northwest, as illustrated by the Dorothea and the Silberfund veins of the Himmelsfürst mine, the Neuglückstern, and the Johannes and David veins of Bescheert Glück. A peculiar development is shown by the Habacht stringers in the Bescheert Glück and Einigkeit mines. They strike northeast and dip northwest at no more than 15° to 40°. They are narrow veins, consisting mainly of rhodochrosite and quartz, argentiferous gray copper and argentiferous galena. In the aggregate 350 of these high-grade lead lodes are known. They have been opened up over a distance of 600-1,000 m. (1,968-3,280 ft.), while a few like the Neue Hohe Birke Stehende are worked for a distance óf 2 km. (1.2 miles.)   A somewhat abnormal character is shown by the vein of the Güte Gottes mine at Scharfenberg near Meissen, discovered in 1225 and worked anew from 1867 to 1898.   According to H. Zinkeisen, the vein

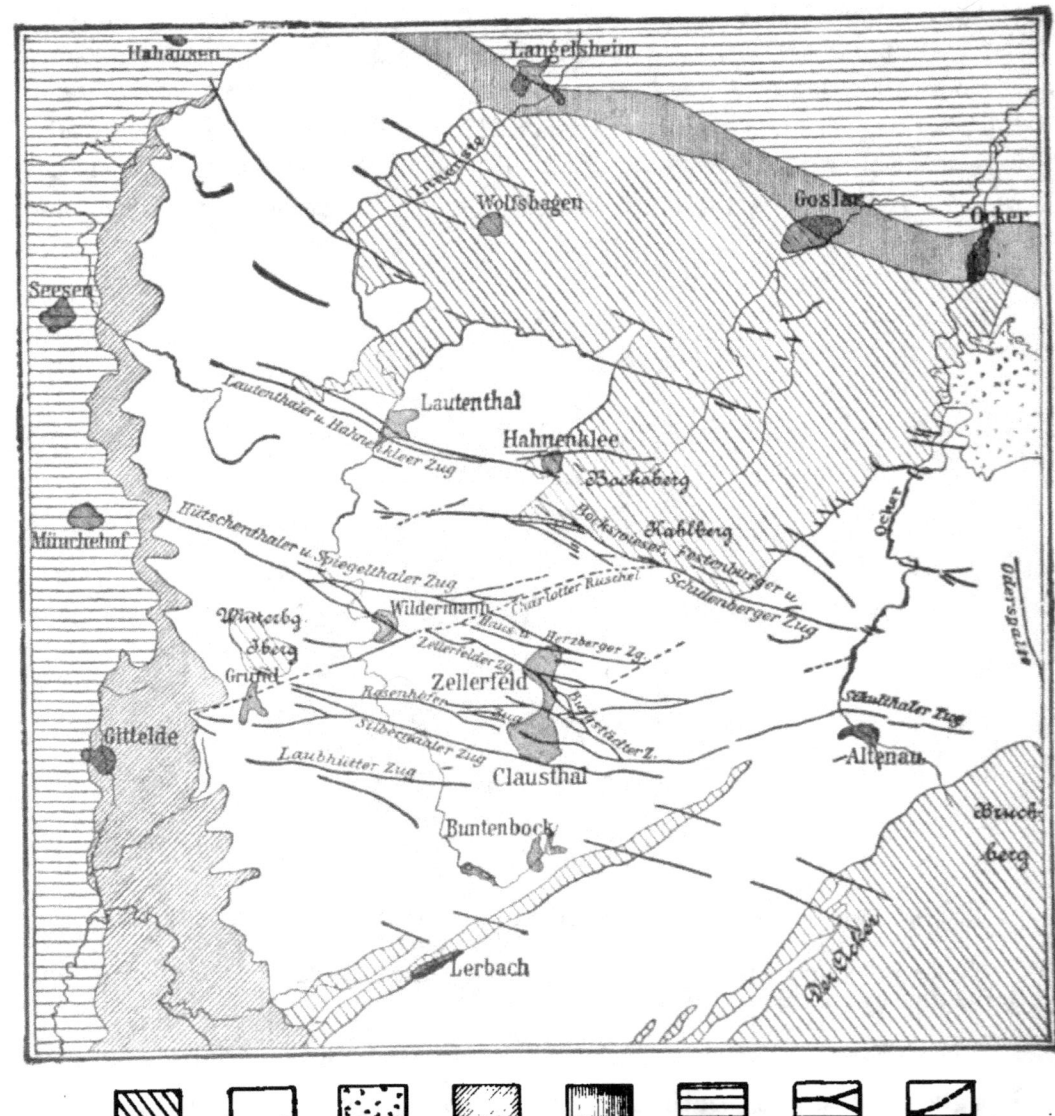

Fig. 156.—General sketch of the lode system of the Upper Harz.
1, Devonian; 2, culm; 3, ochre granite; 4, zechstein foot-hills; 5, Jura-Trias and Lower Cretaceous; 6, Upper Cretaceous and younger formations of the foot-hills; 7, veins; 8, Ruscheln.

fissures occur in the granite-syenite massive of Meissen.   They contain quartz, brownspar, rhodochrosite, celestite and galena, with an average silver content of about 0.20 to 0.30%, and a yellow blende almost as rich in silver, together with gray copper having as high as 8% silver.[1] The veins also include great masses of vein clay and granitic friction débris. The accessory constituents include strontianite, barite, gypsum, pyrite, copper pyrite, very rarely also glass ore (earthy argentite), ruby silver and native silver.

The stringers and veins form a zone of 2 km. (1.2 m.) long, and 600 m. (1,968 ft.) wide, and striking northeast. The thickness of the veins, about 50 of which are known, varies from a few centimeters in the small stringers to 2 meters in the Heinrich Morgengang. All the veins show a propensity to fray out in stringers, and to take an irregular course. They are often dislocated by transverse faults, and by the so-called "Schwebenden," which are horizontal, or nearly horizontal, narrow clay fissures or movement planes, only rarely carrying ore. The filling of the veins and stringers is usually massive. Quartz and brownspar, intimately mingled, contain the ores in scattered particles. More rarely the metallic minerals are found in compact, coarsely crystalline bodies of nearly pure ore, which in such cases are crossed in all directions by veins of hornstone. A banded structure occasionally appears, ordinarily with streaks of blende along the two selvages and with a central often breccia-like part, consisting essentially of brownspar and rhodochrosite.

The Clausthal Veins.

In the Harz, the famous lodes of the Clausthal area are, for the most part, closely related to the carbonspar-lead formation. According to F. Klockmann,[2] the Clausthal district (see general map Fig. 156) comprises an area 18 km. (10.8 mi.) long and 8 km. (4.8 mi.) broad, part of the North German plateau in the region of Clausthal, Grund, Wildemann, Bockswiese and Schulenberg. This plateau, whose elevation is 550-600 m. (1,804-1,968 ft.) above sea level, rising in the Schalke to a height of 764 m. (2,506 ft.) forms a part of the northwestern upper Harz range, which drops down northward and westward with steep scarps to the foot-hills. To the southeast the plateau is bounded by the Bruchberg range, to the north by the mountain wall of the Bocksberg and the Kahleberg. The entire area is formed of the nucleal rocks of the Harz; that is, of Devonian and Lower Carboniferous beds, especially the Culm, whose lower division consists of clay shales and silicious shales, while the upper division is mainly gray-

[1] H. Zinkeisen: ' Ueber die Erzgänge von Güte Gottes zu Scharfenberg.' Freiberger *Jahrb.*, 1890, pp. 40-64. Gives also the older literature of H. Müller.

[2] F. Klockmann: 'Die Erzlagerstätten des Oberharzes im Werke Berg und Hüttenwesen des Oberharzes,' 1895, pp. 43-65.

wacke. Lying unconformably upon the upraised edges of these Paleozoic rocks, close to the west edge of the mountains, are the strata of the Zechstein formation. The Devonian and Culm strata have been squeezed together into northeast folds with associated overturns and overthrusts, with many true fault fissures. Overthrust zones also occur as the so-called *Ruscheln,* lode-like formations, filled with strongly folded or crushed and mashed country rock, called Gangthonschiefer (vein clay slate). In the vicinity of Clausthal, the well-known Faule Ruschel, with nearly east-west strike, crosses the main Burgstadt series and the so-called Charlotte lode. Many veins are true fault fissures, as shown by A. von Groddeck,[1] and, still earlier, by Schmidt. Other fault lodes at Bockswiese displace the Devonian by about 200 m. (656 ft.), as determined by the dislocation of the Culm beds (Fig. 94).

The vein system of Clausthal consists of several series of veins, often convergent and which frequently fray out. The following may be distinguished from north southward:

1. The Gegenthal-Wittenberg series.
2. The Lautenthal-Hahnenkleer series.
3. The Bockswieser-Festenburg-Schulenberg series.
4. The Hütschenthal-Spiegelthal series.
5. The Haus Herzberg series.
6. The Zellerfeld main series.
7. The Burgstädt series.
8. The Rosenhöfer series.
9. The Silbernaaler series.
10. The Laubhütte series.

The length of each series along the strike is considerable. The Schulenberg series has been traced for a distance of 10 km. (6 m.); the Gegenthal-Wittenberg series is probably much larger. Without exception, they all have a strike of from southwest to west, and dip, almost without exception, south at 70 to 80°.

They consist wholly of compound lodes, mostly with a distinct footwall selvage, but passing into the hanging-wall country rock by a gradually diminishing succession of stringers. They are, therefore, not solid fissure fillings, but zones of stringers or zones of sheeting which may be as much as 40 m. (131 ft.) across. A good idea of this stringer formation is afforded by Fig. 157, after Zirkler.[2]

[1] A. v. Groddeck: 'Ueber die Erzgänge des Oberharzes.' *Zeitschr. d. Deutsch geol. Ges.,* 1866, pp. 693-776. Also, 'Geognostischen Durchschnitten durch den Oberharz' Z. f. d. B. H. u. S. im preuss. St., 1873, pp. 1-14.

[2] Zirkler: 'Essener Glückauf,' 1897, Vol. XXXIII, p. 73.

The Rosenhofer series of veins scatters in passing into the Ruscheln, while in the Zellerfeld and Burgstädter series some veins are deflected, while others pass through with unaltered course.

As has been noted, all the veins are probably fault fissures, though this is hard to prove in the case of the lodes occurring in the monotonously similar beds of the Culm.

The vein filling consists of ore, gangue minerals and fragments of coun-

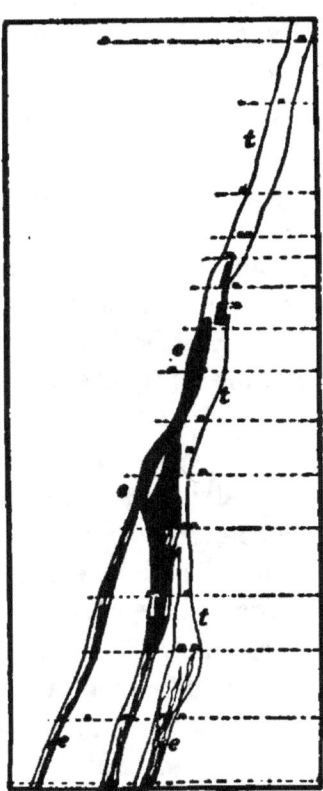

Fig. 157.—Cross-section through the main veins of the Bergmannstrost mine, 70 m. east of the Marien shaft. (Zirkler.)

e, ore-carrying; t, barren veins. The dark places have been worked out

try rock. The predominant ore is argentiferous (0.0—0.3%) galena, with associated zinc-blende, which predominates at Lautenthal, with subordinate copper pyrite, pyrite and marcasite, and rarely tetrahedrite and bournonite. Among the matrices, quartz and calcspar are predominant, except in the Silbernaaler and Rosenhöfer vein series, which are rich in barite; siderite, pearl spar and strontianite also occur. The rock fragments consist of hard sandy slate in angular fragments or of crushed clay slate. The inclusions often far exceed the ore in bulk. In many cases we find an irregularly massive lode structure, and not infrequently 'ring ore' is developed, especially in the lodes of the Ring and Silberschnur mine at Zellerfeld (see Fig. 135-137).

According to A. Lengemann, the silver-lead mining industry began in the Clausthal district about 1220, was abandoned in 1350, but resumed in 1526. In this second period it was at first concentrated around the newly founded mining town of Zellerfeld. Clausthal did not obtain a mining grant until 1554. The greater part of the mining population of the upper Harz came from the upper Erzgebirge in the 16th century. In 1811 the mining academy of Clausthal was founded, a one-year course for prospective miners having been introduced in the Clausthal lyceum as early as 1775.

The mines on the Clausthal plateau (Clausthal, Lautenthal and Grund) produced, in 1898, 196,985 tons of crude ore.

The lodes of Neudorf-Harzgerode, in the eastern Harz of Anhalt, depart even further than those of Clausthal from the typical high-grade (noble) lead deposits.[1] At Neudorf-Harzgerode the plateau of upper Silurian slate is traversed by several vein systems very nearly parallel to one another, among which the Dillenberg lode, with the mines of Pfaffenberg and Meisenberg, is the most important. This double vein consists of a true lead vein and another of spathic iron ore. The two are either directly in contact or are separated only by a narrow layer of rock. The ore vein, which is often split up into stringers, consists of siderite, quartz, calcspar, fluorspar, galena, zinc-blende, gray copper, pyrite, copper pyrite, bournonite, sometimes with the addition of hübnerite, scheelite and wolframite. The silver lead ores of the Neudorf mine contained, towards the end of the seventies, on an average 40% lead and 0.061% silver. The spathic iron ore appears now above, now below, and may attain a thickness up to 4 m. (13 ft.). An abundance of fluorspar and the occasional presence of other minerals, characteristic of tin deposits, distinguishes the Neudorf lodes from the normal type of 'rich' lead veins.

A good Austrian example of this type occurs at Pribram, in Central Bohemia.

The Pribram lode area proper lies in a mountainous region 500-550 m. (1,640-1,804 ft.) above sea level. The prevailing rock is a quartzitic grit (greywacke), which, though it has not yet yielded any fossils, is probably Cambrian. In some beds it becomes conglomeratic; in the vicinity of the Lill shaft intercalations of oolitic limestone occur, as indicated by blocks

---

[1] Heinr. Credner: 'Uebersicht der geogn. Verh. Thüringens und des Harzes,' 1843, p. 123. Kegel: 'Beitrag zur Kenntniss der Neudorf-Harzgeroder Gänge,' *Berg. u. Hutten Zeit.*, 1877, pp. 397-400. C. Blömeke: 'Ueber die Erzlagerstätten d. Harzes,' Vienna, 1885, p. 85. H. Fischer: 'Gutachten über die Anhaltinischen Blei- und Silberwerke,' 1894.

scattered thereabouts. The strata form a trough, as indicated in the section after J. Schmid, Fig. 158. To the southeast below this trough, clay

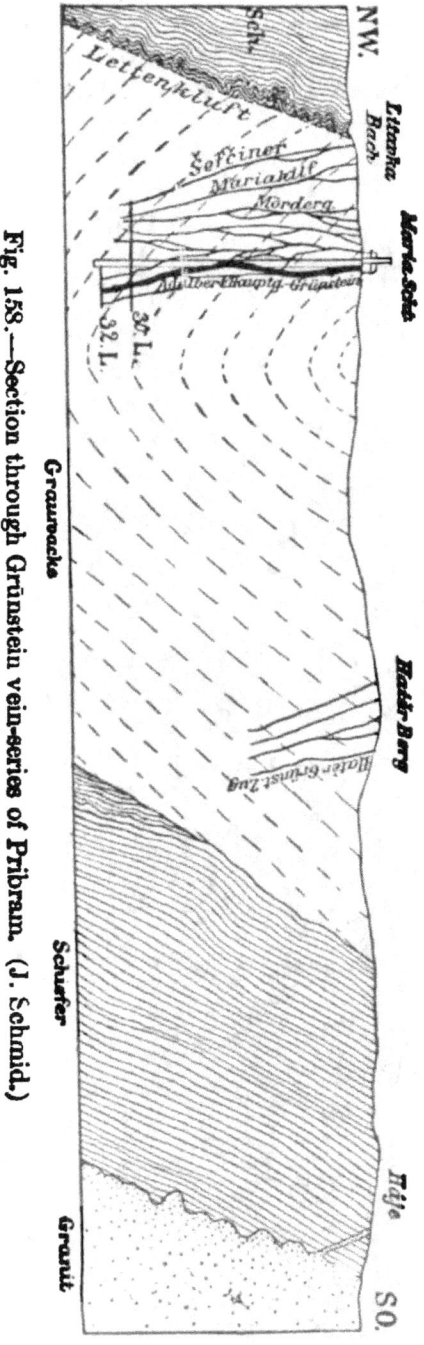

Fig. 158.—Section through Grünstein vein-series of Pribram. (J. Schmid.)

slates, probably of Cambrian age, are traversed by granite and altered by contact metamorphism. The hornfels immediately adjoining the contact is crossed in all directions by fine-grained granite apophyses. The exact relation between the grits and the shales does not appear to have been quite made out. In the section an uncomformity between these two rocks

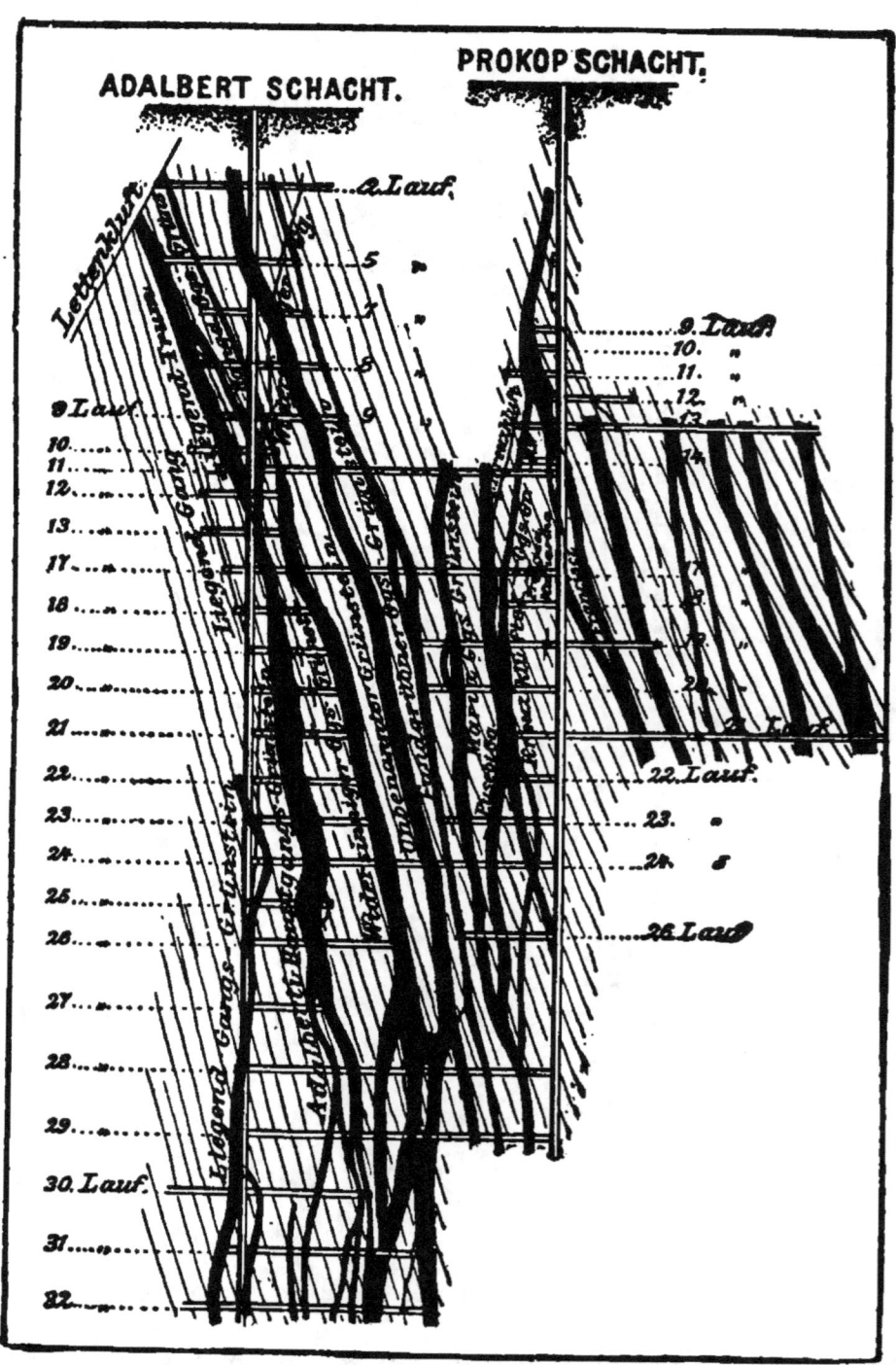

Fig. 159.—Section through Adalbert and Prokop shafts. (J. Schmid.)

is assumed. At the northwest edge of the greywacke trough the stratification has been interrupted by a fault, dipping steeply northwest, and filled with a clayey mass or with highly decomposed material, folded in a most complicated manner, the so-called Lettenkluft (clay fault), most probably an overthrust fault, along which the greywacke beds abut sharply against a second slate zone. The minute folding of the adjoining slate is indicated in the section. Quartzitic graywackes recur further northwest.

Where the ore veins occur, at Bohutin and Birkenberg, the trough of Cambrian grits is traversed by numerous greenstone dikes, often swelling into stock-like forms 1 to 30 m. (3 to 98 ft.) thick. The rocks are mostly diabases, rarely diorites. These dikes of igneous rocks are followed by the lodes, the mineral veins either clinging to the selvages or running through the middle of the dike and parallel to its contact. The section, Fig. 159, based on surveys by J. Schmid, as well as the ground plan in Fig. 160, by the same author, gives a clear representation of this close connection between the veins and the greenstone dikes. Besides the eruptive masses thus far named, a quartz diorite mass encountered in the mining operations west of the Stephan shaft at Bohutin is noteworthy because the mineral veins in passing from the sandy slate into this quartz diorite lose the lead and silver ores and carry antimony ores instead. This eruptive stock is surrounded by a contact zone.

The most important and richest of the Pribram lodes occur in a narrow belt, 4 km. (2.4 miles) long, running southwest and northeast along the Lettenkluft (clay fissure). Where the lodes, issuing from the grit, run into the clay fissure, and up against the greatly folded clay slate, they contract decidedly and become almost barren, or they split up into several insignificant stringers, which are often found to be deflected eastward. Some of them indeed have been followed beyond the 'clay fissure' into the slates, but there they turned out to be much poorer than in the grit, and as a rule carried only spathic iron ore, brownspar and calcspar, as well as some blende. In the granite area, mineral veins are also known, both iron- and lead-bearing ones, as, for example, at Milin and Vrancice, but they are not of economic importance.

Within the graywacke, the veins have, as a rule, the following character: Down to a remarkable depth, usually 60 m. (196 ft.) in the Segen Gottes shaft southwest of Pribram, in the Liegend lode, and in the Nordwest lode even to 270 m. (885 ft.), the ore is altered to a gossan, consisting of brown iron ore, cerussite, pyromorphite, malachite, native silver and secondary silica as chalcedony. At greater depths the veins assume the character of sulphide lead veins, mostly of the character of the spathic lead formation. The veins, according to J. Schmid, consist, in the richer por-

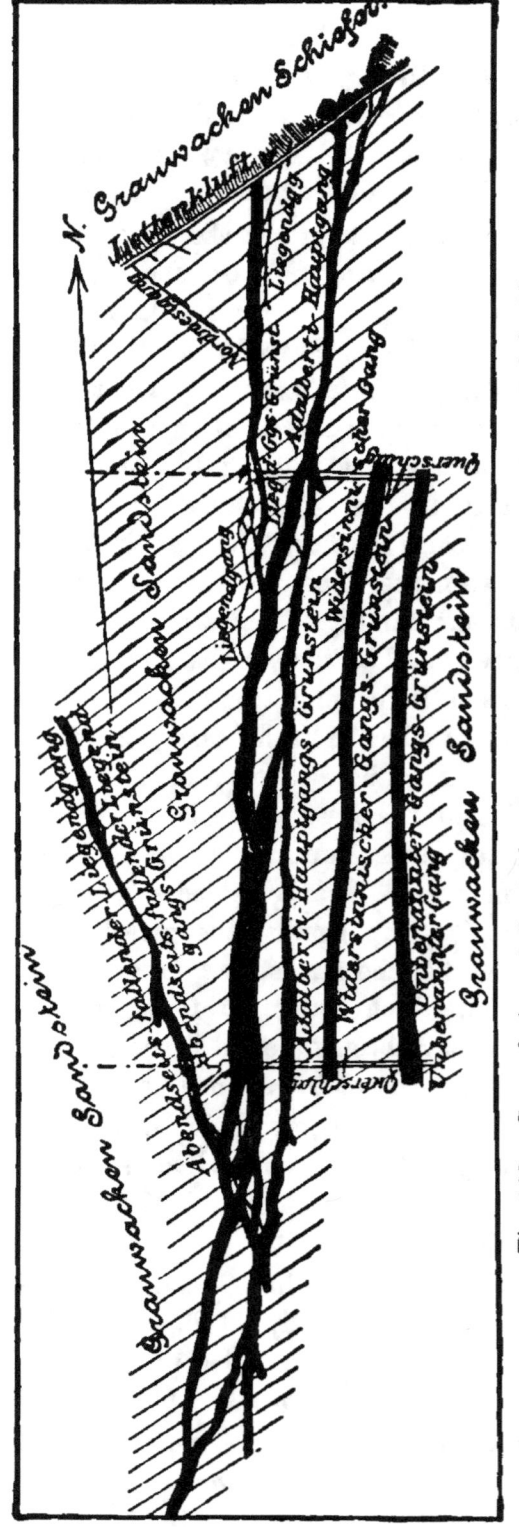

Fig. 160.—Giound plan of the 24th galle y of the Adalbert-Maria mine. (J. Schmid.)

tions, mainly of compact argentiferous galena, zinc-blende, siderite, quartz and calcspar, in the barren portions of only iron spar and calcspar.

The silver content of the galena varies between 0.1 and 0.7%, and it was not possible to ascertain whether there was a decrease in the content with increase in depth. The zinc-blende is generally, though but feebly, argentiferous (0.04—0.06%). Higher percentages are rare. Besides the minerals just named, fragments of the country rock also occur frequently in the midst of the vein filling. The structure of the latter is sometimes massive, banded or with stringers of ore. The illustration of the lode in

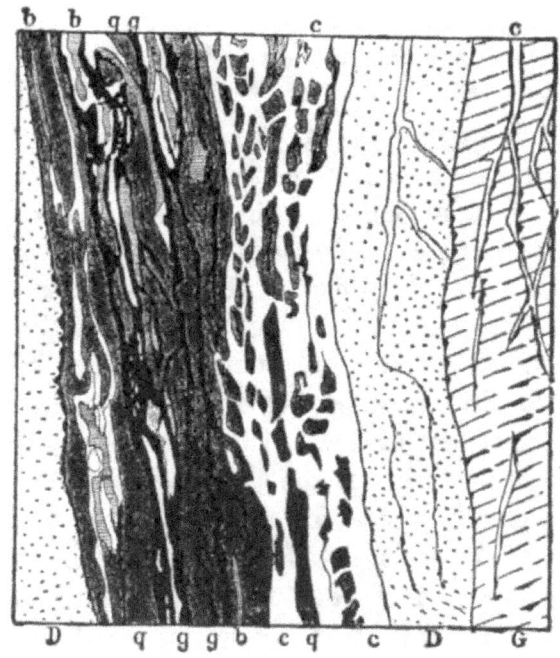

Fig. 161.—The Adalbert main vein at Pribram.   (J. Zadrazil, reproduced by J. Schmid.)
G, graywacke; D, diorite; q, quartz; c, calcspar; g, galena; b, zinc-blende.

Fig. 161, taken from the official report on Pribram, shows the Adalbert main lode, and illustrates a type which is of common occurrence there.

In the Anna mine and elsewhere, especially in the deep workings, another kind of lode filling often occurs, called there Dürrerz (dry ore); approaching the character of the high-grade quartz formation of the Freiberg area. A matrix rich in quartz, but poor in carbonates, carries, besides silver-bearing galena, ruby silver ore, native silver, stephanite, gray copper ore and stibnite, mostly in disseminated grains.

The filling of the fissure often consists of several bands quite sharply divided from one another, a galena band, one or two bands of dry ore and a younger one of calcspar. The dry ores steadily increase in amount with increasing depth.

The thickness of the Pribram veins varies, according to J. Schmid, from nothing up to 8 m. (26 ft.). The most important veins are the main Adalbert vein, the Abendseitsfallende (westward dipping) Liegend vein, the northwest vein, the Adalbert Liegend vein, the Fundgruben vein, the Eusebi vein and the Widersinnige (contrary) vein.

The silver mining industry of Pribram is stated to date back to 843. In recent time the production has vastly increased. In 1898 it amounted to 263,979 tons of crude ore. From this were obtained 20,882 tons of pure ore, with 38,599 kilograms of silver and 4,826.1 tons lead. The ore delivered at the furnace contained on an average 0.185% silver and 23.10% lead. The work has been carried to extraordinary depths, in the Adalbert shaft to 1,099 m. (3,606 feet).[1]

Another undoubted example of this type of deposit occurs in the Austrian Alps, at Metnitz and Zweinitz, in Carinthia.[2]

The silver lead deposits of Mazarrón, in the Spanish province of Murcia, which were mined by the ancient Romans, have become notorious by their dangerous eruptions of carbon dioxide. The basal rocks consist of micaceous and talcose slates, passing into argillaceous, chloritic and amphibole schists, and of crystalline limestones. These are covered by Tertiary strata, which are cut by rhyolite stocks. The north-south mineral veins which cut these stocks at the Cabeza de San Cristobal are thus proved to be of very recent age. As the veins pass into the slate they quickly wedge out. One vein, Las Laguenas, clings close to the contact between the two rocks. The vein filling consists of disintegrated rhyolite, silver-bearing galena, blende, pyrite and siderite. At a depth of 400 to 500 meters the veins become impoverished. The horizontal length is also small, but the veins sometimes reach a thickness of 16 meters (52 feet).

The outbursts of carbon dioxide always take place when slate is struck, and are accompanied by large outflows of warm water at 35° C. (95° F.) The rapid diffusion of the gas and the hurling of rock fragments by these

---

[1] The most important publications include: W. Vogelsang: 'Die Pribramer Erzniederlage.' Cotta's Gangstudien, I, 1850, pp. 305-329. John Grimm: 'Die Erzniederlage von Pribram,' 1855; and several later essays by the same author. Babanek: 'Zur Kenntniss der Pribramer Erzgange.' Oesterr. Z. f. B. u. H., 1878. J. Schmid: 'Bilder von den Erzlagerstätten zu Pribram.' Published by the Imperial Royal Ministry of Agriculture, 1887, with atlas. Contains a bibliography of the subject. F. Posepny: 'Beitrag zur Kenntniss der montangeolog. Verhaltnisse von Pribram.' Archiv f. Prak. Geol., II, Freiberg, 1895, pp. 609-745.

[2] R. Canaval: 'Die Blende und Bleiglanz führenden Gänge bei Metnitz und Zweinits in Kärnthen.' Carinthia, II, No. 4, 1899. D. F. Villasante y Gomez: 'La industria de Mazarrón,' 1892. 'Sur les filons de Mazarrón,' Revista Min., No. 1393-1396, 1902. D. F. Iznardi u. a. Revista Min., 1902, No. 1873.

outbursts caused repeated accidents. An eruption of gas in the Triumfo mine at a depth of 440 m. (1,443 ft.) produced 200 cubic meters of débris and filled the levels with the gas.

As an American example of the carbonspar lead ore formation we will mention the Enterprise mine at Rico, in southwestern Colorado. According to T. A. Rickard,[1] the gently dipping Lower Carboniferous slates, lime-stones, and sandstones of this locality are traversed by two systems of veins. One set has a northeast course and very steep dip; the other strikes north-south, and has a low dip. The fissures of the former are ore-bearing; those of the latter are of importance only as 'cross veins' enriching the others, being themselves merely lean quartz veins, much too poor to pay for working. The ore-veins cannot be followed up beyond a definite horizon, the so-called 'contact.' This contact is between high fissured limestone, overlain by solid but warped shales. Within the 'contact zone' lateral infiltration from the tops of the veins has produced rich deposits of banded ore. Followed downward the mineral veins continue, but their character changes. At 30 to 45 m. (98 to 147 ft.) below the 'contact zone' the ore and rhodochrosite disappear and the veins below this depth contain only barren quartz and crushed country rock. All the veins are fault fissures, with but slight throw and a thickness rarely exceeding a foot. (0.3 m.) Their physical development depends greatly on the nature of the country rock. In the sandstone the veins are formed by a single fissure; in the limestone the fracture splits up into a number of stringers. The filling consists of rhodochrosite and quartz, with galena, zinc-blende, pyrite, copper pyrite, argentite and stephanite. They often show a remarkably well banded structure in which the quartz crusts show comb structure. At one point the veins change from their normal development, as silver veins with carbonate gangue, and occasionally carry native gold, besides native silver. Thus their example represents a transition of the carbonspathic lead ore formation to the silver-gold ore formation.

The ore of the Rico district was discovered as early as 1864, but it was only in 1881 that an active and remunerative mining industry began to develop, at the Enterprise, Rico and Aspen mines.

This class also includes the silver-lead veins occurring in the quartzites of the Coeur d'Alene Mountains in Idaho, whose main gangue is siderite (see later).

As an extreme member of the carbonspathic lead type of vein we have the two pure zinc-blende veins, with calcspar and some hornstone as gangue, which occur in the granite at Merklin, southwest of Pilsen in

---

[1] T. A. Rickard: 'The Enterprise Mine, Rico, Colorado,' *Trans.* of the Am. Inst. Min. Eng., XXVI, 1897, p. 906 *et seq.*

Bohemia. According to E. Rüger they strike northwest, and belong structurally to the lead vein system of Mies. The southwestern vein especially has been extensively mined, though work has recently been suspended. The lode matter contained many angular fragments of decomposed granite embedded in it, and also in places lenticular masses of impure graphite lying across the strike. When the vein was narrow the mineral filling consisted of nothing but zinc-blende; when of greater thickness the filling was calcspar, with blende, in which case the country rock also was found to be impregnated with blende. In the upper portions smithsonite and calamine were found. A very remarkable phenomenon is the fact that the zinc-blende veins on passing into the pre-Cambrain (?) clay slate contain more and more galena, until this mineral finally predominates. Merklin was founded in 1842. The annual production sometimes amounted to 2,000 tons of zinc-blende ores, with a zinc content of about 52%.

11. Barytic Lead Veins.

In this class of veins the gangue consists of predominant barite with fluorspar and quartz or jasper besides calcspar. These minerals are generally intergrown in a remarkably thin-banded structure. The barite especially occurs in finely crystalline crusts (calc-barite), the fluorspar in diverse tints, but mainly green and yellow. The ore minerals, which either form thin crusts or appear sprinkled through the gangue, consist of argentiferous galena, often developed in large flakes, pyrite and marcasite, also zinc-blende, copper pyrite, gray copper and at times rich silver ores.

In the Freiberg mining district (see page 238), the most prominent representatives of this formation are found in the region of Halsbrücke, where the Halsbrücke Spat and the Drei Prinzen Spat called for special mention. These lodes all strike northwest, and ordinarily dip steeply northeast. At Halsbrücke itself they traverse biotite gneiss; farther north they cut through mica schist and granulite. The main veins are distinguished by their thickness, usually 1 to 4 m. (3 to 13 ft.), at times up to 6 m. (19 ft.). They are the youngest of the Freiberg veins, often traversing and dislocating both the pyritic and the high-grade lead veins. On the Halsbrücke Spat, according to H. Müller "besides the normal, so-called soft band, carrying lead ore poor in silver, there is also a so-called hard band, consisting essentially of crystalline and jaspery quartz with thinly laminated barite, fluorspar, argentiferous tetrahedrite, galena, carrying 0.02-0.08% of silver, antimonial silver blende and arsenical silver blende, bournonite and copper pyrite. The two main bands of the lode diverge toward the northwest in the mining field of Kurprinz Friedrich August Erbstolln near Gross·

schirma, and are there worked as separate veins, the soft one under the name of Drei Prinzen Spat (see Fig. 130), the hard vein under the name of Ludwig Spat."

Aside from the main representatives just named, the barytic lead veins of the Freiberg area proper are not particularly rich.. They show a peculiar behavior at points where they cross older veins, especially the pyritic lead veins. At these crossings they carry rich silver ores, 'edler Geschicke,' including especially argentite, ruby silver, acanthite, polybasite, stephanite, native silver and tetrahedrite, ordinarily in a carbonate gangue. In rare cases the veins also carry nickel and cobalt ores. Such rich intersections were especially productive in the Himmelfahrt mine directly outside the town, especially at the crossing of the Neu Hoffnung Flache, the Ludwig Flache and Ludwig Spat, the Abraham Spat and the Friedrich Spat, with the pyritic lead lodes. One square meter of vein surface sometimes contained rich silver ores having a value of several thousand marks.

A very typical development of barytic lead ore was also observed in the Tobias Flache, Hilfe Gottes Morgengang and Friedrich Flache, at the Segen Gottes Erbstolln at Gersdorf, near Rosswein, in Saxony. In these lodes, the crustified or laminated structure was found to be developed in wonderful perfection. Magnificent druses of fluorspar were often found, those of dark wine-yellow color being found in all large collections. This mine, which had good orebodies as late as the fifties, is at present idle.

Another Saxon district showing barytic lead veins is the zone 2-3 km. (1.2-1.8 miles) wide between the towns of Mittweida[1] and Oederar., a distance of about 24 km. (14.4 miles). They are mostly Spat and Flache lodes (see page 132), which here have united into a series. As late as the eighties mining continued in this region, being formerly quite extensive, on the right bank of the Zschopau, in the Alte Hoffnung Erbstolln, near Schönborn. The vein series traverses a body of schists intercalated among the granulites (gneiss, mica schist, biotite gneiss, cordierite gneiss, amphibole schist, alum slate, clay slate and quartzite slate), to a small extent also granulite and granite. The main lode, the Clementine Spat, strikes on an average north 60° west and dips northeast at 60-80°. Its thickness varies ordinarily between 1.5 and 2.5 m. (5-8 ft.), but in exceptional cases rises to 7 m. (22 ft.), where side stringers unite with the main vein. Quartz, fluorspar, barite and calcspar, with galena, pyrite, copper pyrite and gray copper, besides fragments of the country rock, form the main filling of the vein, the structure being unusually massive, but sometimes stratiform. The galena occurs in two varieties: coarse-foliated with

---

[1] H. Müller: 'Die Erzlagerstätten der Sectionen Mittweida, Frankenberg und Schellenberg,' Leipzig, 1881, in *Erlaut. z.* Sect. Frankenberg-Hainichen, pp. 84-120.

0.02-0.03% silver and fine-foliated (feinspeisig) with 0.05-0.10% silver. The high content of the latter is attributed to finely intermingled gray copper. The mining industry of Schönborn may be traced back to the twelfth century (Ermisch). In 1847 it was resumed, but has again been abandoned for a decade.

This mineral zone also includes the veins of Hilfe Gottes near Memmendorf, in the vicinity of Oederan, a typical specimen of which was shown in Fig. 129.

There are also transitions from the barytic lead veins to the barren barite veins, which are worked at several points in the Erzgebirge of Saxony, for example, near Aue.

As further German examples, we mention the lodes in the Münsterthale. in the southern Black Forest, described by A. Schmidt[1], especially the Shindler lode and the Teufelsgrund lode, as noted long ago by B. von Cotta[2]. Some lodes of the Kinzig valley, near Shapbach, also belong to this class.

On the Iberian peninsula this barytic lead type is found in the lode district of Hien de la Encina, in the Sierra de Guadalajara, between Madrid and Zaragoza, in a high mountain country built up of mica schist and gneiss. The gangue consists of barite, quartz and siderite. Among the ores, silver glance and other high-grade silver ores predominate to such extent that the habit of the deposit approaches that of the high-grade silver ore lodes[3].

The same transitions to the high-grade silver ore formations are exhibited by the very rich barytic lead veins in the mining region of Sarrabus, in the southwest part of the island of Sardinia[4]. The region is composed of Lower Silurian slates, abutting on the south against a granite mass, which has altered them by contact metamorphism. The lodes, which for the most part strike east and west, and traverse these schists, are for the most part developed as double veins. One may distinguish in them an older band, consisting of finely crystalline barite, some calcspar, quartz and coarsely foliated galena, poor in silver, associated with a little blende. The younger part of the lode forms a band of reddish, foliated barite. calcspar and fluorspar, with galena and blende both highly argentiferous, with

[1] A. Schmidt: 'Geologie des Müntserthales,' part II. 'Erzgänge und Bergbau,' Heidelberg, 1889.

[2] B. v. Cotta: 'Erzlagerstätten,' Vol. II, p. 177.

[3] B. v. Cotta: 'Erzlagerstätten,' Vol. II, p. 443. Fuchs and De Launay: 'Traité des Gîtes Minéraux,' Vol. II, p. 778.

[4] Traverso: 'Giacimenti a mineral d'argento del Sarrabus,' *Ann.* Mus. Civico, Genova, 1880-81, Vol. XVI, p. 493. De Castro: 'Descriptione et carta geologica de la zona argentifera del Sarrabus,' 1890. Fuchs and De Launay: 'Traité,' 1893, II, pp. 769-776.

brightly glistening fracture surfaces, together with native silver, horn silver, argentite, stephanite, pyrargyrite and proustite. Pyrite, marcasite, arsenopyrite, copper pyrite, molybdenite, as well as cobalt and nickel ores, appear merely as accessories. The lodes were first mined in 1622, but an active industry has existed only since 1870, and yielded rich returns in the seventies.

(*) **High-Grade or Noble (Edle) Silver Veins.**

The lead veins connect by gradual transitions with those of the high-grade silver veins, in which the rich silver minerals, which in the first-named class were only occasionally present, decidedly predominate and impart to the veins their distinguishing character. It is possible to introduce some order into the chaos of phenomena, if the deposits are grouped according to their natural mode of occurrences. If, however, one follows the method adopted for other groups of veins, basing the division solely on the predominant matrix, we should be obliged to separate veins that belong together. We will therefore adopt a combined method, even though it is not consistent. Two groups, both of which are characterized by rich silver ores, will be distinguished by the prevailing gangue. Two other groups we will distinguish by the abundant presence of ores of a certain other kind, laying less stress on the matrix, because it is not always uniformly developed.

On this basis we may distinguish:

A. Veins with rich silver ores only, grouped as:

1. Those with quartz as predominant matrix, the *rich quartz formation*:

2. Those with calcspar as predominant matrix, the *rich calcspar formation;*

B. Veins with silver ores and other characteristic ores, grouped as:

1. Those in which many copper ores enter, with barite as the predominant gangue mineral, but frequently also with quartz and earthy carbonates; the *high-grade silver-copper veins.*

2. Those in which many cobalt ores, together with nickel-bismuth ores and uranium ores, are present, with quartz or carbonates forming the gangue; the *high-grade silver cobalt veins.*

Another group, the silver-gold veins, might have been added. It appeared, however, that the most prominent character of these veins is marked by the presence of gold, not silver, especially from an economic point of view. Hence, the silver-gold veins are not included here, but described with the gold veins proper. From what has been said it is apparent that the various types of rich silver veins admit of no sharp delimitation, their frequent transitions to other very diverse types causing the distinction to become at times very faint. This will appear still more decidedly in the detailed presentation. In characterizing the different groups, advantage may also be taken of their geologic age.

12. High Grade Silver Quartz Veins.

These veins consist of ordinary white or gray quartz or hornstone, frequently showing a brecciated structure, with fragments of the country rock or of an older quartzose filling held together by later quartz. The ore minerals occur sprinkled through this matrix, in rather fine particles rarely united into bunches, nests, streaks or stringers. They are chiefly rich silver sulphides, such as silver glance, ruby silver, native silver, etc., but include argentiferous arsenopyrite and iron pyrite, more rarely argentiferous zinc-blende and galena. The mixture of ores ordinarily contains also a slight percentage of gold. Some departures from this prevailing character and some rarer ingredients will be enumerated and discussed in the several examples that follow.

The silver quartz veins of Freiberg in Saxony are first noted. According to H. Müller[1], the high-grade quartz veins are especially numerous to the north and west of the town, in the region of the upper gneiss (a microgranular biotite gneiss), and of the mica schist. The veins are grouped in an important series in a belt about 9 km. (5.4 miles) wide that extends from the region of the lower Muld valley between Hohentanne and Zella, near Nossen, in a southwest direction, for a distance of 22 km. (13.2 miles) across Bräunsdorf and Oberschöna, to the region of Oederan. The veins strike mostly between north-northeast and east-northeast and dip northwest. Their thickness varies between 0.1 and 1 m. (0.3-3 ft.) Some of them have been opened up to a distance of 2,600 m. (8,528 ft.) and for a depth of 460 m. (1,508 ft.) The general characterization given above applies with the following additions: Alongside of the principal quartzose matrix there are occasionally found subordinate amounts of ankerite, calcite and rhodochrosite. Besides the ores already mentioned, there are also found both ordinary and silver-rich tetrahedrite, miargyrite, stephanite, polybasite, freieslebenite, xanthocone, hydrostilpnite (3 Ag_2S, Sb_2S_3), alabandite, copper-pyrite, marcasite, capillary pyrite, stibnite, berthierite, kermesite, bournonite, zinkenite, jamesonite, etc. Arsenopyrite sometimes contains as much as 0.3% silver, iron pyrite up to 0.5 or even 1.2%, galena and zinc-blende usually still more. The zinc-blende is dark and "vitrified," that is to say, coated with films of argentite or intimately overgrown with granules of argentite, giving it its high silver content. The pyritic and antimonial ores contain 0.5-8 g. per ton in gold. The ore minerals rarely form large compact bodies, being on the contrary for the most part in the form of fine dust-like particles scattered through the quartz. The silver minerals occur mainly as small parasitic crystals in small druses. The exceedingly fine distribution of the silver minerals in the quartzose matrix is exhibited in the reproduction of a thin section shown in Fig. 162, which is also of interest for the information it gives as regards the mode of treatment necessary for such ores. The lode structure is massive or breccia-like.

The most important veins of this series, and the ones most typical of the

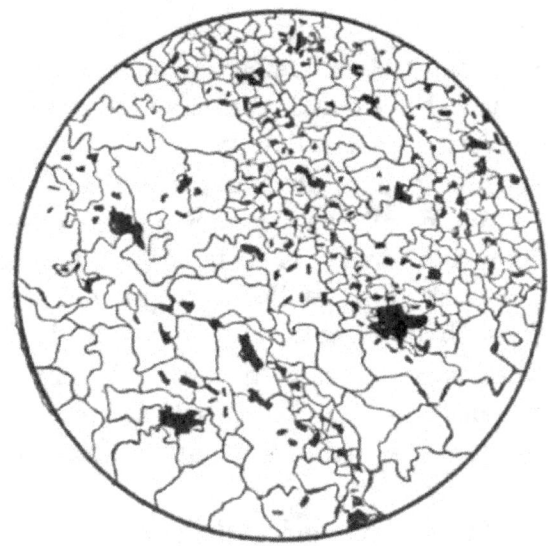

Fig. 162.—Thin section of gray quartz of Freiberg silver quartz veins, with fine particles of argentite and other high-grade silver minerals. (Enlarged 50 times.) (Outlines of quartz grains as shown in polarized light.)

formation, but no longer worked, are the lodes of the so-called 'Schwarze Gebirge' (black rock) of Bräunsdorf, a very irregular block of carbonaceous rock, resembling alum shale lying in the midst of the mica schist. Hence J. C. Freiesleben called the entire formation "the Bräunsdorf formation." In the black rock, according to H. Müller[1], the vast Neue Hoffnung Gottes Stehende vein, the main lode, from 300 m. (984 ft.) downward, becomes broken up into a series of smaller veins, such as the Verlorne Hoffnung Stehende, the Felix Morgengang, Aaron Morgengang, Jupiter Morgengang, Zweifler Stehenden, Augustus Spat, etc. All these veins, from 5 cm.

[1] H. Müller: 'Erzlagerstätten nördlich und nordwestlich bei Freiberg,' in B. v. Cotta's 'Gangstudien,' Vol. I, p. 174.

to 2 m. (0.1-6.5 ft.) thick, break up again into innumerable stringers. This whole network of stringers extends from northeast to southwest through the black rock, but only in rare cases does it send short branches into the adjoining mica schist. The most important ores of the Bräunsdorf lodes were what was there called 'Weisserz' (white ore), that is to say, a highly argentiferous arsenopyrite, ruby silver, miargyrite, native silver, stibnite, kermesite and berthierite. Beside the predominant quartz matrix, rhodochrosite, brownspar and calcspar appeared as accessories, in rare cases also fluorspar, barite, etc.

Somewhat farther to the southeast are the old mines of Oberschöna, one of which, the Zenith, was in operation as late as 1879-1893. Here, too, the typical high-grade silver quartz veins were mined.

The Bräunsdorf lodes are similar to the Peter Stehende vein of the Christbescherung mine near Grossvoigtsberg, which is still mined. Mention should also be made of the veins of Emanuel Erbstolln near Nieder-Reinsberg, Romanus Erbstolln and Vereinigt Feld near Siebenlehn, the last of which is still mined in a small way.

At the Gesegnete Bergmanns Hoffnung mine near Obergruna, which has recently been abandoned, all the veins, but more especially the Helmrich Spat and the Traugott Spat (see Fig. 126), are examples showing a slight variation from the type of the high-grade silver quartz veins. The ores are especially rich in argentiferous blende ('vitrified' blende, that is, coated with films of vitreous or earthy argentite) and galena, with pyrite, and also argentiferous arsenopyrite, which as 'coarse ore' are associated, even in the richer leads, with the 'high-grade ores' constituting the silver ores proper.

The lodes of the neighboring and still flourishing mine of Alte Hoffnung Gottes, near Klein-Voigtsberg, especially the rich main veins, the Peter Stehende and the Christliche Hilfe Stehende, are also not strictly rich silver quartz veins, but transition veins to the rich brownspar formation. Besides quartz, they carry much rhodochrosite, brownspar and calcspar, and besides the rich silver ores, much coarse matrix occurs. This mine is the only one of the veins of the Freiberg district proper which, thanks to its rich ore shoots, continues to pay working to the present day, despite the depreciation of silver.

Somewhat farther off from Freiberg one finds the typical rich quartz vein of the Friedrich August mine near Reichenau in the neighborhood of Frauenstein, which was in operation as late as the middle of the seventies. This, and the Emanuel Erbstolln near Drehfeld, were the mines which revealed the fact, so important in determining the age of the veins, that the rich quartz lodes are traversed by the Permian quartz porphyries.

Rich silver quartz veins also occur in the granulite (gneiss) formation of Rosswein, especially in the fibrous-uralite gabbro of Gersdorf and Wolfsthal, and again in the gneiss area near the Grossdorfhain and Höckendorf near Tharandt (Unverhofft Glück and Edle Krone mines), near Reichstädt and Ammelsdorf, in the vicinity of Dippoldiswalde, at all of which mining has ceased.

Rich silver quartz veins were also worked in former times in the neighboring regions of Bohemia, as at Nicklasberg and Klostergrab in the Bohemian Erzgebirge, at Adamstadt and Rudolstadt, northeast of Budweis.

Among the veins of the Kinzig valley in the Black Forest, this type is represented by those of the Morgen series.[1]

Numerous examples of rich silver quartz veins occurring in countries outside of Europe differ from the original Erzgebirge type in being of much younger age.

Where silver veins are associated with andesites and other Tertiary eruptive masses, as at Pachuca and Real del Monte, and where the ores contain gold even in small amounts, it is difficult to draw a distinction between veins of this class and those of the rich silver-gold ore formation.

Among the many known occurrences of this class those of Mexico probably claim the first rank. Most of the silver veins of that country[2] belong by their mineral composition to this division of the system into which we have grouped ore deposits. With few exceptions the predominant matrix of the Mexican veins is a crystalline quartz, often violet-colored, accompanied by subordinate quantities of calcite and rhodonite. In the zone of unaltered primary ores, they consist essentially of ruby silver ore, both proustite and pyrargyrite, accompanied at increasing depths by more and more gray copper, and finally also zinc-blende and pyrite, so that the richness slowly decreases. As regards the alteration products, there has been developed, directly above the primary ore, first a zone with simple silver sulphides, which usually contained the richest bonanzas, above this a second zone rich in chlorine-, bromine- and iodine-compounds of silver, and finally the gossan with native silver predominating, but usually of moderate richness (see gossan). These silver ore lodes traverse very diverse rocks. The famous Veta Madre of Guanajuato,[3] Mexico,[4] is probably the richest

---

[1] Vogelgesang: 'Geogn.-bergm. Beschr. des Kinzigthaler Bergbaues,' Carlsruhe, 1865, p. 9.

[2] Fuchs and De Launay: 'Traité des Gîtes Minéraux,' 1893, pp. 811-829.

[3] Visited in 1903 by W. H. W., who supplies description.

[4] 'Carta Minera de la Republica Mexicana por Antonio del Castillo,' 1893, Mexico. See also 'The Mineral Deposits of Mexico,' by Aguilera. *Trans.* Am. Inst. Min. Eng., Mexican Volume, 1903. C. Henrich: *Mining Magazine*, New York, 1904, Vol. VI, No. 1, p. 83.

silver-gold vein of the entire world, the production of the district, mainly from this one vein, reaching the enormous total of nearly $1,000,000,000. The vein occupies a well defined fault on the west slopes of a range of steep and barren mountains, composed mainly of well bedded volcanic tuffs with a capping of dacitic rocks and intrusive dikes of rhyolite. The andesitic tuffs, called sandstones by all earlier writers, are of varying colors and textures, mostly well bedded, but showing conclusive evidence in both field relations and in thin section under the microscope, of their origin by ejection, but in part sorted by water. These beds of andesitic tuffs rest upon micaceous schists and slates, cut by a mass of intrusive diorite, which altered the shale about it. The Veta Madre cuts through these earlier rocks, and at the south end through the tuffs, but so far as known does not cut the outflows and dikes of dacitic lava. The vein varies from a few feet to 50 feet across, widening in places to 120 feet. The ores occur in well defined shoots, the greatest—that of the Valenciana mine—being 1,400 feet long, 40 to 50 feet thick, and worked to a depth of 2,100 feet on the dip, or 1,750 feet vertical. The vein strikes northwest, and dips west with the mountain slopes at 45°. The vein filling between and alongside of the ore shoots consists of crushed and much altered country rock. The ore consists of quartz with orthoclase (var. valencianite) dusted and stained with minute specks of rich silver sulphide and rarely with pyrite. Galena is a rarity. The neighboring La Luz vein carries druses with remarkable crystals of apophyllite. The ores have averaged about 50 oz. silver for 350 years, but the rich oxidized and secondary sulphide orebodies run much higher. The ore shoots occur at and just beyond the intersection of the hanging-wall veins, coming in from the west, or La Luz system. At such points the vein is enormously thickened and the bonanzas occur. Earlier writers locate the richest ores at the points where the vein is crossed by ravines, and it is possible that secondary enrichment may have been particularly active at such places. The vein has a generally obscure outcrop, but is to-day traceable for seven miles by a very remarkable line of shafts and buildings.

Mining began at Guanajuato in 1558, and has continued with some interruptions ever since. Though in the last half century the production has come mainly from La Luz system, mining in the Veta Madre area has lately been very successful.

The lodes of Zacatecas and Fresnillo traverse similar conglomeratic porphyry tuffs, clay slates and calcareous shales, alternating with quartzites. At Catorce the veins occur in hornblende porphyrites. As these hornblende porphyrites are intruded in white Jurassic limestones with *Perisphinctes plicatilis,* the post-Jurassic age of the lodes is there established beyond doubt. We will dwell in somewhat greater detail on the famous lodes of Pachuca and Real del Monte, situated north-northeast of the capital, abstracting from the official account published by the Geological Institute of Mexico.[1]

The mining area of Pachuca lies about 150 kilometers north-northeast of the City of Mexico in the region of the Sierra of the same name. Embracing an area of about 20 square kilometers, it is the largest and at the same time the oldest mining district of the entire republic. The Sierra de Pachuca, whose main strike is northeast-southwest, extends over a length of 43 km. and attains its greatest height in the northeastern part in the summit of the Cerro de las Navajas, 3,212 m. (10,536 ft.) The mountain range is traversed by numerous valleys, but only on the east slope are there any rivers that carry water throughout the year; they belong to the drainage area of the Rio del Amayac, a tributary of the Montezuma. According to Aguilera and Ordoñez, the Sierra de Pachuca owes its origin to a period of violent eruption, which apparently occurred in the middle Tertiary as a sequel to great dislocations. The latter led to the upraising of non-fossiliferous sediments, probably belonging to the Cretaceous. The products of those eruptions were, in the order of their age, andesites with their tuffs and breccias, rhyolites and their glassy forms, and, as last member, basalts. Dacites, too, were observed. The andesites, for the most part pyroxene andesites, sometimes diabasic in character, possess by far the greatest extent, and are of special interest, because they form the most widely distributed country rock of the lodes.

The latter may be divided into four groups, each of which embraces a main lode and several lateral veins. They comprise the Vizcaina, Cristo and Analcos lodes and of those of the Santa Gertrudis region. The strike of all these lodes is east-west, the dip, at a high angle, mostly south. The thickness, both in the strike and in depth, is subject to great change. Thus the Vizcaina lode at some points attains a thickness of 8 m., while that of the other parts of the lode varies between 0.3 and 6 m.

The matrix is, in the main, quartz, often developed as chalcedony; more rarely calcite, rhodonite and rhodochrosite occur. This explains why the outcrop of the lodes may often be traced far across the ground in the form of white crests. The lode structure is for the most part markedly banded, and only in isolated cases has a breccia structure been observed. A remarkable feature is a wide zone of decomposition along the selvages. The country rock, which often appears strongly silicified, has been found to be abundantly impregnated with pyrite at all depths. Part of it is also propylitized.

[1] J. G. Aguilera and E. Ordoñez: 'Mineral de Pachuca,' 1897. E. Ordonez y M. Rangel: 'El Real del Monte,' 1899.

According to the richness of the ore, three zones are distinguished from above downward:

> 1. *Metales colorados* (*podridos*) (red ore, rotten).
> 2. *Metales negros* (black) } (a) *Metales de pinta* (spotted ore).
> } (b) *Metales de fuego* (fire ore).

In the first zone, that of the "colored or rotten ore," which owes its name to the prevailing iron and manganese oxides, silver chloride, bromide

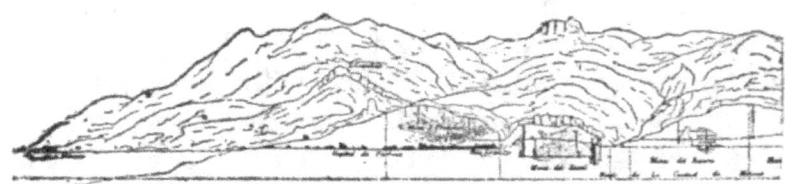

Fig. 162*a*.—Cross-section of veins at Pachuca, Mexico.

and native silver are found, sometimes also native gold. This zone is considerably richer in gold than those which follow.

In the second zone (2a), fine-grained galena, argentite, pyrite and more rarely chalcopyrite occur.

The lowest zone (2b) contains galena, zinc-blende, pyrite, stephanite and polybasite. Some native copper, too, was observed. These ores, which are found but very rarely in definite crystalline forms, occur intimately

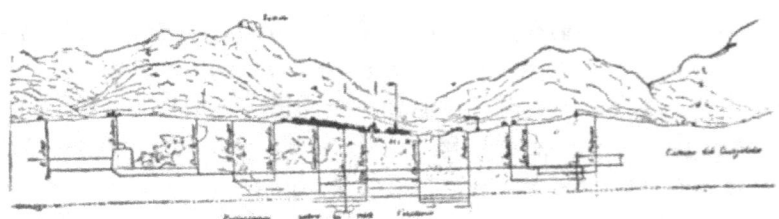

Fig. 162*b*.—Pachuca cross-section, continued.

intergrown with quartz. The presence of zinc-blende is unfavorable and indicates early impoverishment. Analysis has shown that the pyrite is always argentiferous, the percentage rising to 0.05%. The galena, too, is argentiferous, though not to the same degree, while the blende is almost always free from silver. Here, as in most of the silver lodes of Mexico, the occurrence of rich shoots, the bonanzas, is of the greatest importance in mining. It is not known that their occurrence is subject to any law, but they certainly begin only at a depth of 150 m. and seem on the whole to be normal in their course to the strike of the lodes. The size of these bonanzas varies greatly; the greatest, encountered in the San Rafael mine, has a length of 1,000 meters.

Mining at Pachuca extends back to the time when the Aztecs obtained the ore by the aid of fire and stone implements. Immediately after the conquest of Mexico, the Spaniards began to exploit these rich mines, and, down to the eighteenth century, the Spanish Crown possessed the exclusive right of mining. The most famous mine at the beginning of that period was Xacal, which produced $7,000 worth of ore a day. After the war of independence the mines were sold to English companies (1824). After 28 years, however, these companies, burdened with debt, ceased operations, and the work was resumed only later and after the completion of railways, with Mexican capital. The production thereupon increased considerably, the San Rafael y Annexas company obtaining in eight years a gross return of $12,500,000.

Exactly similar conditions are found at Real del Monte, situated somewhat farther northeast of Pachuca. Mining operations were begun in 1578 and were formerly exceedingly remunerative. The work is mainly on three lodes: the Veta Vizcaina, 2 to 15 m. (6.5 to 48 ft.) thick, striking west-northwest and dipping south, extending over from Pachuca and opened up to a depth of 400 m. (1,312 ft.); the Veta Santa Inez, 40 m. (131 ft.) thick; and the Veta Santa Brigida, as much as 10 m. (32.3 ft.) thick, the two latter, striking north-south and dipping more or less steeply west and east, intersect the first named lode. All these veins have lenticular enlargements at close intervals, these ore shoots proving to be very rich in ore. These lenticular bonanzas coincide with those portions of the lode in which the dip is especially steep, almost perpendicular.

In the lower or deeper zones, to which working is now confined, at depths of 300 to 400 m. (984 to 1,312 ft.), the vein filling consists of quartz, often brecciated, and frequently associated with rhodonite and to a subordinate extent with calcite. The ores consist mainly of pyrite, copper pyrite, galena, in part argentiferous, and zinc-blende, together with argentite, polybasite and native silver. Druses studded with crystals of quartz, calcite, manganocalcite and rhodochrosite are common. In the upper part of the veins, on the contrary, the rich silver ores, especially ruby silver, exceed the pyritous ores in amount. The outcrop consists of a gossan rich in manganite, pyrolusite and wad.

Very commonly the country rock is pyritized or transformed into clayey masses, 'lamas,' or impregnated with pyrites.

The lodes of Real del Monte, too, are Tertiary beyond a doubt. The volcanic activity began in this region in the middle Miocene coincident with a strong folding of the Mesozoic strata, the earliest eruptive being pyroxene-andesite. Next followed flows and dikes of rhyolite, in whose train the ore-bringing thermal waters rose. The next and last act of the volcanic period was an eruption of basalt, during which the vein fissures were in many cases fractured and displaced.

The period of greatest prosperity of Real del Monte was in the last two decades of the 17th century. From 1687-1697, La Trinidad mine alone is said to have yielded 40 million pesos. As late as 1759-1771, the mines of Real del Monte yielded to the principal owner at that time, Don P. R. de Terreros, $6,125,000. When mining became confined to the pyritous zone, the profit quickly decreased.

From what has just been said, it is doubtful whether Guanajuato, Pachuca and Real del Monte, with the other Mexican lodes mentioned, should be included in the rich silver quartz formation, since the ores contain from 20 to 30% of their value in gold, and have carbonates in the matrix. On the other hand, the association with rhyolites and andesites, the propylitization of the country rock, the participation of rhodonite as matrix, and the occasional gold content, establish a close similarity with Schemnitz. But for that matter, Schemnitz, too, as we shall see, does not represent a pure type, but constitutes a transitional form between the rich quartz and the silver-gold vein-types.

Another example of this class of veins occurs at Tonopah, Nevada, a district that has astonished the world by the richness of its bonanza ores. According to Spurr, the district is one of Tertiary volcanics, or sedimentary volcanic tuffs. The oldest rock is hornblende andesite, and still later rhyolite and dacite rocks occur, and the whole is cut into a complex system of blocks by faulting. The important veins of the region occur only in the early andesite, but four periods of hot-spring action and vein-formation are recognizable. The veins are linked veins, branching and reuniting both laterally and vertically. They are typical replacement veins formed along zones of fracturing. The primary ore consists of argentite, polybasite, and small amounts of chalcopyrite, galena blende and pyrite, in a gangue of quartz and orthoclase (valencianite). The rich ores occur in roughly defined shoots, with west-east pitch. These veins, though rich, are so cut by faults of different systems as to cause great trouble in working. The Mizpah vein, for example, is not only cut off east, west, and at the bottom by faults of great magnitude, but is broken up by minor fractures. The veins in the rhyolo-dacite rocks are characterized by a lack of persistence and definition. Bunches of high-grade ore sometimes occur in these usually barren veins, but only near andesite masses. The Goldfield veins, 23½ miles south of Tonopah, are similar, showing irregularly branching outcrops of pipes of jaspery quartz, the jackets about pay shoots of very rich gold ores. In the past two years the district has produced ores to the value of $2,000,000.

A very rich development of the formation occurs at some points in Peru, in particular in the lodes of Quespesisa mine near Castrovirreyna in the department of Huancavelica, exploited during the last decades. The matrix there, according to the material furnished to the Freiberg collection by Mr. E. Treptow, is mainly a milk-white to glassy quartz, of distinctly crystalline, often drusy development, often colored dark gray by minute ore inclusions. This is found in association with subordinate amounts of a hornstone-like material composed mainly of chalcedony, only rarely of barite. The ores consist of pyrargyrite, polybasite and other rich silver ores, together with galena, fine-grained zinc-blende, a little chalcopyrite and pyrite. The rich portions contained on an average 2% of silver. The country rock is a strongly decomposed augite andesite.

The famous lodes of the Cerro de Pasco, in the Peruvian province of the same name, known from d'Achiardi's and Fuchs and De Launay's descriptions[1] may also be included with the rich silver quartz lodes, despite some aberrant features. At this place profitable silver mining has been carried on ever since the end of the thirties of the 17th century, despite the altitude, 4,352 m. (14,278 ft.), and the inhospitable climate. As late as 1879, 179 mines were in operation. The veins are of post-Jurassic age. The unaltered ores found in the deeper workings consist of highly argentiferous gray copper associated with galena in a gangue of quartz. In the upper zones, much pyrite and argentiferous copper pyrite are found, and at the very top, in the gossan, the rich secondary ores were found, which are here called *pacos* or *cascajos,* and contain an average content of 500 gm. silver per ton. The primary ore contains much copper and the district will soon be one of the great copper producers of the world.

Among the numerous silver veins of North America, those of Austin, Nevada, which, according to S. F. Emmons[2], occur in granite, and consist of a series of quartz stringers with pyrargyrite, proustite, polybasite, stephanite and tetrahedrite, and associated galena and blende, fit remarkably well into this category. Rhodochrosite and calcite occur together with the predominant quartz of the gangue. Some veins have been worked for stibnite. This occurrence is, therefore, completely analogous to the typical Braünsdorf or rich silver quartz lodes of the Freiberg area.

As an appendix to the discussion of the rich silver-quartz veins, mention may be made of a rare example, decidedly unique in its composition: a rich silver vein with quartz and orthoclase as gangue. According to

[1] Fuchs and De Launay: 'Traité.' II, pp. 829–832.
[2] S. F. Emmons: 'Fortieth Parallel Survey,' Vol. III, p. 349.

W. Lindgren,[1] the Black Jack-Trade Dollar vein occurs in granite, and in the basalt and rhyolite which rest upon it, at Silver City in southern Idaho. Where granite is the country rock, the vein filling contains orthoclase in addition to the more abundant quartz, and is often banded. The ores contain pyrite, copper pyrite and silver glance, enclosed in milk-white orthoclase, intergrown with quartz. This vein also contains druses lined with crystals of feldspar of the type called valencianite by Breithaupt, and first described from the Valenciana silver mine near Guanajuato in Mexico as noted in the description of that property.

13. Rich Silver-Calcite Veins (Rich Calcspar Formation).

Veins of this class consist mainly of calcite, sometimes colored dark by bituminous matter, together with quartz, fluorspar, zeolites and rarely axinite. Rich silver ores occur either finely disseminated through this gangue, or forming small pockets and short stringers. They comprise pyrargyrite, proustite, silver glance and native silver, more rarely dyscrasite (antimonial silver), arsenical silver, pyrostilpnite (fire blende), etc., usually associated with galena, zinc-blende, pyrrhotite, pyrite and native arsenic.

The veins are mostly thin and in large part scattered into stringers. The best known example of this class is seen in the veins at Sanct Andreasberg in the Harz, the following summary being based largely on the account given by F. Klockmann.

The Sanct Andreasberg[2] region is underlain by the Wieder slates, belonging to the so-called Harz formation of the lower Devonian, which contains many limestone layers, and holds numerous intrusive sheets of diabase. A little north of the mining area the Brocken granite forms intrusive bodies surrounded by contact zones of altered slates. The Andreasberg veins lie south of the contact zone of the granitic intrusive masses.

The veins are limited to a narrow wedge of rock, about 3 km. (1.8 miles) long and 1 km. (0.6 m.) wide on the east, but tapering westward. This block is bounded by two zones of faulting, viz.: the so-called Neufanger Ruschel at the north and the Edelleuter Ruschel at the south, as shown in the following sketch, Fig. 163, after F. Klockmann. These faults or "Ruscheln" meet near the Sieber river; eastward the network of

[1] W. Lindgren: 'Orthoclase a Gangue Mineral.' *Amer. Journ. of Sc.* 1898, Vol. V, p. 418. Also in 20th *Annual Rep.*, part III, Geol. Surv., Washington, 1900.

[2] Heinr. Credner: 'Geognost. Verhältn. Thüringens und des Harzes.' 1843. Also, 'Geognost. Beschreibung des Bergwerks-Distrikts von Sct. Andreasborg.' *Zeitschr.* d. d. G. G., 1865. C. Blömecke: 'Die Erzlagerstätten des Harzes.' Vienna, 1885, p. 48 *et seq.* F. Klockmann: *Berg u. Huttenwesen des Oberharzes.* Stuttgart, 1895, pp. 50–57.

veins extends geologically as far as the Oder valley. However, lodes of the same character also occur farther east, at Braunlage.

These delimiting faults are as much as 60 m. (186.8 ft.) wide, and consist of highly altered, leached and crushed rock, with alternating layers of harder rock, less affected by crushing, and consisting of layers of slate traversed by innumerable slips and planes of pressure and of slight movement.

The course of the northern fault corresponds approximately to the strike of the surrounding country rock, and may be conceived of as a system of overthrust planes, after the manner of the Clausthal lodes (see page 247). The Edelleuter Ruschel, on the other hand, cuts the strata at an acute angle, and the fault brings the slates against the diabase, the fault forming the boundary for some distance. This fracture is accompanied by two lesser converging faults, the Silberburger Ruschel and the Abendrother Ruschel. The two main fault lodes dip south 60 to 70°.

The Andreasberg veins are confined to the space between the main faults. Fissures do, indeed, occur beyond these limits, but they are of different character and filled with other ores and gangues. North of the Neufanger Ruschel the veins contain a quartzose red hermatite filling, while south of the Edelleuter Ruschel the veins contain chalcopyrite in a barite filling.

Grouped according to their strike, the Andreasberg veins belong to two divisions. Those of the first group, for example the Fünf Bücher Moses series, the Samson lode, etc., strike northwest transversely to the faults (Ruscheln), and dip steeply northeast. They terminate against the fault-fissures by wedging out and breaking up into stringers. The others run approximately parallel to the faults, that is to say, between east-west and west-northwest, and dip north at 60-80°. They are deflected by the veins of the first group. The Bergmannstrost and the Gnade Gottes lodes are examples of this.

All the veins are in the main simple in structure, without stringers, and of but little thickness, usually between 1 cm. and ½ m. The filling consists mainly of compact, whitish calcspar (older calcspar generation), in which ores occur either as impregnations or in stringers and pockets and consist of galena, zinc-blende, native arsenic, proustite, dyscrasite, arsenical silver, native silver, rarely argentite. Here and there other minerals are seen, namely, antimony, and arsenious nickel (breithauptite and niccolite), cobalt pyrite, linnaeite, fluorspar and barite, as well as axinite as accessories. In numerous druses a second generation of the above named ores and gangues are found, particularly crystals of calcspar and ruby silver, in a great variety of forms with associated fireblende, antimony and numerous zeolites, such as apophyllite, analcite, harmotome, desmine, stilbite, natrolite and sometimes also datolite. The distribution of the ores in veins is very irregular.

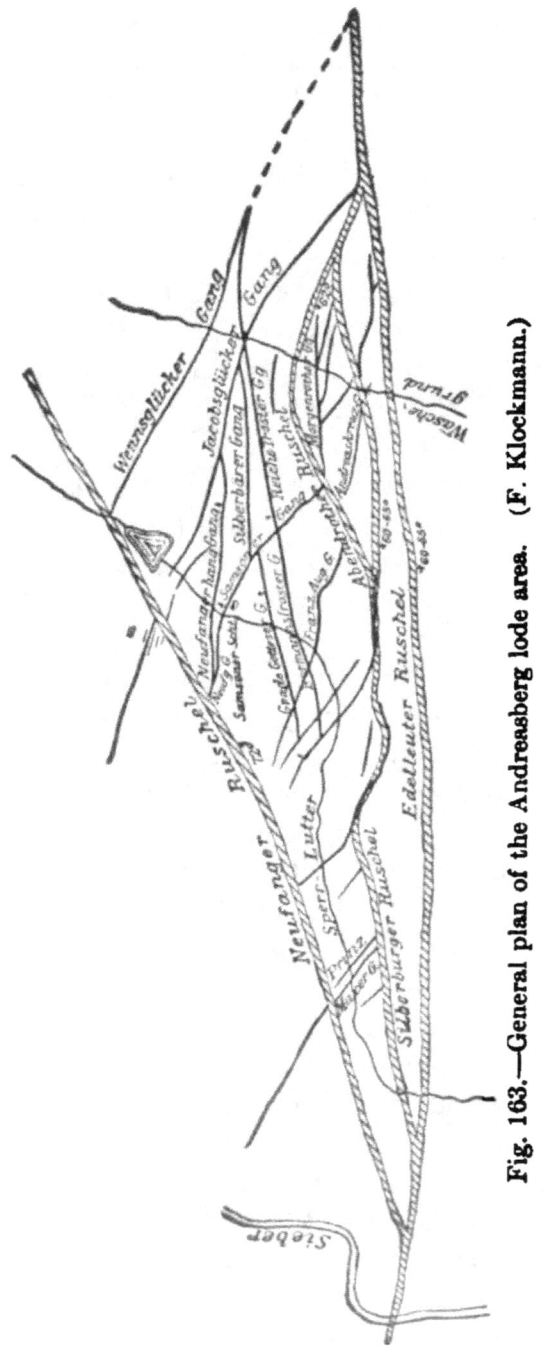

Fig. 163.—General plan of the Andreasberg lode area.   (F. Klockmann.)

The Andreasberg veins were discovered in 1521 by Joachimsthal miners, who opened the first mine, called Sanct Andreaskreuz am Beerberg, so christened, it is said, because two of the lodes intersected like beams of a St. Andrew's cross. The newly founded town was called by the name of the mine. The period of greatest prosperity was from 1565 to 1570. Though abandoned soon after that, work was resumed in 1646 and is in progress to this day.

A second typical representative of the rich calcspar formation is the group of veins at Kongsberg in southern Norway.

Kongsberg[1] lies in the southern part of the Numedal (southwest of Christiania), on both sides of the Laagen Elv. Most of the mines are situated about 6 km. (3.6 miles) west of the town, in a highly mountainous region, near the Jonsknut, 908 m. (2,978 ft.) high. The veins occur in steeply tilted crystalline schists, biotite schists, chlorite, talcose, quartzite and hornblendic schists as well as a garnetiferous biotite gneiss. These rocks are intruded by plutonic rocks near the ore deposits, as, *e. g.*, an olivine gabbro near Vinor, uralite gabbro and Flasergabbro (gabbro altered by pressure), at the Jonsknut, and norite at Skollenberg. In the Armengrube and the Kongensgrube, a dike of augite porphyry, a few decimeters thick, is known, which is cut by the veins and is but slightly displaced. In certain zones the foliated schists are impregnated with pyrite, which imparts to the weathered rock a rusty brown, or yellowish brown color, appearing gray when compared with the color of the normal schists. This led the German miners who founded the Kongsberg mining industry to call such zones 'Fahlbander' (gray bands).

Six bands of this sort are usually distinguished, whose aggregate thickness amounts to 300 m. (984 ft.) The pyrite is most abundant in the mica schist, less in the hornblende schist and is quite scanty in the gneiss. This impregnation with pyrite may be so abundant as to actually form compact layers of pyrite, as is seen in the pyrite pits of the Oberberg in the main fahlband of that locality. The distribution of the sulphides is always very irregular; the fahlbands may often enclose masses of rock entirely free from pyrite. The pyrite consists mainly of ordinary pyrite and magnetic pyrite, but includes copper pyrite, rarely cobalt glance (see the Modum occurrence). Fahlbands are also described as occurring in the gabbros.

[1] Most important publications: Böbert: 'Ueber den Kongsberger Bergbau.' Karsten's *Archiv.*, vol. XII, pp. 267–346, 1833. Report of Commission of Jahre 1865. Th. Kjerulf u. Tell. Dahll: 'Ueber den Erzdistrict Kongsberg.' Christiania, 1860. Th. Scheerer: 'Vorkommen des Silbers zu Kongsberg.' *Berg u. Hütten. Zeit.*, 1866, p. 25. C. F. Andresen: 'Om Gangformationer ved Kongsberg.' 1868. G. Rolland: 'Mém. sur la géologie de Kongsberg.' *Ann. d. Mines.* vol. VII, ser. XI., 1877, p. 301. Chr. A. Münster: 'Kongsberg ertsdistrikt.' Christiania, 1894. P. Krusch. 'Das Kongsberger Erzrevier.' *Zeit. f. Prak. Geol.*, 1896, pp. 93–104. J. H. L. Vogt: 'Ueber die Bildung des ged. Silbers, etc. und ein Versuch zur Erklärung der Edelheit der Kongsberger Gänge an den Fahlbandkreuzen.' *Zeit. f. Prak. Geol.*, 1899, pp. 113–123 and 177–181.

The pyrites of the fahlbands contain merely traces of silver. According to Chr. Munster, the pure pyrrhotite of a fahlband contained 0.0005% silver, besides 0.2% nickel and copper, while the pure pyrite contained only 0.0005% silver.

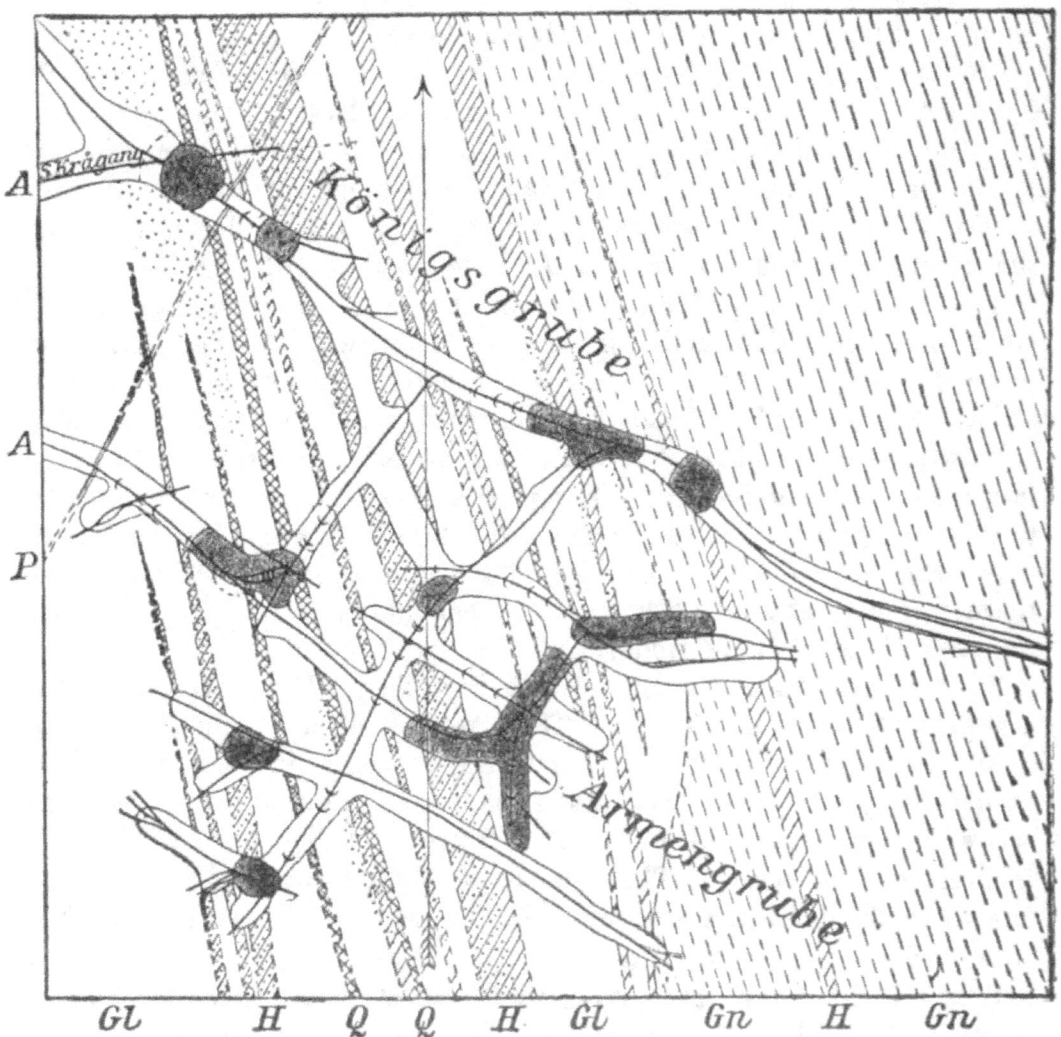

Fig. 164.—Ground plan of the 310 fathom level of the Armen mine and the Kongens mine at Kongsberg. (From the official plans.)

Gl, mica schist; H, hornblende schist; Q, quartzite schist; Gn, gray biotite gneiss (dots indicate development as fahlbands); P, dike of augite porphyry; A, drifts with lodes and orebodies (the latter shaded); (((rich, ((medium, (poor ores.

The Kongsberg veins are distinguished by their steep dip and their decided tendency to stringers. In various parts of the mine, several stringers, usually only 10 to 30 cm. thick, are worked simultaneously. This is shown in Fig. 164, representing a portion of a mine map placed at our disposal by the administration of the mine. The longitudinal section, Fig. 165, also shows the same thing, and represents the distribution of the working drifts, not over a single compact lode, but over a series of stringers. The stringers may show local thickening, and in such cases may enclose druses of as much as 1.25 cubic meter in capacity.

The prevailing strike of the lodes is northwest. The main lodes are con-

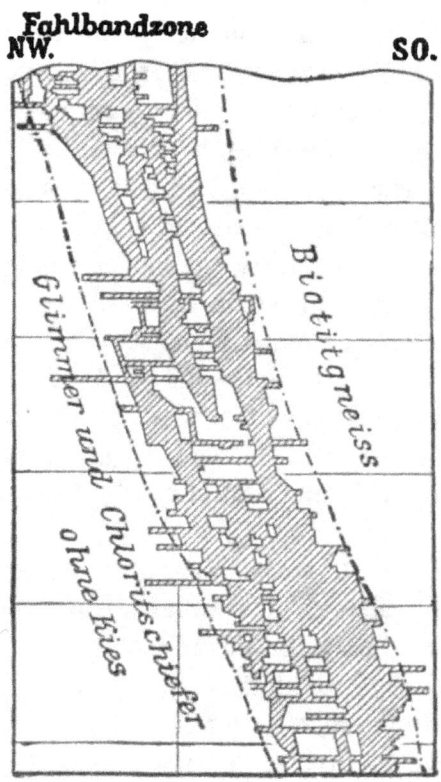

Fig. 165.—Horizontal plan of a series of vein stringers of the Kongens mine at Kongsberg. (J. H. L. Vogt.)

nected by a perfect network of diagonal and transverse stringers. All dip very steeply.

The predominant gangue is calcspar with lesser amounts of fluorspar, quartz and anthracite. Barite, axinite, albite, adularia and zeolites are rarities. The calcspar is usually colored brown by bitumen or coal. The ores consist in the main of native silver and silver glance. The native silver occurs as wires and points forming mosslike patches, as well as in large lumps of odd shape, especially fine specimens of which may be seen in the collection of the Royal Museum of Mineralogy at Copenhagen, collected.

in the old days of mining. As late as 1867 there was found in the main lode of Kongens mine at a depth of 530 m. (1,738 ft.) a lump weighing 500 kilograms (1,110 lb.) consisting of silver with silver glance. J. H. L. Vogt has shown that the native silver is almost always younger than the silver glance, which is often coalesced with it. Sprigs and mosslike aggregates of silver often encrust the silver glance. Vogt thinks there has been an alteration of the argentite into native silver. Alongside of the silver glance is also found some ruby silver ore (proustite). The proportion of silver glance to native silver is about as 1 : 15. The Kongsberg native silver is often alloyed with other metals, especially quicksilver (up to about 1%) and antimony (up to about 0.6%). Silver amalgam, acanthite, hornsilver, and stephanite have also been found. These rich ores are often accompanied by pyrite, sphalerite, galena (with up to 0.05% Ag), pyrrhotite, copper pyrite, sometimes also arsenopyrite and cobalt ores, but pyrite is the only one found as a general constituent of the veins.

The ore shoots of the silver veins are confined chiefly but not entirely to the fahlbands, *i. e.*, where the veins traverse the pyritized gneiss. This local distribution of the ores is apparent from the longitudinal section of the vein series of the Kongens mine shown in Fig. 165. The output of silver in the fahlband zone is as a rule 0.2 to 4 kilograms to the square meter of lode surface, and is only exceptionally as much as 20 kg. (642 oz.) per square meter. At a little distance from the fahlbands, the veins are invariably barren.

In addition to the silver veins, younger, barren calcspar veins are known at Kongsberg; also quartz veins with chalcopyrite and bornite, which sometimes carry electrum (silver and gold). Neither of these two varieties of lodes is of any economic importance.

The Kongsberg lodes were discovered in 1623. The German names of the mines Gottes Hilfe, Armen Grube, Haus Sachsen, together with Kongens (Königs-) Grube, as well as the family names of many of the miners, recall the fact that Germans started this mining industry. In the Kongensgrube the work has now advanced to a vertical depth of 720 m. (2,260 ft.) The mines belongs to the Norwegian crown and continue to produce annually about 5,000 kilograms of silver. From 1623 to 1805 they produced about 543,000 kilograms of fine silver.

Very remarkable veins of this class characterized by rich silver-antimony ores occur east of Broken Hill in the Barrier Range of New South Wales. A vein occurs in the crystalline schists of that region which, wherever it crosses amphibolite intercalations, carries rich silver ores, namely the otherwise rare ores dyscrasite ($Ag_3 Sb_m$), stromeyerite ($CuAgS$) and antimonious silver chloride, besides gray copper, in a gangue of calcite and iron spar[1]. These ores occur in rich bunches associated with clayey material. Cobalt ores play a subordinate part.

14. Rich Silver-Copper Veins.

The veins of this class consist of rich silver and copper ores in a barite gangue, carrying quartz in the deeper levels, with associated zeolites and carbonates. The rich silver and copper ores are generally associated, especially in the deeper levels, with rich gray copper and pyrite-blende ores, and cobalt and nickel ores as mere accessories[2]. The rich silver-copper veins occur most often in basic plagioclase-augite rocks, especially augite porphyries, or at any rate within sediments of Mesozoic age traversed by those rocks. This is especially true of the Chilean occurrences described by W. Moricke[3] at Tres Puntas, Cabeza de Vaca, Los Bordos, Chañarcillo, San Antonio, and Caracoles, all in the region of the Atacama desert, as well as those of Arqueros in the province of Coquimbo.

Only a few of the best known examples can be described in detail.

Chañarcillo[4] lies about 80 km. (48 miles) south of Copiapo, the most important town of northern Chile. The prevailing rock is a gray-blue or dark gray limestone of Upper Jurassic age. This limestone is cut by numerous dikes and intrusive sheets of augite-porphyrite accompanied by contact zones rich in calcium-silicates. These rocks are themselves cut by numerous veins, mostly striking north-northeast. The best known veins are the Corrida colorada, Guias de la Descubridora (a network of stringers in the Lorete mine; Guia = guide-streak), Guias de Carvallo, Mercedes vein, the Candelaria vein, whose strike, by way of exception, is east-northeast. The Corrida colorada has been found to carry rich ore to a depth exceeding 600 m. (1,968 ft.), and has been traced over a distance of 1,800 m. (5,904 ft.) horizontally, with a thickness of 10 m. (32.8 ft.) in the upper regions, only 1 m. (3.2 ft.) in the lower. At the north it is cut off by a great fault. The veins in the upper levels carry native silver, silver chloride, silver bromide and malachite, besides brownspar, calcspar and barite. These ores lie in a clayey gangue colored yellow by ferric hydrate; farther down they carry

[1] G. Smith: *Trans.* Am. Inst. Min. Eng., Vol. XXVI, 1897, p. 69.

[2] On account of the predominance of barite, it might seem natural to use the term "rich barite formation," which would be very appropriate for many cases, especially in Chile, and would moreover be more consistent with preceding categories, but the occurrences at Huanchaca and Huancavelica, though closely related to the others, yet almost or entirely free from barite, would then have no place. Hence the above designation, first used by W. Moricke, was chosen.

[3] W Moricke: 'Die Gold- Silber- und Kupfer erzlagerstätten von Chile.' Freiberg. 1898, p. 27.

[4] Fr. Moesta: 'Ueber das Vorkommen der Chlor-, Brom- und Jodverbindungen des Silbers in der Natur.' Marburg, 1870, p. 15.

native silver, argentite, polybasite, proustite and pyrargyrite; still deeper, increasing amounts of blende, galena, arsenopyrite and a little pyrite. In crossing certain beds (porphyrite sheets), not only are the veins themselves enriched but they have also impregnated these adjoining rocks with valuable ores. The enriched beds adjoining the vein are called *mantos pintadores*. The impregnation is especially strong in the hanging-wall of the lodes and in the upper levels. It is often associated with silicification. Such enrichment may also be observed at the points where the augite-porphyrite dikes (the so-called chorros) cross.

The lodes of Chañarcillo were only discovered in 1832 by the Indian Juan Godoi, and soon became famous because of the fabulous wealth of their outcrop. The mines yielded a very large silver production until very recently.

A similar reputation is enjoyed by the mining district of Caracoles (meaning "snails") in northern Chile, north of Antofagasta in the Atacama desert. Highly fossiliferous Upper Jurassic marls and limestones, whose fossil ammonites gave the name to the mine, are cut by intrusive masses of quartz porphyry and augite-porphyry. The silver veins are said to be much richer in these porphyries than in the limestone. The veins carry ores similar to those at Chañarcillo, in a matrix consisting essentially of barite and calcite. Argentite, proustite, pyrargyrite and highly argentiferous galena are common, while in the upper depths several kinds of silver chloride and bromides occur with native silver.

The veins have only been known since 1870. According to an estimate by Domeyko they produced in the best years 120,000 kilograms of silver.

Similar conditions prevail at Arqueros in the province of Coquimbo. Upper Jurassic limestones, broken through by porphyries, are traversed by barite veins rich in native silver, amalgam, silver chloride, stephanite, gray copper, cobalt pyrite and copper pyrite.

In the adjoining republic of Bolivia, the copper-silver veins also contain lead. The veins of the region of Huanchaca in the province of Potosi are most interesting. Among the various mining fields of this region, Pulacayo, Ubino and Asiento—the first named in particular—have become famous during the last few decades by a large production, and the mining operations, begun by the Spanish conquerors at an early date, and afterward long abandoned, have recently been extensively resumed.

The lodes of Pulcayo[1] occur in trachytic and andesitic rocks. The main lode, striking east-west and 1/3 m. thick, has been opened up to a distance of 1,100 m. (3,608 ft.) in the strike, and to 500 m. (1,640 ft.) in the dip. In

---

[1] A. Gmehling: 'Huanchaca.' *Oesterr. Zeitsch.* f. B. u. H., 1890, p. 281. Fuchs and De Launay: 'Traité,' Vol. III, 1893, p. 852.

the upper depths it is said to have shown a great resemblance to the Chilean examples of the rich silver-copper formation, and, like them, to have been characterized by barite and rich silver minerals, together with gray copper. With increasing depth, quartz with a little calcite took the place of barite. Finally, in the lower workings, according to A. Gmehling, the proportion of barite to quartz was 0.5 : 21. At the same time the rich silver ores were replaced by a mixture of metallic minerals, which, according to the same author, consisted, as stated in the relative order of quantity, of zinc-blende, with 0.2% and more of silver, pyrite, copper pyrite with about 0.3% silver, and galena with an average of 0.6% silver, a mixture whose real value was derived from the presence of tetrahedrite with as much as 10% silver. This gray copper often occurred finely crystallized in druses associated with quartz. In rare instances radial aggregates of stibnite are observed. Mention should also be made of a small content of gold, tin and bismuth in the various ores, while large masses of white lithomarge, with inclusions of gray copper, also occur in the vein filling.

In 1893 the Huanchaca mines produced 281,006,924 kilograms of silver. In 1898 it was only 144,049,443 kilograms, 51,500 tons of ore being exported.

The same type of lode is found in Peru, in the department of Huancavelica, for example, the mines of Morlupa and Julio Caesar, whose ores at some depth consist mainly of a mixture of gray copper, chalcopyrite and galena, in a gangue of quartz and carbonates. Sometimes, as in the case of the Caudalosa mine in the department of Vuitava, there is found a mixture of gray copper, red zinc-blende, famatinite (Cu_3SbS_4), antimony glance and realgar, with quartz and barite.

The occurrence of famatinite associated with enargite (Cu_3AsS_4) also characterizes many veins in the Argentine Republic. Such veins were described by A. W. Stelzner[1] at the Cerro de Famatina in the province de la Rioja. They occur in clay slate broken through by intrusions of dacite and andesite. Rich silver ores occur but very scantily and the ore deposit has the character of a copper formation. Besides enargite, which is decidedly the prevailing ore, the veins also carry famatinite, pyrite, copper pyrite and zinc-blende, with a little proustite, in a gangue composed of quartz, hornstone and barite. The mines are now worked mainly for copper. The Mejicana, Upulongos and Capillitas are the best known.

In the Old World, this type of copper-silver vein, so widely prevalent in South America, presents conditions of a widely different nature. It is probably developed in the remarkable silver mines of Smeinogorsk or

[1] A. W. Stelzner: 'Beiträge zur Geologie der Argentin. Republ.' 1885. Vol. I, pp. 216, 232.

Schlangenberg (snake mountain) in the Altai Mountains of Siberia described by B. von Cotta[1].

The mines lie in a region of Paleozoic rocks, traversed by porphyries and porphyrites. At Snake mountain, there is a vast interbedded ore deposit which expands locally into the form of a stock in the midst of these strata. It is composed essentially of barite and some quartz, and contains rich silver ores, such as native silver, silver chloride, argentite, proustite, miargyrite and highly argentiferous gray copper, as well as some native gold, together with copper ores, such as glance, copper pyrite and secondary copper ores, as well as galena, pyrite and zinc-blende. The footwall consists of a fossiliferous Devonian hornstone, swarming with stringers of ore-bearing barite, and representing perhaps a silicified limestone. The hanging consists of clay slates of the same formation. The mass of the vein, which is still found here and there to be 20 m. (65.6 ft.) thick, and is said to be still thicker at the outcrop, may be divided into five streaks, defined by the richness of the ore; the uppermost consisting of pure barite and containing but little silver; the second, a mixture of baryta carrying silver ores as very fine dust; the third, a mixture of ore-bearing barite and hornstone; the fifth consisting of hornstone traversed by ore-bearing barite veins; and, finally, the hornstone of the footwall, which is devoid of ore. The deposit is traversed by porphyrite dikes, 1-3 m. thick.

The mineral veins of Smeinogorsk, which were worked by the Tschudes in prehistoric times, were rediscovered in 1742, and have since then been imperial crown land, being worked with varying success.

15. Veins of Rich Silver-Cobalt Ores.

In this type of vein, the gangue is not everywhere developed in the same way. In one variety, seen at Joachimsthal, Austria, it consists mainly of quartz, hornstone, calcspar and dolomite; in the other, as at Annaberg, barite and fluorspar are added to the first named minerals. The ore minerals, on the contrary, are everywhere uniform in occurrence, consisting of cobalt, nickel and bismuth minerals, as well as uranium ores, which usually occur associated with the non-carbonate gangues that form an older part of the lode filling; also of rich silver ores, which are wont to represent a younger generation. These veins belong to the older geologic formations. In the Erzgebirge they are found about the borders of granitic

[1] B. von Cotta: 'Schlangenberg-Silbergrube im Altai.' *Berg u. Hutten Zeit.*, 1869. No. 28. Also, 'Der Altai.' Leipzig, 1871, p. 192. C. Griwnak: 'Les gisements de minérais dans l'Altai.' *Journ. d. Min.*, 1873. Vol. II, No. 5, 6, pp. 172-265; 1875, Vol. II, No. 6, pp. 277-311. (In Russian.) G. Mayer: 'Les mines d'argent de l'Altai.' *Compte-rendu* du Conseil de la Soc. Amat. Investig. d'Altai, 1891-1893, p. 31.

intrusive masses of late Paleozoic age, with which they are genetically connected. They are traversed by Tertiary basalts. Analogous veins found at other localities and of similar age are of no economic importance, as compared to the Erzgebirge.

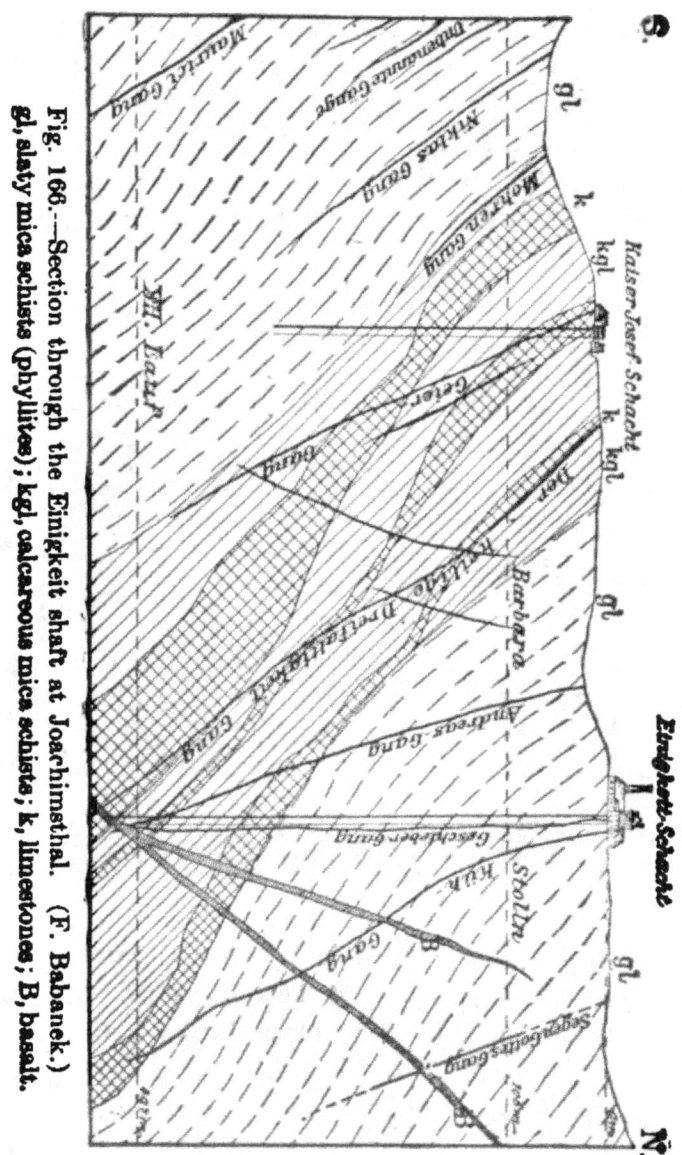

Fig. 166.—Section through the Einigkeit shaft at Joachimsthal. (F. Babanek.) *gl,* slaty mica schists (phyllites); *kgl,* calcareous mica schists; *k,* limestones; *B,* basalt.

For our first example of this formation we will select Joachimsthal, Bohemia, where at the present time mining operations are confined to veins of this type.

The Joachimsthal district lies on the higher southern slope of the Erzgebirge, southwest of its highest elevation, the Keilberg, 1,238 m. high, near the Saxon boundary. The rocks are mica schists, with east-west to westnorthwest foliation and north dip. In the vein area finely crystalline, slaty mica schists predominate, with intercalated layers of calcareous mica schists, crystalline limestones and coarse-fibered mica schists. (See Fig. 166.) Toward the northeast and east gneisses occur, while toward the southwest the schist ends abruptly against granite, the northeast contact of the great Eibenstock-Karlsbad massive of tourmaline granite, which crosses the axis of the Erzgebirge. At its border a contact zone is developed, which, however, does not quite extend into the Joachimsthal mineral vein area. Numerous dikes of quartz porphyry, sometimes quite large and running northwest to north-northwest, traverse both the region northeast of the

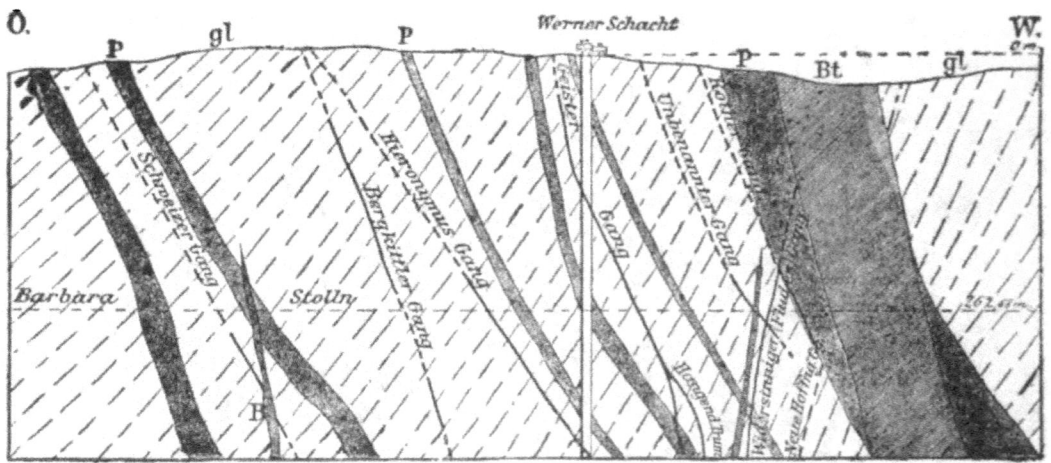

Fig. 167.—Section through the Werner shaft at Joachimstahl. (F. Babenek.)
gl, mica schist; P, quartz porphyry; B, basalt; Bt, basalt tuff (Putzenwacke)

granite, and the mineral area. Dikes of basalt and phonolite, as well as bosses of these rocks, were intruded in Tertiary time. The tremendous earthquake shocks that must have taken place during that period have left their record in wide fissures extending deep into the rocks, which must have remained open for long periods, since volcanic, phonolitic and basaltic material, together with silicified deciduous trees, that is to say, débris washed from above, were found in the Joachimsthal deep levels within the so-called *wacke lodes* in mica schist, particularly in the Barbara drift, at a depth of 262 m. (859 ft.) (See Fig. 167.) This remarkable fact was mentioned as early as 1557 by Mathesius, the preacher miner and reformer of Joachimsthal.

According to their strike, the silver-cobalt veins of Joachimsthal, especially those close to the town on the northwest, fall into these two groups:

(a) Morgengange, striking eastnortheast (7 hours) and dipping north at 60-80°; (b) Mitternachtgange, striking northnortheast to northnorthwest, for the most part, almost exactly north-south (11-1 hours) and dipping at times 45-85° west or east.

Some of the lodes of the first group, for example, the Geyer and Mauritius veins, were very rich in the upper levels, but grew poorer in depth, so that they became of little importance in the modern mining period. On the contrary, some of the veins of the second group have in recent decades been worked by deeper drifts with success, particularly the Geistergang, which showed fine bodies of silver ore, especially in 1853; also the Rothe, Prokopi, Anna, Geschieber, Hildebrand, Junghäuerzecher and the Evangelisten veins.

The width of the veins varies between 15 and 60 cm. and only exceptionally reaches 1-2 m. (3.28 to 6.56 ft.) Stringers are common. Some of the Mitternacht veins reach the surface not as true veins, but as narrow, barren cracks[1].

The filling is not the same in all the veins. Thus in the western Mitternachtgänge it is for the most part a brittle clay, with quartz and hornstone; in the eastern veins it is mostly calcspar and dolomite, while both sets of veins occasionally show a brecciated structure. The ores in these gangues form stringers, branches and pockets that are very spotty. According to G. Laube, these ores may be divided into:

1. Silver ores (native silver, argentite, polybasite, stephanite, tetrahedrite, proustite, pyrargyrite, sternbergite, argentopyrite, besides rittingerite, acanthite and cerargyrite).

2. Nickel ores (niccolite, chloanthite, millerite).

3. Cobalt ores (smaltite, as well as bismuth-cobalt-pyrite and asbolate).

4. Bismuth ores (native bismuth, as well as bismuth glance and bismuth ocher).

5. Arsenic ores (native arsenic, arsenopyrite).

6. Uranium ores (pitchblende).

Galena, zinc-blende, pyrite, marcasite, copper pyrite and bornite only occur subordinately and occasionally. Among these ores, the cobalt and nickel ores appear to be on the whole the older, the silver ores the younger. From the numerous sections of the lode published by F. Babenek, we have chosen the one shown in Fig. 168, which exhibits the characteristic distribution of ores and gangues.

[1] Compare with W. P. Jenney: 'Mineral Crest,' *Trans.* Am. Inst. Min. Eng., Vol. XXXVII, p. 46; discussion, p. 1060.

Near the lode fissures, the country rock frequently has been impregnated with extremely finely divided ore particles. This explains the small percentage of metals shown by F. Sandberger and A. Seifert to exist in various rocks of Joachimsthal, especially copper, cobalt, nickel and arsenic. In like manner the presence of minute granules of uranium pitchblende in a scapolite-mica-schist of that locality, first discovered by F. Sandberger and conclusively confirmed by F. Babenek and A. Seifert,

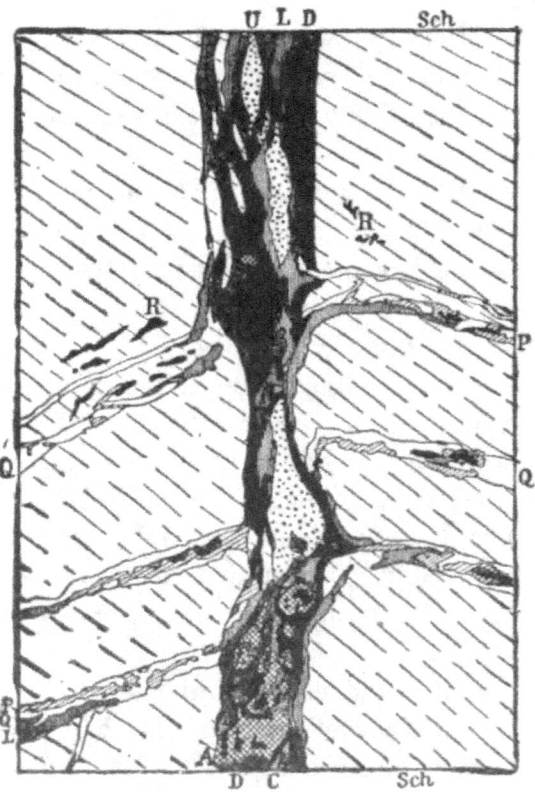

Fig. 168.—Cross-section of the Hildebrand lodes at Joachimsthal. (Babanek-Nemecek.)
Sch, schist; D, dolomite spar; C, calcite; Q, quartz; L, vein-clay; R, pyrargyrite; U, pitchblende; A, argentiferous arsenic ore; P, pyrite. The vein crosses three old lodes with quartz filling.

by means of large-scale ore concentration experiments, is most naturally explained by an infiltration from the lode fissures. Their presence as primary constituents seems to be gainsaid by their very unequal distribution in the rock. In the experiment just noted, the dressing of 6,358 kg. from a richer zone gave 226 kg. of concentrates with a content of 0.3% of uranic binoxide.

The veins traverse dikes of quartz porphyry, and in their turn are cut across by dikes of basalt and the *wacke* veins. However, since these *wackes* sometimes contain some interspersed argentite (earthy silver glance) where they cross the ore veins it is inferred that at the time of the eruption of the younger volcanic rocks the vein formation had not yet been quite completed. It may also be mentioned here that in 1864, according to Fr. Weselsky, a spring with a temperature of 23° Reaumur (84° F.) was tapped on the Geschieber lode at a depth of 531 m. (1741 ft.)

At the acute-angled lode crossings there has been an enrichment of the ore. The veins are richer in the porphyry and to the east in the limestone intercalation than in the schist.

The Joachimstahl mining industry began, according to G. Laube, at the end of the 15th or in the early years of the 16th century. As early as 1517 a mining settlement existed in the valley, and as early as 1518 the first "Joachimsthaler" were minted, this coin being now known as the thaler. In 1520 the settlement obtained the privilege of a free mining town. The total output for the first 44 years is put at 40 tons of gold, that is to say, over four million gulden, reckoning silver at the value prevailing at that time. After 1545 the industry suffered a great decline, but acquired new vigor when the cobalt and bismuth ores became valuable. In recent time the industry has languished because of the depreciation of silver. During the last decades, special attention has been paid to the extraction of uranium, for which a government factory has been set up at Joachimsthal. In 1898, 50.9 tons uranium ore was obtained[1].

Conditions similar to those of the Joachimsthal lodes were found to prevail on the Saxon side of the Erzgebirge in the silver-cobalt veins of the vicinity of Annaberg, described in the detailed monograph of H. Müller[2].

In this vein area the rock consists mainly of gray gneiss (the "bimicaceous principal gneiss" of the geologic map). The district forms the southwest part of the Annaberg-Marienberg gneiss dome. In this dome, at the southwest periphery of the area, at Buchholz, a granite stock is exposed, from which a tongue of fine-grained granite (porphyritic microgranite) runs through the mineralized district northeastward in the form of a dike. Similar smaller dikes of this rock have been met with at

[1] Some of the more important publications upon Joachimstahl are: G. Laube: 'Aus der Vergangenheit Joachimsthals,' Prague, 1873. Also, 'Geologic des böhmischen Erzgebirges,' Prague, 1876, p. 176–192. F. Babenek: 'Ueber die Erzführung der Joachimsthaler Gange.' *Oesterr.* f. B. u. H., 1884, pp. 1, 21, 61. Also, 'Die Uranhaltigen Skapolith-Glimmerschiefer von Joachimsthal,' 1889, p. 343. Also, Geologic. bergm. Karte mit Profilen und Bildern von den Erzgangen in Joachimsthal.' Published by the Imperial Royal Ministry of Agriculture, 1891.

[2] H. Müller: 'Die Erzgänge des Annaberger Bergrevieres.' Geolog. Survey. Leipzig, 1894.

several points above ground and underground. At the northeast edge of the field, at Wiesenbad, in the valley of the Zschopau, the erosion of the river has laid bare an intrusive stock of granite, so that this rock may be presumed to be rather widely distributed in the deeper levels of the mining area. Furthermore, mining operations very frequently encounter narrow dikes of lamprophyre (fine-grained mica syenites and mica diorites), at times also basalt dikes. The latter are genetically connected with the sheet of leucite basalt, which resting on a substratum of fluviatile clays, gravels and sands of the Oligocene, forms the tomb-shaped Pöhlberg immediately southeast of Annaberg.

More than 300 veins traverse the rocks near the town and in the district about it. They may be divided according to age into two groups. Of the older, it may suffice to mention briefly the tin lodes of the former Alte Thiele Fundgrube near Buchholz and the pyrite blende lead veins, which, because of their unusual richness in copper-pyrite and other copper ores (see page 238), were worked in the Sanct Briccius, Heilige Dreifaltigkeit, Weinkeller, Rothe Pfütze and Spanier mines on the east slope of the Pöhlberg and at several other points. The strike of these veins is usually about east-west. They have always proved to be younger than the lamprophyres, but older than the basalts.

On the other hand, the younger group consists of the silver-cobalt veins, in which we are here especially interested, and of examples of the iron and manganese veins which are developed farther southwest, in the region of Scheibenberg.

The silver-cobalt veins are the most important ore deposits of the Annaberg field. Besides those of the Annaberg district itself, the only one which we will here consider, there are others at Schmalzgrube and Steinbach in the Pressnitz valley, at Jöhstadt, at Bärenstein and Weipert, and at Oberwiesenthal, where they form a transition to the great Gottesgabe-Joachimsthal mining field, and a sixth area near Scheibenberg. In all these fields, mining is no longer carried on.

The silver-cobalt veins occur in two principal series, whose directions intersect at nearly right angles. Most of them strike northnorthwest, a less number strike eastnortheast to east-west. Only a few have a diagonal strike of north-south to northnortheast. The northnorthwest veins belong exclusively to the silver-cobalt class; the veins of other courses are in part of different characters and age. The veins of the northnorthwest series have usually a steep dip, those of the east-west series have a dip of but 45-60° in different directions. Most of the veins have been followed for a horizontal distance of 800 m. (2,624 ft.), some much farther, the Treue Freundschaft Stehende at least 2 km. (1.2 miles). On the other hand, mining operations on these lodes have penetrated to no great depth, the deepest, about 400 m. (1,312 ft.), being in the Erstneuglück Flachen and the Heynitz Flachen. The thickness varies between 10 and 20 cm., in exceptional cases amounting to 2 meters (6.5 feet).

The vein filling consists mainly of barite, fluorspar, quartz and brown-spar with various cobalt, nickel and bismuth ores, especially chloanthite, smaltite, gersdorffite, annabergite (nickel-ocher) and native bismuth, rarely bismuthinite, rich silver ores, especially pyrargyrite, proustite, argentite, earthy silver glance, native silver, silver chloride, and finally pyrite. The subordinate gangue minerals include hornstone, chalcedony, amethyst, calcspar, aragonite, kaolin, lithomarge, gypsum; among the ores, copper pyrite, galena, zinc-blende, marcasite, gray copper, siderite, uranium pitchblende, uranochalcite, gummite, zippeite, native arsenic, with others of minor interest.

An interesting feature of the vein is the occurrence in former times of great masses of silver chloride; for example, in the 16th century, in the Himmlisch Heer vein, masses of pure horn-silver were found as much as 20 pounds in weight.

The veins are frequently filled with fragments of country rock and the brittle inter-vein rock and attrition clays. Argentiferous vein clays have sometimes been mined. The prevailing structure of the vein filling is irregularly massive.

According to H. Muller, the result of more than 200 observations showed as a rule a definite succession in the deposition of the minerals of these Annaberg silver-cobalt veins, which he summarized as follows:

V. Decomposition products such as annabergite and cobalt efflorescence.
IV. Rich silver ores and native arsenic.
III. Calcspar and uranium pitchblende.
II. Brownspar and cobalt-nickel-bismuth ores.
I. Barite, fluorspar and quartz.
I and II together always form the bulk of the filling.

As already stated, the silver-cobalt veins cut through the tin-bearing and pyritic lead veins, as well as the dikes of microgranite and lamprophyre. The basic dykes are often accompanied by silver-cobalt veins. Thus the Heynitz Flache vein follows a lamprophyre dike 0.1-3 m. thick, for almost 600 m. along its strike and to a depth of 340 m. It follows first one wall, then the other, and in places crosses the dike. On the other hand, both this and other examples of this type of vein are cut across by basalt. The basalts encountered in the mines not only form true dikes, but also occur as vertical chimneys or pipes of round cross-section. As rich silver ores occasionally occur interspersed in fissures in the midst of the basalt mass near vein crossings, as at Oberwiesenthal, it is inferred that chemical transpositions of the metallic compounds took place even later on.

The various relations of veins and eruptive dikes are illustrated by the accompanying Fig. 169.

An important feature in the mining industry of Annaberg is the presence of the so-called "Schwebenden," that is to say, sheets of decomposed rock ordinarily parallel to the foliation of the gneiss, but sometimes also transverse to it. These sheets are from 1 cm. to 2 meters thick, are ordinarily colored blackish by fine earthy, carbonaceous substances, and often contain finely interspersed pyrite and sometimes copper pyrite. As indicated by their name, they are distinguished by a flat dip. The "obere Schwebende," of the Markus Rohling Fundgrube at Schreckenberg, dips in

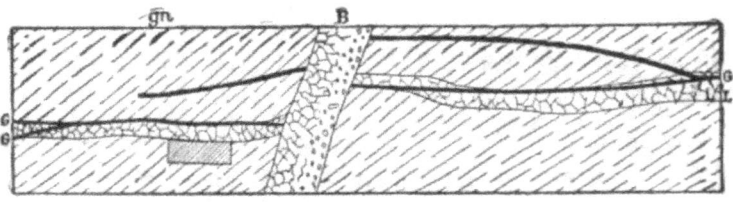

Fig. 169.—Ground plan of a drift in the Neu Unverhofft Glück at the Luxbach Annaberg. (H. Müller.)

gn, gneiss, micaceous; L, lamprophyre (fine-grained mica syenite); GG, lodes of the rich silver-cobalt ore formation; B, basalt, with fragments of gneiss and lamprophyre on one side.

various directions at 15-20°. Where the silver-cobalt veins cross these Schwebenden, they often show an extensive enrichment, as appears by the mine section, Fig. 170. According to H. Müller, at least half the entire output of silver and cobalt is derived from ore bunches found at these crossings, especially in the Bäuerin, Heilig Kreuz, Zehntausend Ritter, Himmlisch Heer, Galiläische Wirthschaft, König David and Markus Röhling mines. Besides these, there are others at the intersections of various mineral veins, and the junctions of various stringers are marked by the occurrence of rich ores.

The Annaberg veins are said to have been first opened for copper mining in the first half of the 15th century at the town of Geyersdorf, near the Pöhlberg. On October 27, 1492, a prospector discovered rich silver veins at the Schreckenberg, near Frohnau. By 1496 the "Neue Stadt am Schreckenberge" was formed by the great influx of foreigners. In 1497 the town received the name of Sanct Annaberg. In 1495 silver mines were opened in the beech forest of the Schottenberg, which led in 1501 to the foundation of Sanct Katharinenberg in Buchholz, subsequently abbreviated to Bucholz. According to H. Müller, during the greatest period of prosperity of the Annaberg-Buchholz mining industry, from 1496-1600, a total of approximately 1,352,900 marks of silver and 48,460 centners of copper were obtained, having a total value of about 24,300,000 marks ($6,000,000) in present coinage. The industry declined from 1560 until towards the middle of the 17th century, when the utilization of the cobalt ores caused a gradual revival of mining activity. Cobalt mining was particularly active from 1701 to 1850, during which time 25,524 centners of cobalt ore were

Fig. 170.—Longitudinal Section of the Heynitz Flache vein, Markus Röhling mine, Annaberg. (H. Müller.)

B, intrusive mass of basalt; Schw., "Schwebende"; M, various Morgengänge.

produced. The silver produced as a by-product during this time amounted to 26,945 kilograms. Since 1850 the mining industry has been at a low ebb.

Silver-cobalt veins as well as simple cobalt veins occur also at Schneeberg proper, as described farther on.

The same is true of the mining field of Johanngeorgenstadt. At the Fastenberg, the Eibenstock tourmaline granite is concealed by a cover of phyllite, which it has altered by contact metamorphism into spotted slate and andalusite-mica rock. These contact rocks are traversed by many dikes of fine-grained granite. Numerous mineral veins form a dense network in this rock, some belonging to the tin-cobalt and some to the silver-cobalt formation, as well as pure cobaltic veins, all showing but little uniformity in strike. The veins are cut off by a vast network of iron ore stringers, the so-called Faule (rotten lode). The filling of the silver-cobalt veins of this locality is similar to that at Annaberg. They are often rich in uranium pitchblende.

Johanngeorgenstadt, the youngest mining town of the Erzgebirge, was founded as late as 1654 by the Protestants who were driven out of Bohemia, but mining began a few years earlier. It still continues, and produced in 1901 nine tons of bismuth and uranium ores.

The group of the silver-cobalt ore lodes also includes the deposits of Wittichen (Kinzig valley) and of Wolfach in the Black Forest, described by Vogelgesang[1] and F. Sandberger[2].

(ʃ) Gold Veins.[3]

The gold veins are exceedingly varied in their composition, but may be grouped according to the prevailing gangue into: (1) Gold quartz veins; (2) Quartz-calcite veins, which contain other carbonates and rarely barite also; (3) Quartz-fluorite veins. The gold quartz veins contain a variety of gold-bearing sulphides and other minerals besides free gold. The predominance of a particular sulphide gives a special character to the particular example and thus various sub-divisions of the group may be distinguished. This differentiation of separate vein-types may be of practical interest, because of a corresponding difference in metallurgical treatment. Of course, these sub-classes are connected by transitions. The most important of them are the following:

Gold Quartz Veins (the Gold Quartz Formation).

(a) Pyritic gold-quartz veins in which pyrite prevails or at any rate predominates as compared to the other sulphides.

(b) Cupriferous-gold-quartz veins in which copper ore is prominent.

(c) Antimonial gold-quartz veins in which stibnite is abundant.

(d) Arsenical gold-quartz veins in which arsenopyrite is the predominant sulphide.

---

[1] Vogelgesang: 'Geogn.-bergm. Beschr. des Kinzigthaler Bergbaues.' Carlsruhe, 1865, p. 19.

[2] F. Sandberger: *Neues Jahrb. f. Min.*, 1868, p. 385; also the volume for 1869, p. 290.

[3] A general discussion of the occurrence of gold in veins is given by Fuchs and De Launay: 'Traité,' II, 1893, pp. 890–893. ('Généralités sur les filons d'or.') E. Cumenge and F. Robellaz: 'L'or dans la nature.' Paris, 1898. Ed. Suess: 'Die Zukunft des Goldes.' 1877.

(e) Cobaltiferous-gold-quartz veins in which smaltite holds the gold.

Silver-Gold Veins.

The second main group, in which the gangue consists of quartz, together with earthy carbonates and sometimes barite, is also sharply defined by the fact that both gold and silver ores occur together, and gold-bearing and silver-bearing tellurides also occur very often.

Fluoritic Gold Quartz Veins.

The prevalence of fluorspar intermixed with the quartz and the presence of tellurides marks this group.

We will now give a detailed description of the different groups and cite examples of them.

16. GOLD QUARTZ VEINS.

These are simple or compound lodes, mostly of massive structure, and with the particles of ore irregularly distributed, or more rarely showing a banded structure, the individual thin or thick-sheeted layers being in such case divided by scales or films of ferric hydrate with free gold, or by talc, sericite, chlorite and other decomposition products. As shown by W. Lindgren, a *ribbon structure* is often present, having been produced by secondary movements and the formation of cracks, parallel to the strike, in the vein mass. This is proven by the pressed and crushed pyrite particles lying along the plates, and also by the forking of the joints between the quartz bands. In lodes that are rich in sulphides, however, a primary banding also occurs occasionally, in which case the ores are strongly concentrated along special planes parallel to the selvage.

The nature and habit of the quartz, the only essential gangue, varies greatly, and it is not possible to formulate any set of characteristics for the gold-bearing variety of this mineral, that may be applicable in all regions. In general it is a micro-crystalline, whitish, faintly translucent mass, appearing opaque to the naked eye, tinted grayish by finely divided sulphides in the deeper parts of the veins, and yellowish or rusty colored from the oxidation of those sulphides in the upper portions. The luster varies from glassy to greasy.

On examining a thin section of this with high powers of the microscope (see Fig. 171) the various individual grains show under polarized light a very irregular outline, which is but rarely formed even in part by crystal planes. Many specimens exhibit the marks of a strong secondary pressure, and the aggregations are often traversed by delicate, dark streaks representing slip or fracture planes, being minute fissures of displacement filled with extremely fine grains of crushed quartz, and ordinarily impregnated and colored with various ores or the products of their decomposition (see Fig. 171 c). Posepny[1], who described such minute shear planes in the quartz from the Tauern gold veins, explained these fissures as the result of a shrinkage of the mass in passing from an opal-like condition into a crystalline quartz, analogous to the process advocated by Fuchs and Breithaupt[2]

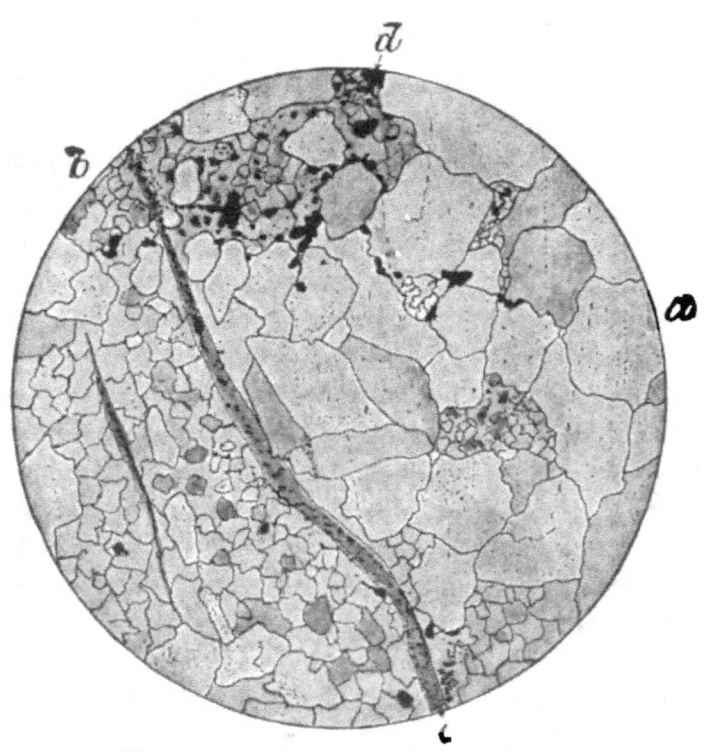

Fig. 171.—Gold quartz from Sheba mine.

a, large quartz crystals with enclosed liquids; b, small quartz grains formed by secondary crushing; c, brownish gliding zone, consisting of finely ground quartz ferric hydrate and gold particles; d, black spots are free gold.

on a large scale to explain the so-called Quarzbrockenfels of the Erzgebirge. It appears unlikely that this explanation will maintain itself.

Individual crystals of the gold-quartz are usually rich in liquid inclusions. As the bubbles disappear in most cases on heating the sections to 30° C., and reappear on subsequent cooling, it may be inferred that the liquid is carbonic acid. This, however, is not true in all cases. From a

[1] F. Posepny: 'Goldbergbaue der Hohen Tauern,' *Archiv f. Prak. Geol.*, I, Freiberg, 1880, p. 46.

[2] A. Breithaupt: 'Paragenesis,' 1849, p. 9.

chemical analysis by G. Steiger, Lindgren[1], for example, inferred that such liquid inclusions consisted of sulphates of calcium and alkalies, mixed with a small amount of chlorides. It is remarkable that the lines into which these inclusions are so often grouped sometimes pass through several adjoining quartz individuals without regard to the contours. W. M. Courtis[2] tried to establish definite relations between the microscopic structure of the gold quartz and the gold content of the lodes concerned, but was unable to find any rule of practical value.

Through the leaching out of the sulphides in the gossan, the gold quartz becomes porous. Such occurrences are sometimes erroneously designated as silicious sinter.

True sinter is probably deposited by the silicious water of some gold-bearing veins or lode-like deposits, though deposited only in the uppermost regions. The pure white, porous and even foam-like silicious sinters, suggesting pumice stone, seen in the gold deposit of Mount Morgan in Queensland and at some other Australian localities, are occurrences of this sort, which may without hesitation be compared with the deposits of some geysers, an assumption which is confirmed by the amount of water shown to exist in them.[3] Opal, too, has been described from gold quartz lodes, for example, in California. Its presence in veins need not be considered surprising, since opal, accompanied by gold-bearing pyrite and stibnite, has been observed in process of deposition by the spring waters of Steamboat Springs (see under 'Thermal theory') close below the surface of the earth, at the Monarch Geyser in the Yellowstone Park, and Boulder Hot Springs, Montana, and it is known that the thermal springs of Taupo in New Zealand produce a slightly auriferous silicious sinter.

Sometimes chalcedony is observed, for example in the gold ore veins of Donnybrook[4] in Western Australia. Here the chalcedony-quartz aggregates formed at first were afterward leached out in such way that a loose gold-bearing quartz dust remained.

For that matter, besides quartz, carbonates are not unknown in lodes of this group, as for example at Nevada City and in Grass Valley in California, and Idaho. However, they always play merely a subordinate part.

The native gold of quartz lodes is usually interspersed in the quartz grains in exceedingly fine scales, dust-like particles or small crystals, or

[1] W. Lindgren: 'Gold Quartz Veins of Nevada City,' etc., 17th *Ann. Rep.* U. S. Geol. Survey, 1896, p. 130.

[2] W. M. Courtis: 'Gold Quartz.' *Trans.* Am. Inst. Min. Eng., Vol. XVIII, 1890, p. 639.

[3] Weed, Walter Harvey: 'A Gold-bearing Hot Spring Deposit.' *Am. Journ. Sci.*, 3d Series, 1891, Vol. XLII, pp. 166–169

[4] F. Beyschlag and P. Krusch: 'Die Goldgänge von Donnybrook in W. A.,' *Zeit. f. Prak. Geol.*, 1900, pp. 169-174.

lies wedged in between the surfaces of the quartz grains. More rarely is it found as dendrites, sheets, tangles of wires or crystalline lumps. The latter may in exceptional cases, as in that of a specimen from Monumental Mine, Sierra County, California, attain a weight of 43.08 kilograms. (See discussion of gold nuggets under the head of placer gravels, farther on.) Such nuggets are always jagged in outline and usually coated with crystals of gold. A special form is exhibited by the aggregations of gold derived from the decomposition of tellurides. They form dark yellow moss-like or spongy crusts, which, because of their color, are popularly termed "mustard gold."

At times the gold is so exceedingly finely divided as to suggest that it is contained as a silicate in the quartz of some lodes. This assumption is unnecessary, since the fineness of the flakes may far exceed our ordinary visual power. J. A. Edman,[1] using very strong magnifying powers, was able to detect, in quartz, particles of gold dust whose diameter was only from 1/40 to 1/480 of a millimeter.

All the gold of this type of veins contains silver. The degree of fineness differs greatly. In California, for example, in the case of gold dust it is 0.850 to 0.870, while in larger nuggets it is up to 0.950.

The sulphides, which often occur in great abundance in the lodes, and are rarely entirely absent, are developed either in finely interspersed crystals and granules, or in compact aggregates. The most widely distributed sulphide is iron pyrite, which is sometimes, but rarely, replaced by pyrrhotite. Associated with the pyrite there is also copper pyrite, galena, zinc-blende, arsenopyrite, stibnite, as well as molybdenite, all of which are apt to contain gold and silver.

The lodes often enclose fragments of country rock, which in such cases are usually more or less strongly impregnated with gold, auriferous pyrite and other ore minerals, as is also the case with a zone of the adjoining country rock along the vein-walls.

Gold quartz veins occur in all kinds of rocks, but are most frequent in crystalline schists. They most often occur in regions where the older schistose rocks are broken through by granitic, dioritic and diabasic rocks and are genetically connected with such intrusions as will be shown in describing the various examples.

The genetic connection of many gold quartz lodes with granitic intrusive masses has been brought into prominence by late researches. They cluster about the borders of the granite areas in Montana, and elsewhere, even the ancient veins of the Appalachian gold belt showing this association. It has been shown that some of these lodes show a close relationship to pegmatite

[1] Cited by E. Cumenge, *op. cit.*, p. 40.

veins. J. E. Spurr[1] has shown that gold quartz veins contain a great number of such accessory minerals as are also characteristic of pegmatite veins, although the amount is trifling compared to the gangue proper, the quartz. Thus among the 62 mineral species enumerated by G. F. Becker[2] from the gold veins of the southern Appalachians, there were, among others, tourmaline, cassiterite, apatite, orthoclase, albite, garnet and scheelite. In the Fortymile and Birch Creek districts of the Yukon, Spurr observed what he considered to be transitions from gold quartz lodes to aplites, that is to say, finely crystalline granites, poor in mica (alaskite), which often passed into pegmatites. These transitions take place very gradually by an increasing abundance of feldspar. This phenomenon only became intelligible after the work of W. O. Crosby, M. L. Fuller, W. C. Brögger and others had afforded us a better understanding of the probable conditions under which pegmatites are segregated. These conditions must have been intermediate between those under which an acid eruptive rock congealed, and those under which an ordinary quartz vein was segregated. As the water and various gaseous compounds become more and more concentrated in a residual "mother liquor" during the crystallization of the magmas, these residual solutions, charged with silicic acid, penetrate from the contact into fissures and deposit vein quartz, together with metallic and nonmetallic compounds, which were previously uniformly distributed in the molten magma, but gradually retreated into this residual water.

The reader should compare these deductions with what was previously said (page 14) on the processes of magmatic differentiation.

Independent of the above mentioned American authors, E. Hussak expressed similar views in regard to the deposits at Passagem (see later).

From this point of view, the genesis of such gold quartz veins is analogous in nature, though of very different material, to that of tin veins, whose genesis is discussed farther on. The statements just made also apply particularly to the cuprous gold quartz veins characterized by the presence of tourmaline.

I. *Pyritic Gold Quartz Veins.*[3]

A typical development of this class is found in the famous "gold belt" of

[1] J. E. Spurr: 'Geology of the Yukon Gold District, Alaska.' 18th *Ann. Rep.*, U. S. Geol. Surv., III, 1896–97, p. 298, *et seq.*

[2] G. F. Becker: 'Goldfields of the Southern Appalachians.' 16th *Ann. Rep.*, p. 274.

[3] M. Laur: 'Du gisement et de l'exploitation de l'or en Californie.' *Ann. d Mines*, 1863, Ser. VII, Vol. III. J. D. Whitney: 'Geol. Surv. of California,' 1865, Vol. I. H. W. Turner: 'Notes on the Gold Ores of California.' *Amer. Journ. of Sc.* 1894, pp. 467–473. 'Geolog. Atlas of the United States,' Californian folios, 1894, etc., W. Lindgren: 'The Gold-Silver-Veins of Ophir, California.' 14th *Ann. Rept.* U. S. Geol. Surv., 1894, pp. 249–284. *Ibid.*: 'Characteristic features of California Gold-'

California. This very extensive mineralized area lies on the west flank of the Sierra Nevada, sloping gently toward the great longitudinal valleys of the Sacramento and the San Joaquin rivers.

This great mountain range is essentially composed of slates of Carboniferous (Calaveras formation) and Jura-Triassic (Mariposa formation) age, which are everywhere broken through by granite rocks and altered by contact-metamorphism. Black slates with limestone intercalations and diabasic intruded masses, as well as sandstones, prevail in both the two formations. The beds appear overturned, probably in consequence of an overfold in an easterly direction, and a steep easterly dip prevails. The beds strike north, or toward the axis of the range, which runs northnorthwest. After some older folding and faulting, the principal uplift of the range began in late Tertiary time.

The gold veins are in the main parallel with the older folds, that is, parallel to the axis of the range. The main so-called 'mother lode' is a zone of veins 70 miles long, $6\frac{1}{2}$ miles wide, extending from Mariposa County (Mount Ophir) to beyond Placerville, and often following the contact zones of intrusive plutonic masses. It is, in fact, a collection of parallel quartz veins, often interrupted, which either coincide in strike with the black clay slates of the Mariposa formation, or cross them at a very acute angle. The dip is usually east at 50 to 70°, and follows in the main the dip of the slates. The thickness of the larger veins is often 10 m. (32.8 ft.) or more, but occasionally decreases to 1 m. (3.3 ft.) North of the Lawrence mine the mother lode breaks up into stringers, and finally resolves itself into a tangled network of small, often highly auriferous quartz stringers, associated with a definite slate zone (seam diggings). Other important veins occur independently of the mother lode, in the contact zone between the Calaveras slates and the granite at Grizzly Flat, southeast of Placerville, others to the north at Grass Valley and Nevada City. Some of the lodes follow for some distance the contact between the slates and intercalations of altered diabase rocks. The veins are sometimes displaced by cross faults, as observed by Laur.

Outside of the real gold belt of the Sierra Nevada, California contains various other gold veins of less importance, for example in the **Coast Range** at San Diego.

The workable ore occurs in pay shoots in true fissure veins, best developed in schists, and absent in serpentine. Very large outcrops of white quartz are low grade, lenticular, and pinch out. (Ransome.)

quartz veins,' 1895, *Bull.* of the Geol. Soc. of Am., Vol. VI, p. 221. *Ibid.*: 'The Gold-Quartz-Veins of Nevada City and Grass Valley Distr., Calif.' 17th *Ann. Rept.* U. S. Geol. Surv., 1896. pp. 13-262. B. Knochenhauer: 'Der Goldbergbau Californiens' ,Leipzig, 1897. F. L. Ransome : Gold Belt,' folio, U. S. Geol. Surv.

In all these lodes, the only gangue is quartz, which in the richer occurrences tends to separate into thin layers (ribboned quartz), while in the poorer ones it is more massive. In this quartz the gold is usually interspersed as very fine, hardly perceptible particles, more rarely in larger aggregations. The fineness of this gold is as a rule 850-870. The gold is always accompanied by gold-bearing pyrite, occasionally also by zincblende, galena, copper pyrite, arsenopyrite and tellurides, rarely by pyrrhotite, molybdenite, tetrahedrite, cinnabar and antimonial-lead ores. The concentrates contain 120-140 gm. (3½ to 4 oz.) gold to the ton, rarely more. The average amount of gold in the total amount of ore extracted is assumed by B. Knochenhauer to be 15-20 gm. (0.4 to 0.6 oz.) per ton. In the gossan, which extends to a depth of 40 m. (131 ft.), rarely to 60 m. (196 ft.), much higher values were found, 125-160 g. (3.7 to 4.7 oz.) per ton being not infrequent.

Though the amount of gold gradually decreased below the gossan, yet in the undecomposed zone itself no decided impoverishment or even barrenness has yet been observed in the Californian lodes. The shafts of Kennedy mine near Jackson City encountered workable gold ore even at 600 m. (1,968 ft.) to 655 m. (2,150 ft.) The distribution of the gold in the lodes, it is true, is not uniform. Ore shoots occur in overlapping lenses and ordinarily replace each other in such fashion that one wedges out downward, while a new one wedges in somewhat to the side.

The origin of these lodes is thought by Lindgren, one of the best posted authorities, to be independent of the country rock, and hence the veins cannot have been formed by lateral secretion as that theory is understood in its narrower acceptation, but rather by rising thermal waters, charged with silica, various carbonates and carbonic acid. This hypothesis is further strengthened by a study of the alteration of the vein walls and the included material. This rock is altered with the formation of carbonates and partial sericitization.

The auriferous wealth of California was discovered in 1848. The production continues to be considerable; in 1903 the total production was $16,363,000, of which $1,475,749 was produced by dredges working in the river bottoms, while $872,812 was the result of hydraulic mining and $905,679 of drift mining.

During the last 50 years, California produced gold to the value of $1,251,696,817, an annual average of $25,031,936. In 1898 the gold production was 22,418 kilograms, $15,906,478; for 1903 it is in round numbers $16,363,000, or including the $498,412 commercial value of silver alloyed with the gold, it is $16,861,412.

Gold veins of this type are also found in various other parts of the United

States, as in Idaho[1], Oregon, southern Alaska[2], the Yukon[3] gold district, and in part also in the southern Appalachians[4]. A special description of these areas, treated in a masterly way in the works cited below, would exceed the limits of space of the present work.

In South America an important mining district containing this kind of gold vein occurs at Callao, on the right bank of the Yuruari, south of the Orinoco delta, in the so-called Caratal district of Venezuela. The lodes of this region are remarkable in that the pyrite is subordinate in amount to the free gold, which is probably due to a specially deep secondary decomposition of the pyrites. The region exported from 1866 to 1886 the grand total of 55,862 kilograms of gold[5].

Pyritous gold quartz lodes are also worked in Dutch Guiana (Surinam) and in British Guiana.

A grand development of the pyritous gold quartz formation is witnessed in Australasia[6], both on the continent and on the islands of New Zealand and Tasmania. Out of a great number of examples, we can only select a few.

A large and rapidly rising gold production has recently drawn attention to the gold fields of West Australia, whose veins are mostly of the class under discussion.

The gold field of that region, which covers an area greater than Germany, occurs in the arid tableland of Western Australia, and owing to its lack of water and wood, presents great difficulties to mining enterprises. The country consists of upturned clay slates, quartzites, mica schists, phyllites, talcose schists and chloritic schists, with many diorite and diabase dikes, which have often been transformed by pressure into schistose amphibolites. In the Coolgardie district, east-northeast of the port of Perth, at Kalgoorlie and elsewhere, these rocks are traversed by many zones of decomposition, mostly striking north-northwest or north-northeast and penetrated by numerous auriferous quartz stringers. Simple gold quartz lodes, often developed as bedded or lenticular veins, sometimes traverse the country rock. While the stringer zones do not show at all on the surface, the simple quartz lodes, called 'reefs,' often form long and high quartz crests, a prominent

[1] Lindgren: 'Mining Districts of Idaho Basin,' 18th *Ann. Rep.* U. S. Geol. Survey, 1898. 'Blue Mts. of Oregon,' 22d *Ann. Rep.*, 1902.

[2] G. F. Becker: 'Recon. Gold Fields Southern Alaska,' 18th *Ann. Rep.* U. S. Geol. Survey, 1898, III, pp. 1–86.

[3] J. E. Spurr: 'Geol. of Yukon Gold Dist., Alaska,' 18th *Ann. Rep.* U. S. Geol. Survey, 1898, III, pp. 87–392.

[4] G. F. Becker: 'Gold Fields of Southern Appalachians,' 16th *Ann. Rep.* U. S. Geol. Survey, 1896, III.

[5] C. Le Neve Foster: 'On the Caratal Goldfield.' *Quart. Journ.* Geol. Soc., XXV, 1869, p. 236. Fuchs and De Launay: 'Traité,' II, pp. 896–902.

[6] K. Schmeisser: 'Die Goldfelder Australiens.' Berlin, 1897. Wolff: 'Das Australische Gold, seine Lagerstätten und Associationen.' *Zeitschr.* d. d. G. G., 1877.

feature in the landscape. The gold is free in some veins, as in the Great Boulder main reef at Kalgoorlie, and also occurs in calaverite and other tellurides of various kinds, and is everywhere associated with pyrite. Besides this, some arsenopyrite, galena, copper pyrite and some rarer ores are mentioned. The telluride lodes, according to Pittman, accompany eruptive dikes of porphyries now altered to schists. They form perhaps a special type and should not be included in the gold-quartz veins. The amount of gold in the quartz lodes averages, according to K. Schmeisser[1], 30-60 gm. per ton ($20-$40), and in the quartz stringers of the compound lodes and alteration zones 30-120 gm.; in the rest of the filling of these shear zones it reaches at most 30 gm. per ton. In exceptional cases, some mines for a while recorded a gold content of the ore treated in their stamp batteries of 90-470 grams per ton. The gold content decreases almost everywhere from the zone of oxidation downward. According to H. C. Hoover, at a depth of but 30 m. (98 ft.) it has declined to one-half, or even to one-fourth of the amount found at the top.

The remarkable residual placer gold deposits of this region, called cements, are discussed under gold-bearing gravels.

The gold production of West Australia in 1898 was 29,218 kilograms of fine gold. In 1902 it was 1,819,308 oz. or $82,454,344.

Most of the gold veins of Victoria belong to the pyritic gold-quartz type, with transitions to the arsenical type, the most important districts being Ballarat, Beechworth, Bendigo (Sandhurst), Maryborough, Castlemaine, Ararat and Gippsland. The veins occur mostly in Silurian slates and in many cases are closely related to dioritic eruptive masses. We have already described the remarkable saddle lodes of Bendigo in the Sandhurst district. This district also contains the largest deep mines of Australia. One shaft on these gold quartz lodes has, according to Lindgren, reached, in 1904, a depth of 4,000 ft.[2] Mention has already been made of the 'ladder' lodes of Waverley.

In New South Wales[3], veins of this class occur in the Wyalong goldfield, and in Queensland at Charters Towers[4]. In the latter area the pyrite is in part replaced by auriferous pyrrhotite. This same colony also contains the

[1] K. Schmeisser: 'Die Goldfelder Australiens.' Berlin, 1897. p. 43. A. Gmehling: 'Beitrag zur Kenntniss der Westaustralischen Goldfelder.' *Oesterr. Z. f. B. u. H.*, 1898, p. 161, and 'Ueber die Golderzlagerstätten von Coolgardie.' Ebenda, 1897, p. 425. Sloet van Oidruitenborgh: 'Technical observations upon the Coolgardie Goldfields.' *Mining Journal*, 1897, p. 819. P. Krusch: 'Zeit f. Prak. Geol.,' 1901, pp. 211-217.

[2] 'Gold Veins of Victoria,' *The Engineering and Mining Journal*, March 9, 1905.

[3] E. F. Pittman: 'On the Geol. Structure of the Wyalong Goldfields.' *Records* Geol. Survey, N. S. W., 1894, p. 107.

[4] R. L. Jack: *Report* on the Geology and Mineral Resources of the District between Charters Towers Goldfields and the Coast, 1879.

Crocodile goldfield, 42 km. (25.2 miles) southwest of Rockhampton, with the remarkable deposit of Mount Morgan, which deserves a few remarks.

Mount Morgan, "the greatest gold deposit of the earth"[1], rises 152 m. (498 ft.) above Linda (Mundic) Creek and 1,225 ft. above sea level. It is an obtusely conical hill, by far the greater part of which consisted of workable gold ore. Its summit has long been removed by open-cuts, and its deeper parts have been explored in all directions by drifts and stopes extending even below the bottom of the valley. Down to a depth of 90 m. (295 ft.) the ore is very uniform, consisting of bluish-gray quartz, silicious red hematite, colored light red to bluish black, brown hematite and manganese iron ore, together with white, vesicular, often foam-like, silicious sinter, kaolin and ocherous earth. Below 90 m. (295 ft.) this gossan changes into a gold-bearing and pyrite-bearing quartzite. The average gold content decreased constantly downward, toward the end very slowly, from about 115 grams to 40 grams per ton. The production of this single mine in 1896 was some 95,000 tons of ore with 4,560 kilograms of gold. In the year ending May, 1902, 213,907 tons of ore were treated, yielding 147,628 oz. of gold.

The form of the deposit is stock-like, tapering downward, and it is traversed by several dikes of dolerite, rhyolite and felsite which, according to T. A. Rickard, are genetically associated with the impregnation of the entire rock mass with gold-bearing quartz. On the other hand, the occurrence of silicious sinter, which seemed to indicate a deposition of dissolved silicic acid near or on the earth's surface, induced R. L. Jack to interpret Mount Morgan as a geyser formation[2]. It seems premature to pronounce in favor of either view, especially as the argument for the former does not seem to include a definite statement as to the petrographic character formerly possessed by the main rock mass of the mountain, which is now completely silicified and mineralized.

This class also includes most of the veins of the African goldfields. We will mention but one example, the famous Sheba mines[3] near Eureka City in the Transvaal, whose ores created universal astonishment by their remarkable richness, up to 250 grams per ton, when the work was as yet confined to the gossan. These mines lie in the DeKaap goldfield between Delagoa bay and the capital of the former Transvaal republic.

---

[1] R. L. Jack: 'Reports of the Queensland Geolog. Survey,' 1884, and 'Mt. Morgan Gold Deposits,' 1892. T. A. Rickard: 'The Mt. Morgan Mine.' *Trans.* Am. Inst. Min. Eng., XX, 1891, p. 133. K. Schmeisser: *op. cit.,* pp. 81–84.

[2] W. H. Weed: 'A Gold-bearing Hot Spring Deposit.' *Amer. Jour. Sci.,* 3d series, Vol. XLII, pp. 166–169.

[3] P. R. Krause: 'Kurze Schilderung der Grubenbezirke von Pilgrim's Rest.' *Zeit. f. Prak. Geol.,* 1897, p. 22.

Mention may also be made of the gold deposits in Matabeleland and Mashonaland, the ancient Semitic mining settlements of Zimbabwe[1], and, lastly, of the lodes exploited by the ancient Egyptians in the desert that lies between the Nile and the Red Sea[2].

Asia, in the Ural and the Amur country, presents numerous deposits of this kind. Among the Ural deposits belonging to this class the veins of the areas east of the mountains proper, at Cheleabinsk, where the great Siberian railway begins, are worthy of mention. The rock of this locality, according to A. Karpinsky[3], is a highly dynamo-metamorphosed granite, in part a hornblende granite. This is traversed by numerous gold-bearing veins, some striking northeast, some northwest. The vein filling consists of completely decomposed country rock and of quartz. The latter usually forms solid vein stringers 0.2-0.7 m. (0.6-0.22 ft.) thick, exceptionally 2 m. (6.6 ft.), or may also pervade the decayed granite in a network of veins. In the upper part of the vein the quartz contains free gold, but at a depth of 30-40 m. auriferous pyrite and arsenopyrite make their appearance. The gold content varies as a rule between 2.0 and 10.4 grams per ton. However, enrichments with 30 grams per ton and above have been encountered. Some of the veins, like those of California, are built up very regularly out of quartz layers 2-5 cm. thick. These veins are especially notable because of their gold-bearing pyrites and the presence of ocher enclosing free gold. Such a lode, when its dip is very low, as in the Ivanovsky vein, which dips north-northwest at 15°, is at first often taken for a bed.

Among the Siberian gold veins of this kind, we will merely mention that of Onon[4], west-southwest of Nerchinsk in Transbaikalia. The lodes there are associated with aplites traversing Paleozoic schists.

Of European gold quartz veins of this type, the Hohe Tauern veins of the eastern Alps are best known. The mines are still worked in Goldberg of Rauris and at the Rathhausberg, near Gastein[5]. Gneisses and other crystalline schists of the central Alps are traversed by narrow quartz veins, usually so thin as to be called 'leaves.' These contain extremely fine

[1] R. Rugg: 'Matabeleland, its Goldfields,' etc. London, 1891. *Idem.*: 'New map of Matabeleland, Mashona and Tati Goldfields,' London, 1891. A. G. Sawyer: 'The Goldfields of Mashonaland,' with 21 maps and plates. London, 1894. Th. Bent: 'The Ruined Cities of Masjonaland,' London, 1896. C. Peters: 'Im Goldlande des Alterthums.' Munich, 1902.

[2] Ch. A. Alford: Reference in *Zeit f. Prak. Geol.*, 1902, p. 9.

[3] 'Guide des excursions du VII Congres. Geolog. International,' 1897, V, p. 30.

[4] E. D. Levat: 'L'Or en Siberie Orientale.' I, p. 64.

[5] B. v. Cotta: 'Erzlagerstätten.' II, pp. 318–324. F. Posepny: 'Goldbergbau der Hohen Tauern.' *Archiv f. Prak. Geol.*, I, 1880, p. 1 *et seq.* R. Canaval: 'Das Bergbau-Terrain in den Hohen Tauern.' Klagenfurt, 1896. With bibliography. P. Krusch: 'Die Goldlagerstätten in den Hohen Tauern.' *Zeit. f. Prak. Geol.*, 1897, p. 77 *et seq.*

particles of free gold, with pyrite and a number of other sulphide ores, such as copper pyrite, arsenopyrite, galena, zinc-blende and stibnite, and very rarely molybdenite and various rich silver ores. Because of this highly varied mixture of ores, in which, however, pyrite greatly predominates, the veins are not so truly typical of the class as the Californian or Australian gold quartz lodes. They form a transition from the pyritic-gold-quartz veins to the pyritous lead formation. A feature of special importance in these Tauern 'leaves' is the strong impregnation of the wall-rock with ore, which in some cases has led to the development of banded masses resembling beds, as at the Heinzenberg, near Zell, in Tyrol. The production of the Rathhausberg in 1898 was only 76.7 tons of gold-bearing concentrates.

A more typical development of the pyritic gold-quartz vein is seen in the Piedmont Alps, as for example at Brusson[1].

In Bohemia, mention must be made of Eule-Jilova[2], between the valleys of the Sazava and the Libre. Besides pyrite and free gold, the ores also carry arsenopyrite. The mining industry of Eule is extremely ancient, dating back to the end of the 8th century. This place may have been the greatest producer of the former golden wealth of Bohemia. In 1898 the work at Jilova had almost entirely stopped and only 38 tons of gold ore were produced. On the contrary, the Borkowitz mines in the Kuttenberg district produced during the same year 2,288.4 tons of gold ore. Work has also been recently resumed on pyritous gold ore lodes at Mount Roudny, east of Wotitz.

II. *Cupriferous Gold-Quartz Veins.*

In this type, the quartz gangue contains not only free gold and auriferous pyrite, but also a good deal of copper pyrite and other copper ores. Tourmaline is very often a characteristic accompaniment. The lodes are generally connected with acid eruptive rocks, especially granites.

The best known example of this type of vein is found in the vicinity of Berezovsk (Berjosowsk)[3], near Ekaterinburg, on the east slope of the Ural mountains.

Berezovsk lies in a low-rolling region, covered with pine and willow, and underlain by a muscovite granite schist and mica schists, strongly affected by dynamo-metamorphism. and deeply decomposed. These talcose, chlorite

---

[1] C. Schmidt: 'Geol. Gutachten,' Berne, 1900.

[2] F. Posepny: 'Goldvorkommen in Böhmen,' *Archiv f. Prak. Geol.*, II, 1895, p. 79.

[3] B. von Cotta: 'Erzlagerstätten,' II, pp. 554-556. F. Posepny: 'Golddistricte von Berezov und Mias im Ural,' *Archiv. f. Prak. Geol.*, II, pp. 499-598. A. Karpinsky: in 'Guide des excursions du VII. Congrès Géolog. Internat.,' 1897, V, p. 42. With bibliography.

and clay schists, all of which in the decomposed, reddish-colored condition, are called 'krassik,' are traversed by numerous vertical dikes, 2-40 m. (6.6 ft. to 131 ft.) thick, running north and south. The dike rock is a micro-granite called beresite, and this rock is also decomposed down to great depth. These dikes have been exposed by deep open-cuts and extensive galleries and shafts. At right angles to the strike of these beresites, extending from wall to wall, and sometimes for short distances past the dike wall, there are numerous veinlets or stringers of gold-bearing quartz, which are seldom over a few cm. (3 inches) thick, but in exceptional cases as much as 1 m. (See Fig. 121.)

Besides free gold, these quartz veins carry a variety of sulphide ores and the products of their decomposition; according to B. von Cotta these sulphides consist of pyrite, some aikinite ($PbCuBiS_3$), gray copper, copper pyrite and compact galena, as well as red lead ore, melanochroite, vauquelinite, pyromorphite, vanadinite, cerussite, anglesite, limonite, and bismuth ocher. Very frequently the quartz also contains long needles of a gray-green tourmaline, running at right angles to the walls, as well as small crystals or radiating balls of pyrophyllite, and sometimes magnesite. A very remarkable phenomenon is the presence of pseudomorphs of a chrome-bearing tourmaline after pyrite, which are occasionally found in these gold quartz stringers, as illustrated by specimens in the collection of the Moscow University.

Besides free gold, these veins also contain gold in the pyrite. According to A. Karpinsky, it varies between 2.5 and 30, locally even 250 g. per ton. The average may be assumed at about 13 g. per ton.

The micro-granite dikes (aplite veins) carrying the gold quartz stringers may be genetically connected with the granite massive of Lake Shartash, which is not far away. The chips of this perfectly fresh rock knocked off in quarrying it for use in stairways and steps, door jambs, etc., are credibly stated to contain as much as 1 gm. (66 cents) of gold per ton. If thus the intrusive stock proper is auriferous, it need not be wondered at that the subsequent extrusions from the same magma hearth in the form of aplites also brought with them a gold content, which was concentrated in the quartz stringers. These quartz stringers must, it would seem, be accepted as having the same origin as pegmatite. See other statements on this subject on page 297.

The gold veins of Berezovsk were discovered in 1745, and the mines are still being worked.

The cupriferous-gold-quartz class of veins also includes most of the gold veins of Scandinavia. On Bömmel Island, on the west coast of Norway, between Bergen and Stavanger, veins of this class have been described by

Th. Scheerer and afterward by H. Reusch,[1] and have been mined since 1884 by an English company. The rocks consist mainly of a saussurite gabbro, broken through by quartz porphyry and dikes of a diorite altered by dynamo-metamorphism. The gold quartz veins occur in these rocks, especially in the quartz porphyry. Their thickness may amount to as much as 1 m., strike northeast, dip 25-45° southwest. The vein filling consists of quartz, some calcite and chlorite, with free gold, copper pyrite, pyrite, galena and at times tetradymite (bismuth telluride) and very rarely native silver. The gold content of the ore, according to H. Louis, varies between 7 and 28 g. per ton.

The gold-bearing tourmaliniferous copper veins in the Thelemark area of southern Norway have already been mentioned. The gold veins of Eidsvold, too, 75 km. (45 miles) north of Christiania, should also be mentioned here.

In Sweden the cupriferous gold quartz veins are represented by the lodes of Adelfors in Småland, as well as by gold quartz stringers in the pyrite stock of Falun (*q. v.*).

Very typical examples of the cupriferous gold quartz veins have been reported by W. Möricke[2] in Chile. They represent an extreme variation of our group of the tourmaline-bearing copper formation. Besides quartz with tourmaline, they contain pyrite, free gold, copper pyrite and other copper ores. They are also apt to be associated with granites and other acid eruptive rocks. As examples may be mentioned the veins of Guanaco, Andacollo and Los Sauces in Chile.

In Australia also gold-quartz veins rich in copper ores seem to be rather common, especially in New South Wales, judging by the material in our collection. Whether they carry tourmaline or in other ways form a well characterized group, has not yet been ascertained.

On the Philippine Islands, according to oral statements by F. V. Voit, this class is represented by the veins of Mambulao and Paracale. A Transvaal example on the Malmani river is a vein in dolomite in the Marico District[3].

A unique type of gold vein occurs at Rossland, B. C., characterized by Lindgren[4] as biotitic gold copper veins. They are well marked fissure veins, contained in granular intrusive rocks, ranging from diorites to monzonites and even syenites. The ore minerals are pyrrhotite, chalcopyrite and a little arsenopyrite, all holding gold, but usually not free, or amenable to amal-

---

[1] H. Reusch: 'Bömmelöen og Karmöen, Cristiania,' 1888, p. 392. Phillips-Louis: Ore Deposits,' 1896, p. 519.

[2] W Möricke: 'Die Gold-, Silber- und Kupfererzlagerstätten in Chile,' 1897.

[3] G. A. F. Molengraaff: 'Géol. de la Rep. Sud Africaine,' 1901, p. 40.

[4] W. Lindgren: 'Metasomatic Processes,' *Trans.* Am. Inst. Min. Eng., 1900, p. 69.

gamation. The chief gangue mineral is biotite, with a little quartz, calcite, muscovite, amphibole, chlorite, tourmaline and garnet. The veins are replacement deposits along fissures, the hornblende of the rock altering to biotite, and the feldspar replaced by pyrrhotite and chalcopyrite. The process indicates dynamic metamorphism rather than ordinary vein-forming processes.

III. *Antimonial Gold-Quartz Veins.*

In place of pyrite this class contains stibnite as a characteristic companion of the free gold, the antimony glance itself being gold-bearing. As subordinate minerals of these veins pyrite still continues to appear with arsenopyrite, rarely galena, blende and copper pyrite, and, together with quartz, there may be a slight amount of carbon spars.

Good examples of the class are seen in the mineral district of Krásnáhora (or Schönberg) and Milesov (or Mileschau) in central Bohemia, southwest of Prague, described by F. Posepny[1]. An intrusive stock of granite intercalated between schists, is traversed by a few porphyry dikes and numerous ones of lamprophyre. Most of the veins accompany these dikes. Besides quartz as the principal gangue, they carry some calcite, and as ore minerals, stibnite with a gold content of 100-133 g. ($66.00 to $87.88) per ton, arsenopyrite and pyrite, together with about 300-400 g. ($198-$264) per ton in native gold in scales and botryoidal aggregations in the quartz; also antimony, ochre and pyrostibite. The stibnite often constitutes the larger part of the filling. Masses with a thickness of as much as 1 m. (3.28 ft.) are said to have been found.

At Mileschau a thriving gold mining industry was carried on as early as the fourteenth century. Of late years the mines have been worked for stibnite. Incidentally, however, in the production of 661.5 tons of antimony ore in 1898, 254.3 tons of gold quartz and 180.4 tons of gold-bearing pyritic concentrates were produced at Schönberg and Proutkowitz.

Another typical example of this class occurs, according to B. von Cotta[2], at Magurka, in Hungary, on the north slope of a granite range, 1,200-1,800 m. high, separating the Comitat of Sohl from Liptau. The granite is there traversed by quartz veins varying from a few cm. (or inches) to 4 m. (13 ft.) thick. The veins contain stibnite and free gold, together with balls of granite. The accessory minerals are galena, zinc-blende, pyrite, copper pyrite, brownspar and calcspar. Ordinarily the quartz occurs along the walls of the vein, while the stibnite is found in the middle of the vein.

This class also includes the veins of the well-known district of Brand-

[1] R. Helmhacker: 'Der Antimonbergbau Milesó bei Krásná Hora,' 1874. F. Posepny: 'Goldvorkommen, Böhmens,' *Archiv. f. Prak. Geol.*, 1895, p. 165.

[2] B. von Cotta: *Berg u. Hutten. Zeit.*, 1861, p. 123.

holz, near Goldkronach[1], in the Fichtelgebirge. At this place Cambrian sericite schists are cut by quartz veins, containing subordinate quantities of brownspar, more rarely calcspar, and occasionally barite, and carrying stibnite, gold- and silver-bearing pyrite and arsenopyrite and free gold, besides very small amounts of striated galena, brown blende, native antimony, antimony ocher, stiblite, heteromorphite and bournonite. The quartz is often thoroughly impregnated with minutely disseminated ores. The veins often wedge out and are for some distance replaced by a simple fissure, alongside of which the adjoining rock is silicified and impregnated with arsenopyrite and iron pyrite. Mining, which dates back to the fourteenth century, was resumed after the transfer of the principality of Bayreuth to Prussia, about 1800, in part under the management of A. von Humboldt, but was again abandoned in 1861.

Antimonial gold quartz veins occasionally occur in gold ore districts outside of Europe; for example, at several points in Australia, in the Selati goldfield and in the Murchison range in the Transvaal.

In the last-named mountains, hornblende schist and quartzite schist are traversed by a series, almost 35 miles long, of veins which generally follow the strike of the schists in the form of beds, and contain quartz with auriferous antimony glance, in part also with gold-bearing pyrite and copper pyrite. The Invicta, Free State and Gravelotte mines are operated on such veins.

IV. *The Arsenical Gold-Quartz Veins.*

In this class free gold is accompanied by arsenopyrite as the most prominent of the ore minerals in a gangue of quartz.

Such veins have been worked since 1881 at Santa Cruz, in the state of Santa Barbara, Honduras[2]. The principal veins, 2 m. thick on an average, consist of quartz, gold-bearing arsenical pyrite and mere traces of galena. The ore contained 25-30 gm. per ton.

Occurrences of similar kind are scattered all over North America, being mostly connected with the pyritous gold quartz veins by transitions. The veins in the Huronian schists at Marmora, Ontario[3], Canada, are typical examples.

The peculiar arsenical gold ore deposits, regarded by many as beds, at Passagem in Brazil, are peculiar, as will be seen by a brief description, taken from E. Hussak.

[1] C. W. Gümbel: 'Fichtelgebirge,' 1879, pp. 386-393. F. v. Sandberger: 'Ueber die Erzlagerstätte von Goldkronach bei Berneck,' *Sitzber.* d. math-.phys. Cl. d. k. bayer. Ak. d. W., 1894, Vol. XXIV, part II.

[2] Fuchs and De Launay: 'Traité,' II, p. 942.

[3] R. P. Rothwell: 'The Gold Bearing Mispickel veins of Marmora, Ontario, Canada,' *Trans.* of the Amer. Inst. of Min. Eng., 1881.

Passagem[1] lies in the province of Minas Geraes, 7 km. (4.2 miles) east of Ouro Preto. Next to Morro Velho it is to-day the most productive gold mine of Brazil, and has been worked since the end of the last century.

According to Hussak, the most important deposit of the locality is a bedded vein of quartz, carrying tourmaline and auriferous arsenopyrite, with subordinate amounts of pyrite and pyrrhotite in patches. It dips southeast at an angle of but 18-20°, and is entirely conformable with the enclosing quartzite schist. The latter is part of a series of rocks made of hematite mica schist and of mica schists and itabirite[2].

The matrix is here and there replaced by lenticular masses of sericite-quartzite schists. Near the underlying mica schist, which contains stauro-

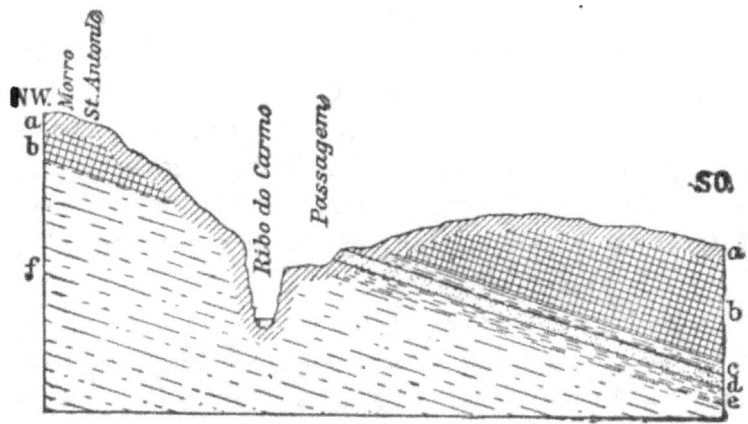

Fig. 172. Cross-section through the quartz bed of Passagem. (M. P. Ferrand.)
a, canga (weathered cover); b, itabirite; c, crypto-crystalline schist; d, quartz vein and quartzites; e, contact quartzites; f, mica schist.

lite and cyanite, a selvage zone or streak of black graphitic schist occurs, and this material is also occasionally seen near the finely crystalline quartz-amphibole-schist, that forms the hanging-wall (see Fig. 172).

In this quartz vein, aggregations are very often found of a finely felted quartz-tourmaline rock, called carvoeira, in which andalusite also occurs at times. Other aggregations consist of garnet with biotite, pyrite and cyanite, as well as of a greenish oligoclase-albite.

The gold occurs either in beautiful crystals or merely in jagged or scaly flakes, filling the interspaces of the tourmaline rock, especially where the latter is rich in arsenopyrite. The gold carries a considerable amount of

[1] M. P. Ferrand: 'I.'or a Minas Geraes,' Vol. II, pt. 1. Ouro Preto, 1894. E. Hussak: 'Der goldführende kiesige Quarzlagergang von Passagem, *Zeit. f. Prak. Geol.*, 1898, p. 345.

[2] Itabirite is regarded by some geologists as a mantle of residual sand cemented by limonite deposited by surface water, and merely a result of weathering. See 'Harrison's Report on British Guiana Laterites.'—W.

bismuth, which apparently is present in the form of an alloy. The highly auriferous arsenopyrite-tourmaline aggregations also carry small crystals of monazite and xenotime.

E. Hussak is inclined to consider the vein as of intrusive nature, and to regard it as an ultra-acid granite apophysis. In support of this view, he points to the presence of zircon, monazite, xenotim and albite, minerals that are characteristic of granite, and which occur as accessories in the quartz vein, and also to the presence of a large granite mass only 1 km. away. (Both Lindgren and Emmons maintain, however, that Hussak's facts prove it to be a normal fissure vein. See 'Genesis Ore Deposits,' p. 758.) The minerals tourmaline, andalusite, staurolite, cyanite, garnet, biotite, hercynite (iron spinel), which are also found in the vein, he regards as contact minerals, resulting from thermal metamorphism of the quartzite schist. His interpretation of this remarkable deposit is akin to the views of North American investigators on the relationship of certain gold quartz veins with pegmatites, mentioned on page 297. In many respects the veins in diorite and granite of Meadow Lake, California, resemble the Brazilian deposits just described. According to Lindgren,[1] these veins contain a quartz and tourmaline gangue carrying pyrite, arsenopyrite, pyrrhotite, blende and copper, besides gold.

Among the gold ore deposits in the Ural, the arsenical gold quartz formation includes the veins of Kotchkar[2], repeatedly described in recent times, situated in the steppe immediately adjoining the east slope of the mountains, in the Government of Orenburg. The region is underlain by sheared and decomposed granite. The vein filling consists of crushed and leached rock with cavernous quartz, particularly porous in the richer portions and carrying free gold, much arsenopyrite, somewhat smaller quantities of pyrite, and still more rarely manganite, copper pyrite, galena and stibnite. The arsenopyrite also impregnates the adjoining granite, which contains from 2.5 to 375 gm. gold per ton. The ore averages 8 to 10 gm. ($5.20 to $6.60) per ton. In the Transvaal the bedded veins of Kromdraai, north of Krugersdorp, belong to this class[3].

V. *The Cobaltiferous Gold-Quartz Veins.*

This class has been established to include the unique deposits which differ decidedly from all the gold-bearing types of veins thus far treated,

[1] *Am. Jour. Sci.*, Vol. XLVI, Sept., 1893.

[2] N. Wyssotsky: 'Les Gisements d'Or du système de Kotchkar dans l'Oural du Sud,' 'Guide des excursions du VII Congrès Géol. Intern.,' 1897, St. Petersburg. H. B. C. Nitze and C. W. Purington: 'The Kotchkar Gold Mines,' etc., *Trans. Amer. Inst. Min. Eng.*, Vol. XXVIII, 1899. p. 24. Chr. T. Nissen: 'Einiges über das Goldvorkommen,' etc., bei Kotschkar, *Berg u. Hütten. Zeit.*, 1900, p. 121.

[3] G. A. F. Molengraaff: 'Géol. de la Rep. Sud. Africaine,' 1901, p. 34.

especially in the fact that the quartzy matrix is small in amount compared to the ores. The real carrier of the gold is in this case the cobalt mineral, smaltite.

According to H. Oehmichen[1] and Dörffel, this unusual occurrence of gold is found in the Middelburg district, northern Transvaal, along the Kruis river. The Lydenburg shales, part of the Cape formation, contain at this place intercalated eruptive sheets of aplite-like rocks and diabase, which are traversed by an east-west vein 2-3 cm. thick, dipping 60-70° south, whose chief and almost only filling is smaltite, the ore carrying 60-150 g. of gold per ton. The ore contains between 7 and 8% cobalt, with 0.5 to 1% of nickel. The aplite wall rock is sericitized and is also impregnated with ore.

A second and similar vein occurs 5 km. farther west and is also in diabase. This, the Laatste-Drift vein, has a gangue of auriferous quartz mixed with kaolin which holds smaltite and copper ores in bunches sometimes containing molybdenite. Erythrite, limonite and scorodite are secondary minerals found in the outcrop. The ores carry from 100-250 g. of gold per ton, of which only a very small part is native, while 90% of the gold is contained in the chalcopyrite. The smaltite contains almost no gold. As the specimens from these veins that have thus far come into our hands differ so much from all other types of gold quartz veins we have thought it necessary to place these deposits in a group by themselves. They form a transition from the gold ore veins to the true cobalt veins, and, in fact, pure, non-auriferous cobalt veins are also known to occur alongside of the gold-bearing ores at this locality.

17. THE SILVER-GOLD VEINS.

The veins of this class are of widespread distribution and include the greatest gold-producing deposits of the world[2]. The matrix of veins of this class is mostly quartz, but also includes lesser amounts of various carbonates, especially calcspar and rhodochrosite, and rarely small amounts of barite. The ores contain gold in various forms, viz.: native gold, auriferous pyrites and other sulphides, and very frequently various tellurides, especially nagyagite (the sulpho-telluride of lead, containing 6-13% gold), sylvanite, a gold-silver telluride and silver telluride. On the other hand, various rich silver minerals occur, such as argentite,

[1] H. Oehmichen: 'Goldhaltige Kobaltgänge in Transvaal,' *Zeit. f. Prak. Geol.*, 1899, pp. 271-274.
[2] See 'Concentration des Metallgehaltes zu Erzlagerstätten,' *Zeit. f. Prak. Geol.*, 1898, p. 416.

stephanite, proustite, pyrargyrite, etc., together with argentiferous galena, zinc-blende and more rarely copper pyrite, gray copper, native arsenic, bournonite, stibnite, realgar and orpiment.

All the examples of the various classes of veins thus far mentioned show a close geologic connection with Tertiary or late Mesozoic eruptive masses, especially andesites and trachytes, dacites and rhyolites. Their formation seems to be conceivable as an aftermath of the eruption of these magmas, eruptions whose only survival or lingering remnant of igneous activity is found in thermal springs and gas exhalations, genetically connected with the eruptions.

The transformation of the above named eruptive rocks into propylite seems to have been a concomitant phenomenon of the origin of those veins.

The silver-gold deposits of Transylvania have been carefully studied for a long time[1].

The deposits occur mainly in the mountain region rising from the flat Tertiary land north of the Maros and drained by the head waters of the Aranyos (that is to say, gold river), Szamos and Körös. The true nucleus of these "Transylvanian ore-mountains" consists of crystalline schists, which, however, are only seen outcropping in large areas in the northeast part of the region. Elsewhere they are covered and concealed by Permian, Triassic, Jurassic and Cretaceous strata, mainly limestones, and also by Tertiary beds. These rocks are cut and broken through by intrusive masses of andesites and dacites, trachytes and rhyolites, as well as basalts. The gold veins are connected with these igneous outbreaks, especially with stocks of the propylites just mentioned, which, as will be explained farther on, probably represent later andesites. The gold veins traverse both eruptive masses and the adjoining Tertiary strata.

Some of the gold deposits of Transylvania are especially characterized by an abundance of tellurides, especially at Nagyág, Offenbanya, Faczebanya and Fericiel. The telluride-bearing deposits of this country will be first

[1] General works upon this District: Frh. F. von Richtofen: 'Studien aus den ungarisch-siebenbürgischen Trachytgebirgen,' *Jahrb.* d. k. k. geol. Reichsanst, Vienna, 1860, pp. 153-277. B. von Cotta: 'Ueber Erzlagerstätten Ungarns und Siebenbürgens,' Gangstudien, IV, 1, 1862 (with E. v. Fellenberg). F. von Hauer and G. Stache: 'Geologie Siebenbürgens,' 1863. F. Posepny: 'Zur Geologie des siebenbürgischen Erzgebirges,' *Jahrb.* d. k. k. geol. Reichsanst, 1688, I, pp. 53-56. *Ibid.* 'Allgemeines Bild der Erzführung im siebenbürgischen Bergbaudistricte,' *Jahrb.* d. k. k. geol. Reichsanst, 1868, II, pp. 7-32. C. Doelter: 'Aus dem siebenbürgischen Erzgebirge,' *Jahrb.* d. k. k. geol. Reichsanst, 1874, I. *Ibid.* 'Die Trachyte des siebenb. Erzgeb.' Tschermaks *Mineral. Mittheil,* 1874, pp. 13-30. *Ibid.* 'Ueber das Vorkommen von Propylit und Andesit in Siebenbürgen,' same journal, II, 1880, p. 1. P. T. Weisz: 'Der Bergbau in den siebenbürgischen Landestheilen,' Also, *Jahrb.* d. k. ungar. geol. Landesanst., Vol. IX, part 6, 1891. Semper: 'Beiträge zur Kenntniss der Goldlagerstätten des Siebenbürgischen Erzgebirges,' *Abhandl.* d. k. preuss. geol. Landesanst. New series. Part 33, Berlin, 1900. L. Remenyik: 'Les Mines de Métaux de Hongrie à l'exposition Universelle,' Budapest, 1900.

described and afterwards those without tellurides. We begin with Nagyág.

The mining town of Nagyág[1], in the Hunyadi Comitat, lies in a narrow gorge in the midst of a magnificent mountain region, near Mount Hajto, 1,047 m. (3,335 ft.) high. The surrounding heights all consist of andesitic and trachytic rocks, part of the eruptive chain of the Transylvanian Erzgebirge, which also contains various other mines of ancient and modern times, such as Coranda, Magura, Füzesd, Boicza, Trestia, Porcura, Zrdaholz and Ruda. On the valley bottom, and in the deep mine workings, there are exposures of the sedimentary rocks, broken through by the eruptive masses. These stratified rocks consist of yellowish sandstones and conglomerates, as well as gray and reddish clays, all of Mediterranean age, and with highly disturbed stratification. Large detached blocks of these sediments are enclosed by the andesites in the mining area. At still greater depth, phyllite-like schists are suspected to exist, being indicated by the inclusions of such phyllites in the andesites. The stratified rocks but very rarely, however, form vein walls, the deposits being almost entirely in Tertiary eruptive rocks. Two types of these eruptives outcrop at Nagyág. Their age is regarded as Miocene, since fragments of them are found in the Sarmatian (upper Miocene) limestones of Vormaga.

The neighboring eminence of Mount Calvary is formed of a quartz-free hornblende-andesite. In the immediate vicinity of the veins, on the other hand, the prevailing rock is a quartz andesite or dacite with biotite and hornblende, as well as augite, in greatly varying quantity. The latest monograph on Nagyág, by B. von Inkey, following Szabo's nomenclature, designates both rock types as trachytes.

The Nagyág dacite has been affected by three different kinds of alteration: (1) Normal weathering or decomposition on the surface, (2) by a propylitization or alteration into 'greenstone trachyte,' with formation of secondary chlorite and of carbonates from augite and hornblende, with accompanying introduction of gold-bearing pyrite, (3) by kaolinization. The two last named processes, certainly the kaolinization, seem to be genetically connected with the formation of the mineral veins. (See 'Thermal metamorphism.') The propylitization is most intense and widespread in the middle and lower parts of the eruptive stock, a place that is also the site of the ore deposits. The kaolinization of the rocks is still more

---

[1] Frhr. von Hingenau: 'Geol. bergmänn. Skizze des Bergwerkes Nagyág,' *Jahrb.* d. k. k. geol. Reichsanst., 1857, pp. 82-143. B. von Cotta: 'Ueber Erzlagerstätten Ungarns und Siebenbürgens.' Gangstudien, Vol. IV, part 1, 1862. F. von Hauer and G. Stache: 'Geologie Siebenbürgens,' 1863. H. Höfer: 'Beiträge zur Kenntn. der Trachyte und der Erzneiderlage von Nagyág,' *Jahrb.* d. k. k. geol. Reichsanst., 1886, p. 1. G. vom Rath: 'Vöröspatak und Nagyág,' *Sitzb.* der niederrh. Ges. f. Natur- und Heilk. in Bonn, March, 1876, and March, 1879. B. von Inkey: 'Nagyág und seine Erzlagerstätten,' Budapest, 1885.

intimately connected with the veins, as is apparent from the cross-section, Fig. 173.

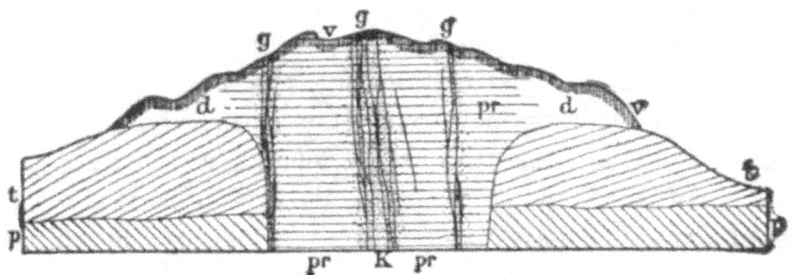

Fig. 173.—Diagrammatic cross-section through the veins of Nagyág. (B. von lnkey.)
p, phyllite; t, tertiary sediments of the Mediterranean stage; d, dacite; pr, propylite; g, vein streaks; k, kaolinized dacite; v, superficial weathered cover.

Another peculiarity of the Nagyág district that needs mention is the presence of 'barren' veins (Glauchgangue), a feature which is also known in other Hungarian and Servian ore deposits in propylitic rocks, as for example at Vorospatak, where they are called 'glamm.' The 'barren'

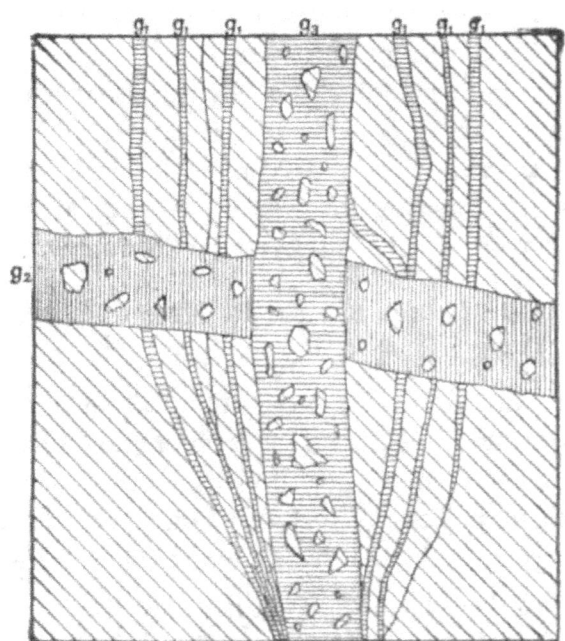

Fig. 174.—Horizontal plan of barren veins at II Longin at the 75 fathom level at Nagyág. (B. von Inkey.)
g_1, oldest barren veins ; g_2, second barren vein, dark 20 cm. thick; g_3, youngest barren vein, light gray, 15 cm. thick.

veins of Nagyág cut through the propylite and the blocks of Tertiary rocks enclosed by it. The fissures vary from mere cracks up to 10 or even 20 m. in thickness. They are filled with a dark sandy or clayey mass, containing angular or sub-angular fragments of the adjoining rock. These 'barren' veins cut one another and are thus shown to be of different ages. Fig. 174, after B. von Inkey, shows as many as three sets of these Glauch-gange differing in age. The included fragments can often be shown to be torn from portions of the wall rock 10 m. (32.8) to even 20 m. (65.6 ft.) distant. This displacement appears to be even greater since bits of phyllite are not infrequently encountered, which seem to have been brought up from a greater depth. The 'barren' veins are explained in various ways. Some authors regard them as fault fissures, filled with friction breccias and the finer products of attrition. This view is, however, opposed by the fact that very many 'barren' veins show no dislocation whatever in the adjoining rock. F. Posepny for some time thought of an in-washing of the material from above. This is hardly conceivable, however, in view of the fineness and manifold ramification of the network of the 'barren' veins. H. Hofer and others supposed them to be decomposed and crushed dikes of eruptive rock, but this is not confirmed by the petrographic character of the finer material composing the filling. B. von Inkey, who seems to have devoted much attention to these phenomena, notes "that the débris material in the Glauch-gange breccia is evidently derived from the fissure walls, while the matrix, which is a mixture of fine sand with clay, was originally a watery mud, whose solid particles may be derived partly from the dust particles produced by the comminution of the adjoining rock, and in part represent triturated material of deeper rocks. The water which made mud of this dust is supposed to be derived from the ground-water of the rocks of the Mediterranean stage. When afterward a mighty eruption fissured these strata and forced great wedges of a hard rock between them, the subter-ranean water circulation was altered, the passage of the ground-water was checked and dammed back, and large accumulations of water or even of thin liquid mud, were formed beneath the trachyte cover and between the eruptive wedges." He next imagines that the mass of the eruptive stock, through the yielding of its sedimentary base and its lateral supports, be-came fissured, and that thereupon those masses of mud were forced by high pressure from below into these collapse-fissures[1].

The mineral veins found within the stock of propylitized dacite, at Nagyág, form a tangled network of fissures showing great variety in thick-ness and direction. Most of the fissures, however, strike about north-south. Minute veinlets and veins ½ m. thick fray out, gather into clusters or cross one another in the most diverse manner, though apparently all be-longing to the same period of formation. According to B. von Inkey, the veins are not contraction fissures, but are due to external orogenic causes.

[1] B. von Inkey: A. a. O., p. 149.

Classified according to the nature of their ores, Freiherr von Hingenau at an early date was able to distinguish three areas, each characterized by its own peculiar distribution of the prevailing vein minerals, a conclusion confirmed in general by H. Höfer and von Inkey. These three districts are the following:

(a) The region of telluride-carbonate veins, with calcite, siderite, dolomite and rhodochrosite, some quartz and hornstone with alabandite (Mn S), nagyagite and other tellurides of gold, also some tetrahedrite and pyrite; the latter, it is true, merely as an impregnation in the adjoining rock. This is in the southeast part of the district, mainly below Mount Szekeremb, extending as far as the Anastasia cleft.

(b) The telluride quartz veins or rich quartz area whose veins carry gray, non-crystalline, often cellular quartz, with auriferous sylvanite, krennerite, gray copper and rarely free gold. This area is in the northeast parts of the district and near Mount Hatjó.

(c) The area of the sulphide ores, which pass into the two other classes by gradual transitions. It is marked by the prevalence of galena, zinc-blende and pyrite in a gangue of carbonates, in which argentiferous tetrahedrite occasionally appears.

The subordinate minerals of the Nagyág veins include gypsum, which sometimes encloses flakes of native gold, native arsenic, bournonite, stibnite, plumose jamesonite, realgar and orpiment, all of which occur mainly in the upper parts of the veins.

An enrichment of the veins is generally found to occur when strings of pyrite cross the fissures, and when the veins themselves form intersections, or when in contact with the 'barren' veins (Glauchgange), which in such cases they either traverse or follow as a network of stringers. When the ore-bearing veins ramify in a large mass of barren vein matter the result is a "Trümerstöcke" (stringer stock), as it is called, such as the Philipp and the Adam stock. Neither a diminution of the gold content or a wedging out in depth has thus far been observed at Nagyág.

The Nagyág goldfield was the last of the Hungarian-Servian mineral districts of the region to be discovered, being found in 1745. Mining operations began in 1748 and have continued in a flourishing condition ever since, the mines being in part the property of the crown and of the state treasury. It may be mentioned that it was in the Nagyág gold ores that the element tellurium was discovered in 1782 by Muller von Reichenstein.

At Offenbanya[1], which is situated in the valley of the Aranyos, on the

---

[1] J. Grimm: 'Die Erzniederlage und der Bergbau zu Offenbánya in Siebenbürgen,' *Jahrb. d. k. k. Montan-Akademien*, XVI, 1867. F. Posepny: 'Ueber den inneren Bau der Offenbányaer Bergbaugegend,' *Verh. d. k. k. geol. Reichsanst.*, 1875, IV, p. 70.

north border of the Transylvanian Erzgebirge, two classes of deposits are distinguished: (1) Contact aeposits, occurring at the contact between intercalated beds of crystalline limestone in the garnetiferous mica schist and Tertiary dacites and hornblende andesites; (2) mineral veins.

The contact deposits locally called lead stocks are formed of pyrite, zinc-blende, arsenopyrite and galena imbedded in a calcareous matrix, and the deposit shows no connection with the gold veins. They were formerly mined by the state, but are now of no importance.

The mineral veins occur in a stock of dacite. Where they pass into a mass of friction breccia, developed locally along the contact between the eruptive rock and the mica schists and limestones, most of the veins wedge out. They are of very small thickness, mostly only 5 to 25 mm., and can only be followed 130 m. in the strike.

Some of them carry only native gold, some only telluride ores, others carry both at the same time. The matrix consists of quartz and calcspar. The more quartz the fissures carry, the richer they are in tellurides (sylvanite and nagyagite). The subordinate accompaniments of the tellurides are pyrite, zinc-blende, realgar, more rarely proustite, bournonite, tetrahedrite, marcasite and arsenopyrite. In druses, crystals of quartz, calcspar, barite, aragonite, albite and gypsum are found. In certain veins, free from tellurides, wulfenite, pyromorphite, as well as stibnite, were also found.

Most of these fissures have a general east-west course. The vein wall-rock is thoroughly decomposed and impregnated with small crystals of auriferous pyrite, so that zones of this altered rock, 5-8 m. (16.4 ft. to 23.2 ft.) broad, are mined as gold ore, and where several fissures meet or come close together great masses of rock have been impregnated with gold, the deposit resembling the so-called gold quartz stock of Nagyág.

Since mining has been resumed the work has been confined mainly to one such stock, the so-called Kreisova stock. This consists of a crushed or brecciated dacite in which the veins consist of drusy quartz with pyrite and silver ores, gray copper, copper pyrite and galena. The amount of silver and gold is stated to be very considerable.

Verespatak occupies the first rank, among the telluride bearing districts, not so much because of its present production as because of its historic fame.[1]

[1] F. von Hauer: 'Der Goldbergbau von Vöröspatak,' *Jahrb.* d. k. k. geol. Reichsanst., 1851, IV, p. 64. J. Grimm: 'Einige Bem. über die geogn. und bergbaul. Verh. von Vöröspatak,' *op cit.*, 1852, III, p. 54. F. Posepny: 'Einige Resultate meiner bisherigen Studien über Verespatak,' etc. *Verhandl.* d. k. k. geol. Reichsanst., 1867, p. 99, 1870, p. 95 and 1875, p. 97. G. vom Rath: 'Vöröspatak und Nagyág,' *Sitzb.* der niederrhein. Ges. f. Natur- und Heilkunde in Bonn, March, 1876. F. J. Kremnitzki: 'Beobachtungen über das Auftreten des Goldes im Verespataker Erzreviere,' Földtani Közlöny, 1888, pp. 517-520. Also the works noted on p. 331; particularly the work of Semper, pp. 125-172.

This place is situated in a deep valley about 10 km. (6 miles) northeast of Abrudbanya. B. von Cotta called it the Eldorado of Transylvania, and even to-day 172 small companies are busy in this region, and the waters from 5 reservoirs drive over 6,000 stamps. Back of the town and to the south of it lies the famous Csetatye, a bald, rocky ridge, completely honeycombed by mining works, which have been carried on from Roman times, A. D. 106-276, down to the present day.

The Verespatak mineral area is an isolated mass consisting mainly of eruptive rocks lying in the midst of the monotonously uniform Carpathian sandstone (Cretaceous and Eocene). This mass consists of dacite, rhyolite, and rhyolite breccias, the latter often silicified. It forms the two mountain stocks of Kirnick and Boi. The great Csetatye mentioned above consists in part of this dacite, which carries large dihexahedrons of quartz, in part of the silicified rhyolite breccia. Younger Tertiary sediments, namely conglomerates, sandstones and tuffs with rhyolitic material in them, are also found there.

A very remarkable feature of the district is the occurrence of the so-called 'glamm' veins, which form a network of fissures throughout the district. Their filling consists of a clayey mass containing rounded rock-fragments, including pieces of coal of considerable size, trachyte fragments and bits of mica schist and phyllite. These masses are impregnated with pyrite and limonite, and seem to have been pressed into the fissures with great force, apparently from below. Semper thinks them intrusions of volcanic mud.

The ore-bearing veins occur partly in the eruptive masses and partly in the above mentioned sediments. In the Carpathian sandstone the veins contain practically only quartz with free gold. Elsewhere the quartz and free gold are accompanied by banded rhodonite, calcite, calcspar, rhodochrosite, gold-bearing pyrite, zinc-blende, as well as galena, gray copper and seldom adular spar. The gold occurs in crystals intergrown with rhodonite and quartz. In the heart of the Csetatye ridge the gold-bearing fissures form a stock of rich stringer veins, called Katronza (petticoat), whose extraction has left great excavations. Similar stringer vein stocks were worked as early as Roman times. They commonly consist of eruptive breccia impregnated and cemented by gold-bearing pyrite and quartz.

A similar deposit is found at Bucsum, about 10 km. southeast of Abrudbanya, in the mines of the Concordia company.

The Vulkoy Korabia and Botes[1], mining districts about 15 km. north of Zalathna, which are similar, were also known to the Romans. The veins occur partly in the trachyte, partly in the Carpathian sandstone. Their

[1] Fuchs and de Launay: 'Traité,' II, p. 932. Semper, *op. cit.*, pp. 178-181.

filling consists mainly of quartz with fine sprinklings of free gold, besides gold-bearing pyrite, and a little chalcopyrite, gray copper and galena. The gangue carries hessite.

The largest gold mines that are being worked at the present time in Transylvania are situated near the town of Brad, the Twelve Apostles[1] and Muszari mines, and at Boicza, all of which are located in the western part of the gold area.

An extraordinary excitement was recently created by the great success of the Geisslinger Industrial Company in the Muszari mine near Brad[2], which in 1895 produced 732 kilograms of crude gold (gold-silver). The entire workings lie in a massive of andesitic and dacitic rocks. Two systems of veins intersect at right angles with very rich orebodies at the crossings, as, for example, that between the Clara and Karpin veins, from which, in 1891, 53 kilograms of crude gold were taken. The vein filling at this place consisted of quartz, calcspar, pyrite and some light-colored zinc-blende. The Clara vein sends out several lateral stringers, which gather and ramify in the Clara stock. This stock, varying up to 7 m. thick, is interspersed with druses containing quartz, pyrite and free gold. The ores in this "treasure chest of Muszari" contained in places as much as 1,000 gm. per ton. This space, which is enclosed by an iron grating, yielded about 600 kg. of gold in 1¾ years.

The other veins of this mine have a thickness of 2 cm. to 2 m., and carry a gangue of quartz and calcspar each in varying amount with gold-bearing pyrite, free gold, sphalerite, chalcopyrite, marcasite, arsenopyrite and some galena. The latter does not seem to have a favorable effect on the gold contents. The amount of gold in the veins varies from 9 to 30 gm. per ton.

The last Hungarian locality to be mentioned as illustrating this class of deposits is that of Boicza, near Déva, in the district of Brad, whose mines are of great importance[3]. The veins are found in a stock forming the Szvregyel mine, which is about 1 km. from the place itself. The core of this mountain is formed of a rock which is regarded by Primics as a melaphyre, whose age lies between lower Triassic and upper Jurassic. This makes it older than the Tithonian limestone which apparently overlies it in many places. According to other authors, the rock is much younger, belonging to the andesitic group. This older eruptive mass is broken through by a spreading stock of dacite, especially broad in the upper part of the moun-

See Semper, *op. cit.*, pp. 86-109.

[2] According to a communication received from H. Oehmichen.

[3] Primics György· 'Die Geologie des Csetráser Erzrevieres' (Hungarian), Budapest, 1896. L. Venator: 'Monogr. über das Gold- und Silberbergwerk Rudolfi in Boicza,' Hermannstadt, 1900.

tain. This dacite encloses numerous fragments of the melaphyre and the Tithonian limestone along the contact zone.

The gold veins occur chiefly along the contact between the two eruptive rocks, but in part also within the melaphyre. They strike northwest and dip 75 to 85° in two different directions. Two systems may be distinguished, whose principal veins show a tendency to unite into a single vast fissure, the so-called stock, in the southeast part of the mineralized district. The two main fissures, Schuhaida and its companion, the Kreuzschlager vein, show payable ore for a length of 800 m. (2,624 ft.) The veins are not thick, varying between 0 and 1 m., the average being 0.35 m.

According to L. Venator, the filling, which differs considerably from vein to vein, consists mainly of quartz, in which gold-silver, calcspar, barite, rhodonite, amethyst, pyrite, chalcopyrite, galena, sphalerite, argentite and proustite occur. In most cf the fissures, pyrite predominates over the other ore minerals, though in some veins galena and zinc-blende are the chief ores. The gold (which contains much silver) occurs in microscopic grains scattered through the ore and it is only in the very richest ore that it is visible to the naked eye. At times the gold is associated with pyrite, more rarely with zinc-blende. The vein junctions and intersections are here followed by distortion and not by ore shoots and enrichments, these being, however, sometimes found at some distance beyond the intersections. The ore 'stock' is due to the meeting of many of the veins. It attains 40 m. in longitudinal and 30 m. in transverse diameter, at the 110 m. level. Its filling consists sometimes of fragments of melaphyre and dacite, cemented by an ore-bearing material, which at some points is very rich.

As is known from various archeologic discoveries, mining operations were carried on at Boicza as early as Roman times. About the year 1444 the work was resumed, but not until 1884 did modern mining begin. In 1898, 212,436 kg. of fine gold and 171,700 kg. of silver were produced.

This brief summary of the Dacian gold districts has shown that less than half of them can be assigned to the typical silver-gold-veins, since the majority of them just described contain tellurides. All of them are, however, of the same age, are genetically dependent on masses of Tertiary eruptive rocks of the andesite and trachyte families, and show so many other geologic characteristics common to all that they can not well be separated in the classification adopted.

A second region containing similar deposits is that of Schemnitz[1] (in Hungarian, Selmeczbanya) in the Erzgebirge of lower Hungary.

[1] Beudant: 'Voyage minéralogique et géologique en Hongrie,' Paris, 1822. Frhr. J. von Richthofen: 'Studien aus den ungarischsiebenbürigschen Trachytgebirgen,'

This long-famed mining region, whose mining work probably dates back to the eighth century, and which, as the seat of an old mining academy, has always had keen intellectual men watching its development, is, geologically, one of the best-studied mining areas of the globe. The older pioneer works of Esmark, Beudant and Pettko were followed by the more recent ones of von Richthofen, von Cotta, Lipold, Szabo, Cezell, Cseh and H. Böckh, of which we are only able to cite some of the more important. According to the recent work of Böckh, the oldest formation in the vic.nity consists of Triassic slates which, along their contact with diorite, are altered to gneiss-like hornstone and mica schists. The Trias also contains limestones and quartzite. After this come Eocene Nummulitic shales. The series of eruptions began between the time of the lower and upper Mediterranean stage. These eruptions were the cause of the vein and ore formation of the district. At first pyroxene andesite, then diorite and granodiorite (quartz diorite), next aplite, then biotite and amphibole-andesite and finally rhyolite. Much later in the Pliocene period basalt eruptions occurred. The rhyolites are predominant and show very diverse modifications, besides forming obsidians, perlites and pumice stones. Furthermore, it happens that just in the region of the ore veins the various trachytic and andesitic rocks have experienced a very widespread hydatometamorphic alteration into propylites, locally called greenstones or greenstone-porphyries. The propylite is, however, no longer regarded as an independent rock of older age than the porphyries, as at one time maintained by F. von Richthofen. The rocks carry much pyrite and are silicified near the veins. Besides andesite flows there are also tuff-like accumulations, trachyte breccias and trachyte conglomerates, which sometimes enclose fragments of coal torn off from Tertiary strata broken through by the eruptive rocks.

The lodes occur in exceedingly great numbers. The strike of most of them is between north-south and northeast and the dip is steep, more frequently southeast than northwest. The vein fissures traverse not only the trachytic and andesitic rocks, but in their southwest continuation they cut Miocene strata. They are for the most part of considerable thickness, usually being compound lodes without clearly defined walls. They enclose a good deal of material from the adjoining wall rock, which is decomposed

Jahrb. d. k. k. geol. Reichsanst., Vol. XI, 1860, pp. 153-277. B. von Cotta: 'Ueber Erzlagerstätten Ungarns und Siebenbürgens,' 1862, pp. 28-41. M. V. Lipold: 'Der Bergbau von Schemnitz in Ungarn,' *Jahrb.* d. k. k. geol. Reichsanst., 1867, Vol. XVII, pp. 317-458; with bibliography. J. W. Judd: 'On the Ancient Volcano of Schemnitz,' *Quart. Journ.* Geol. Soc., August. 1876. 'Die Erzgänge von Schemnitz und dessen Umgebung,' official map, 1883, written by von J. Szabó, L. Cseh and G. Gezell. J. Szabo: 'Geschichte der Geologie von Schemnitz,' announcement, 1885. *Idem:* 'Monographie der Umgebung von Schemnitz.' H. Böckh: 'Ueber das Altersverh der in Umg. von Selmeczbánya vork Eruptivgesteine,' *Földtani Közlony,* Vol. XXXI, 1901.

into clay or is silicified. The prevailing gangue consists of quartz, amethyst and hornstone; other minerals, usually of later formation, include calcspar, brownspar, rhodochrosite, siderite, barite and gypsum. The most common ore minerals are argentiferous galena, zinc-blende, chalcopyrite and auriferous pyrite, which occur scattered through a gangue of quartz and jasper-like material, this ore being called zinopel. Besides the zinopel, there are found rich silver ores, such as native silver, argentite, stephanite, polybasite and pyrargyrite, at times also native gold. Rarer occurrences are, among others, stibnite, hübnerite, tetrahedrite, marcasite, pyrrhotite, copper glance, cinnabar, fluorspar, diaspore and adular spar.

Most of the veins occur (1) in the andesites and rhyolites and (2) in diorite. Among the lodes in the andesite rock we may mention the Grüner, Stefan, Johann, Spitaler and Biber veins, while of the second group, which

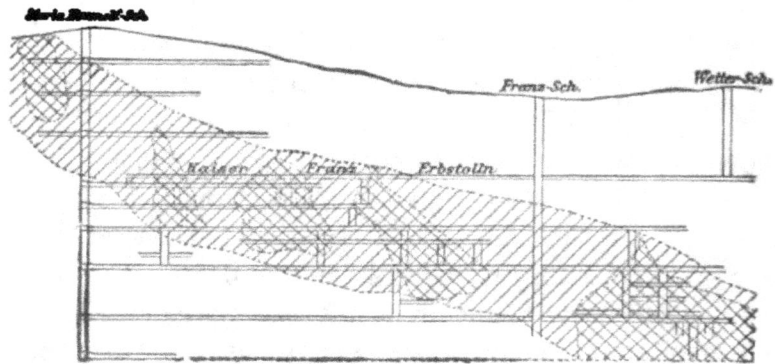

Fig. 175.—Horizontal section of the Grüner lode at Schemnitz. Showing the ore shoots forming a continuous zone of enrichment. (G. Faller.)

is developed at Hodritsch, west-northwest of Schemnitz, we may mention the Allerheiligen, Josefi, Schöpfer and the Brennerstolln lodes.

As representatives of the first group, the Grüner and the Spitaler veins may be characterized in a few words.

The Grüner vein has been traced for a distance of 1.5 km. (0.9 mile). The course is northeast, the dip 70 to 80° southeast, and the thickness 2-12 m. (6.5-39 ft.). The larger part of this great fissure is usually filled with much-crushed and pyritized rhyolite, which is traversed by stringers of ore forming compound veins. The ore is characterized by the great predominance of the true silver ores over galena and blende. The most abundant ore mineral is stephanite with associated argentite, polybasite and a little native silver, all of which, intergrown with pyrite crystals, occur in a grayish densely crystalline or drusy transparent quartz mixed with carbonates. The pay ore occurs in flat ore shoots or columns, whose horizontal extent is but 40 m. (131 ft.), but which extend downward much farther. These shoots have a flat pitch southwest as shown in the sketch (Fig. 175). Outside of the ore shoots the vein is much lower grade, though not barren.

The Spitaler vein is the most important of the entire area. It has been traced for a distance of over 8 km. (4.8 miles) and may have a total length of over 12 km. (7.2 miles). The course is mainly north-northeast, the dip south-southeast at 32 to 70°, being steeper with increasing depth. The vein is very thick, averaging 40-50 m. (131-164 ft.), which, however, does not mean that this is a single vein fissure. On the contrary, it is a lode formed of a system or cluster of veins consisting of single veins and stringers, very close together, but separated by layers of barren material. At points where the stringers converge, the lode attains a thickness of as much as 5 m. The gangue is principally quartz, with some rhodochrosite, calcite, brownspar and barite. The ore throughout the greater part of the lode consists of auriferous and argentiferous galena, blende, copper pyrite and pyrite, with a little free gold and rarely some cinnabar. In the southwestern region, the lode assumes quite a different character, rich auriferous silver ores being predominant.

The lodes of the second group cut the diorite at Hodritsch. They contain rich silver ores in a gangue of quartz and carbonates. The adjoining rock is strongly impregnated with pyrites.

The Schemnitz mines are the oldest in central Europe. According to some authors, the Quadi at "Vania," in this vicinity, had mines as early as the beginning of our era. About 745 the mines passed into the possession of the Slavs, who continued to work them. Their rights were left undisturbed by the Magyars, who came into power at the beginning of the 10th century. The names of the mines, moreover, indicate that there were numerous immigrations of Germans, especially Saxons from Meissen. They infused new life into the operations and the Germans monopolized the mines under Andrew II. Under Bela IV (1235-1270) the town was granted municipal and mining rights under the name of Schemnitz (Sebnitz). The first chair of the Mining Academy was established in the year 1763[1].

The Schemnitz lodes are similar to those of Kremnitz[2] (Körmöczbánya), situated farther north. The latter also are associated with propylitized

---

[1] 'Gedenkbuch zur hundertjährigen Gründung der k. ung. Berg- und Forstakademie in Schemnitz,' 1871.

[2] F. Windakiewicz: 'Gold- und Silberbergbau zu Kremnitz in Ungarn,' *Jahrb.* d. k. k. geol. Reichsanst., 1866, Vol. XVI, p. 217.

trachytic rocks. The ores occur in wide lodes composed of stringers and not in a single fissure. The ores and vein stone are exactly like those of the Schemnitz lodes.

A third Hungarian district distinguished by silver-gold veins is found in the Wihorlet-Gutin mountains, along the boundary line between the Marmaros, Szatmar and Ugocsa comitates. This mountainous tract, which rises steeply to a height of 1,500 m. above the great Hungarian plains to the west, consists of trachytes, rhyolites and dacites in part of propylitic habit. In the Nagybánya region, at Kreuzberg, 1 km. north of the town, at Veresviz (meaning red water) 3.5 km. northwest, at Felsöbánya on the east, and farther off at Turcz, Negy-Tárna, Visk, Illoba, Laposbánya and Miszabánya and at Borpatak and Kapnik, these rocks are traversed by lodes quite similar to those described at Schemnitz. For further details the reader is referred to G. Szellemy's comprehensive work[1].

These silver-gold deposits of Transylvania and lower Hungary, which have been so briefly summarized or merely mentioned, all form a geologic unit. They are associated with vast eruptions of trachyte and andesite, which took place along the inner side of the mountains formed by the great Carpathian anticlinal uplift. Thus they are a final consequence of the shrinkage of the earth's crust. The molten matter rose along a basin formed by the one-sided nature of the mountain-folding, forming intrusive and extrusive masses. These eruptions were followed by the formation of fissures through which the metallic materials from the still hot masses of the interior could be conveyed in solution and deposited in the upper part of the earth's crust. Entirely similar phenomena occur in western North America.

The Great Basin of Utah and Nevada, lying between the Sierra Nevada and the Rocky Mountains, was the theatre of tremendous volcanic activity in Tertiary time. Immense quantities of lava were poured out from the interior of the earth, forming lava flows many miles in length, or spreading out in great sheets and forming monotonous plains. Craters are recognizable at many points, the vents being surrounded by vast tracts of land covered with the ejecta of the volcanoes, either as loose lapilli and cinders or firmly cemented. Erosion and denudation have dissected these volcanoes, exposing the cores as great 'stocks,' whose rocks mostly belong to the andesite family. These volcanic centers and their immediate environs are the seat of numerous vein-like deposits of silver-gold ores. The greatest and most famous is the Comstock lode, situated in the Washoe district in Nevada.

---

[1] Gevza Szellemy: 'Die Erzlagerstätten von Nagybánya in Ungarn,' *Zeit. f. Prak. Geol.*, 1894, pp. 265-271, 449-457; 1895, pp. 17-25.

The Comstock Lode.—This famous deposit has been the object of careful and repeated examination by various scientific observers[1].

The geologic conditions may be generalized as follows: A huge stock, whose highest point forms Mount Davidson, consists of pyroxene andesite, whose upper part is of normal andesitic habit, and contains porphyritic crystals in a partly glassy groundmass. In depth this rock gradually as-

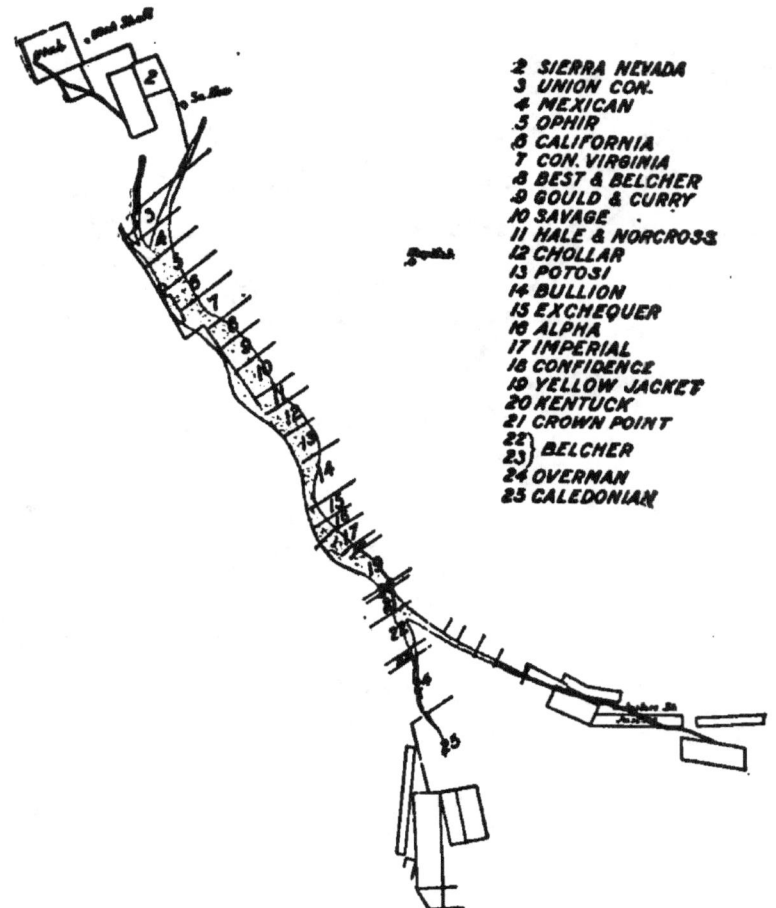

2 SIERRA NEVADA
3 UNION CON.
4 MEXICAN
5 OPHIR
6 CALIFORNIA
7 CON. VIRGINIA
8 BEST & BELCHER
9 GOULD & CURRY
10 SAVAGE
11 HALE & NORCROSS
12 CHOLLAR
13 POTOSI
14 BULLION
15 EXCHEQUER
16 ALPHA
17 IMPERIAL
18 CONFIDENCE
19 YELLOW JACKET
20 KENTUCK
21 CROWN POINT
22
23 } BELCHER
24 OVERMAN
25 CALEDONIAN

Fig. 176.—Outcrop of the Comstock lode, Nevada, showing location of the mining claims.

sumes a granular structure so that the crystalline rocks thus developed, which are badly changed by a secondary propylitic alteration, apparently

[1] F. v. Richthofen: 'The Comstock Lode,' San Francisco, 1866. Clarence King: 'U. S. Geol. Explor. of the 40th Parallel,' Vol. III, *Min. Ind.*, Washington, 1870. J. A. Church: 'The Comstock Lode, its formation and history,' *Trans.* Am. Inst. Min. Eng., New York, 1879. G. F. Becker: 'Geol. of the Comstock Lode,' etc., U. S. Geol. Surv., *Monogr.* III, 1882. E. Lord: 'Comstock Mining and Miners,' U. S. Geol. Surv., *Monogr.* IV, 1883. A. Hague and J. P. Iddings: 'On the development of crystallization in the igneous rocks of Washoe,' *Bull.* U. S. Geol. Surv., No. 17, 1885.

differ so much from the andesite that they were formerly described as diabase or granular diorite. Farther east pyroxene-andesite is intruded by a huge mass of mica-hornblende-andesite, formerly called porphyritic diorite. On the east flank of Mt. Davidson this rock has overflowed the pyroxene-andesite, and like it, passes downward into granular-crystalline varieties, the deep-seated facies being formerly called mica diorite. This mica-horn-

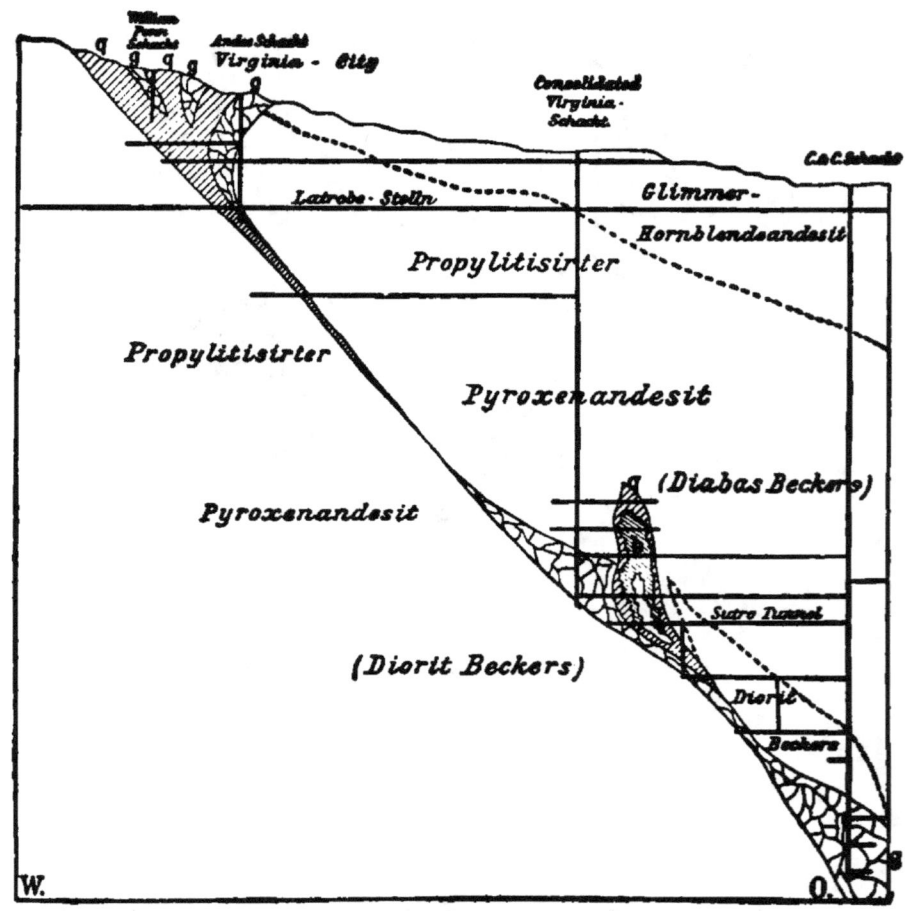

Fig. 177.—Cross-section through the Comstock lode. (Becker.)
g, lode body; q, quartzose lode mass; b, excavation in the bonanza.

blende-andesite has also undergone partial propylitization. The study of Becker's thin section by Hague and Iddings convinced them that the various Comstock rocks discriminated by older authors and assumed to be partly pre-Tertiary and partly Tertiary, may be referred to the two Tertiary andesite magmas mentioned, which were possibly intruded at approximately the same time. Furthermore, in the mines of that locality a great number of younger dike-like or stock-like intrusive masses of mica-hornblende-ande-

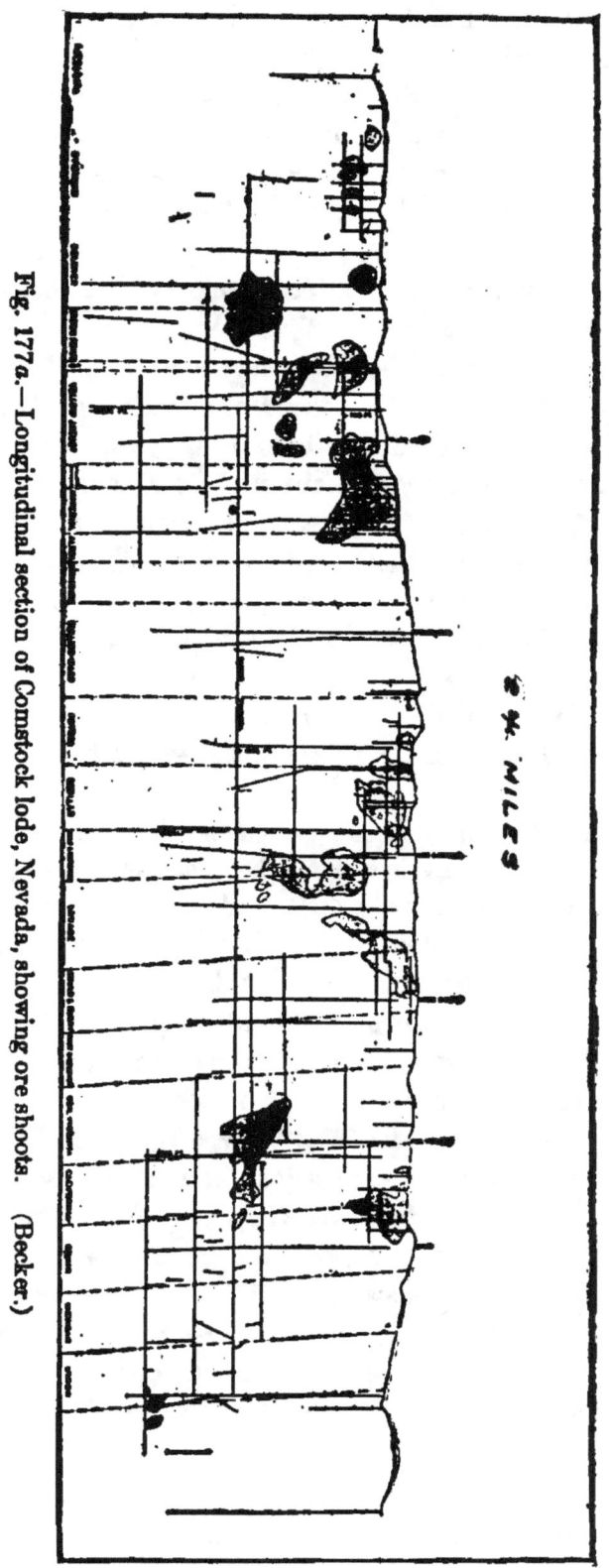

Fig. 177a.—Longitudinal section of Comstock lode, Nevada, showing ore shoots. (Becker.)

site, dacite and rhyolite (formerly called quartz porphyry) and of basalt have been met with.

The Comstock lode has a north-northeast course, and dips at 45 to 50° east-southeast. The lode occurs approximately but not always exactly at the boundary between the altered pyroxene and the mica-hornblende andesites just mentioned. It is a zone of faulting of greatly varying thickness. The upper part of the lode near the outcrop is exceedingly wide, but consists mainly of completely altered country rock, often decomposed into clay. In this, the main body of the lode, there is observed a hanging-wall vein and a foot-wall vein, the latter very large and of very irregular shape. In the cross-section, Fig. 177, taken from Becker's atlas, the orebodies are seen to be great lenses, dividing into two parts, and very irregular in outline.

These lenses and ore shoots have in part a course transverse to that of the lode itself, and sometimes extend far beyond the lode wall. F. Posepny[1] has very wisely suggested that this may be due to cross fissures.

These orebodies consist mainly of quartz. If we consider that the quartz orebodies and the decomposed rock form a unit, the entire lode has an average thickness of 60 to 100 m. (196 to 328 ft), and extends for a distance of nearly 2 miles. The vein divides at each end into several branches; including these its outcrop has been traced more than four miles (7 km.).

The quartz ore shoots just mentioned usually carry considerable gold and silver, and are but rarely barren. The ores include stephanite, polybasite, argentite, native gold, some galena and sphalerite, all of which occur in minute particles distributed through the friable quartz. The ore minerals also occur concentrated into very rich bonanzas. But small amounts of calcspar, gypsum and zeolites are found mixed with the quartz. Near the outcrop the bonanzas were rich in 'horn silver' (silver chloride), as, for example, the one encountered in 1874, which was 360 m. (1,180 ft.) across. The bonanza ores averaged 0.001 to 0.5% of gold, and 0.05 to 1.78% silver. Notwithstanding the large dimensions and rich contents of the various bonanzas, it must not be forgotten that they formed but a very small part of the entire mass of the lode, amounting to only about 1:600, according to Burthe. Nor do they show any definite rule or arrangement in the lode. At any rate this lode was, to quote Ed. Suess[2], "the greatest accumulation of precious metal that man ever laid hands on." From 1859 to 1889 the Comstock lode produced 4,820 tons of silver and 214 tons of gold, of a

---

[1] F. Posepny: 'Genesis der Erzlagerstätten,' *Jahrb.* d. k. k. österr. Bergak. **1895,** p. 124.

[2] E. Suess: 'Die Zukunft des Silbers,' Vienna, 1892.

total value of $340,000,000 (J. Vogt). In brief, the history of the lode is as follows: As early as 1856 gold gravels were discovered and washed near the lode. The lode itself was discovered in 1857 by the Grosh brothers. It was first worked, however, by others in 1859, for free gold only. The silver ore was not recognized for a while, but soon after its discovery it became the main object of mining. The towns of Virginia City and Gold Hill rose like magic on the flanks of the desert! Deeper and deeper the work penetrated until great difficulty was encountered in keeping the mines dry. In 1878 the famous Sutro tunnel completed draining the mines to its level. With increasing depth, however, mining became more and more difficult, owing to the increasing heat, until in 1877, in the shaft of the Savage mine at a depth of about 900 m. (2,950 ft.), a spring with a temperature of 69.4° C. (157° F.) was tapped[1]. This put an end to any downward advance.

The Veta Madre, of Guanajuato, Mexico, is very similar to the Comstock in its structure and ores, belonging to the same type. Its production has far exceeded that of the Comstock, but has been spread out over a period of 300 years. It is described on page 266 as a silver vein.

A somewhat more pyritic example of the silver-gold formation, also associated with andesitic rocks. is represented by the Smuggler[2] vein, near Telluride, in the San Juan Mountains, in Colorado. While the Comstock Lode was characterized by rich bonanzas distributed in a capricious way, the Smuggler lode shows great uniformity and constancy of values, with some variation of ratio of gold and silver contents.

The Smuggler lode fills a fault fissure in augite andesite and the underlying andesite breccias. It has a thickness of 1 to 1.5 m. (3.2 to 4.9 ft.), sometimes up to 3 m., and is traceable a distance of 3 k. (2 miles). The vein filling consists of quartz, with a little calcite and barite. Quite uniformly interspersed throughout the quartzose gangue one finds pyrite, chalcopyrite, galena and zinc-blende, together with rich silver ores, especially proustite and polybasite, as well as native gold. For a distance of three-fifths of a mile not one spot was found unremunerative. The ores average 0.04% of silver and 0.0016% of gold. From north to south the gold content gradually increases, while the silver content decreases until at its south end the Smuggler is a true gold vein.

Silver-gold veins are also found in Custer County, Colorado, and have

---

[1] J. A. Church: 'The heat of the Comstock mines,' *Trans.* Am. Inst. of Min. Eng., 1879, p. 45.

[2] J. A. Porter: 'The Smuggler Union Mines, Telluride,' *Trans.* Am. Inst. of Min. Eng., 1896. Ch. Wells Purington: *Preliminary Rep.* on the Mining Industries of the Telluride Quadrangle, Col., 18th *Ann. Rep.* U. S. Geol. Surv., II, pp. 751-848.

been described by Emmons and Cross[1]. The ores contain besides free gold and auriferous tellurides, various rich silver ores, as well as sulphides of lead, zinc and iron. The gangue, however, differs from the deposits described in that barite predominates. These occurrences, too, are associated with recent volcanic eruptive centers.

Of these deposits the Bassick mine is the most interesting. The bedrock consists of gneisses intruded by granites, as well as by dikes of syenite and peridotite. This rock complex is covered by younger eruptive rocks, lavas, breccias and tuffs. The eruptive series begins with andesites. Next follow in succession diorites, dacites, rhyolites, younger andesites and trachytes. Finally, the andesitic agglomerations of Mount Bassick were accumulated,

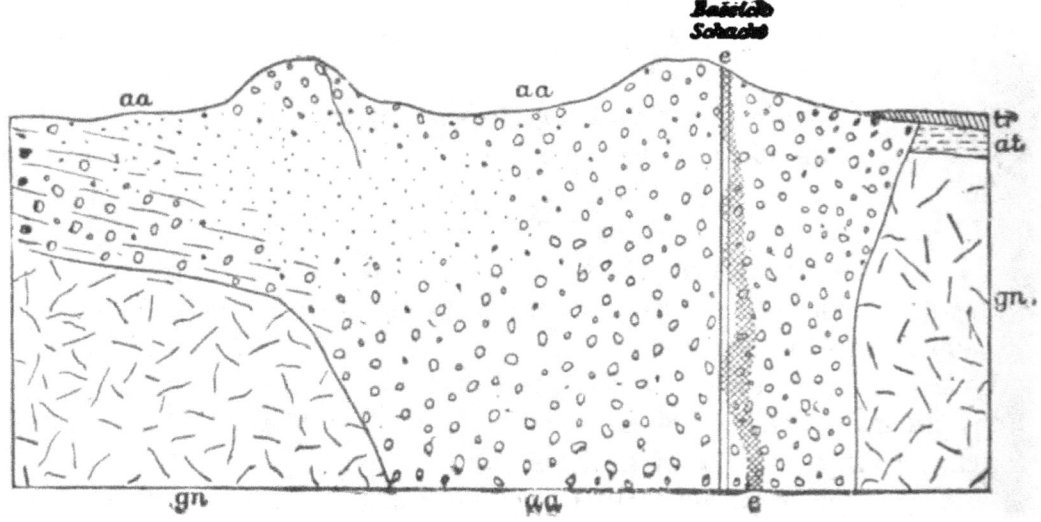

Fig. 178.—Section through the Bassick shaft.   (S. F. Emmons.)
gn, gneiss; at, andesite tuff; tr, trachyte; aa, andesite agglomerate; e, ore shoot.

probably in consequence of vast volcanic explosions which broke up the older rocks. At some points there were subsequent eruptions of mica-dacite and glassy basalt. (See Fig. 178.)

A very remarkable ore deposit occurs in this Bassick agglomerate. Its form is hardly veinlike, but, mineralogically and genetically, it belongs to this class of the system. The deposit forms an upright stock of elliptical cross-section with diameters of 8 m. (25 ft.) and 30 m. (98 ft.) in the midst of andesite breccias. It has been followed to a depth of 1,200 ft. Not far from it a second deposit was found. In these 'pipes' the andesite

[1] Whitman Cross: 'Geology of Silver Cliff and the Rosita Hills, Colorado.'   S. F. Emmons: 'The Mines of Custer County, Colorado,' 17th *Ann. Report* U. S. Geolog. Surv., II, 1896, pp. 269-470.   *Idem.*: 'Some Mines of Rosita and Silver Cliff, Colorado,' *Trans.* of the Am. Inst. Min. Eng., 1897, Vol. XXVI, p. 773.

fragments, which vary from 1 cm. to a meter across, are surrounded by concentric layers of ore, varying between 1 cm. and 60 cm. in thickness. These encrusted fragments are impregnated with pyrite and are much more decomposed than those of the surrounding rock. The crusts of ore consist (from within outward) of (1) a blackish mixture of sulphides of zinc, antimony and lead with 0.2% silver and 0.005 to 0.01% gold; (2) a brighter thin layer, richer in lead, varying up to 0.7% of silver and 0.35% of gold; (3) a layer of crystalline zinc-blende 5 to 50 millimeters thick with 0.2 to 0.4% of silver and 0.5 to 0.16% of gold; sometimes also (4) a crust up to 2 centimeters thick of silver-bearing and gold-bearing copper pyrite; (5) a thin layer of pyrite. The free interspaces are filled with kaolin, but sometimes also contain quartz, gray copper, gold-telluride and silver-telluride, or secondary ores such as hydrosilicate of zinc and smithsonite.

The ore shoot is not sharply divided from the surrounding barren breccias, but in the latter there are fissures intersecting in such way that the ore column lies between them.

Emmons states that the vehicle for the metallic compounds was not vapor and gas which condensed on the fragments of the breccia, but solutions containing such gases, and he regards their rise as the last phase of the dying activity of this volcanic throat.[1] The John Jay mine near Jimtown, in Colorado, is worthy of mention in this place on account of the native tellurium found in masses weighing as much as 25 pounds.

Examples of the present group are also found among the gold deposits of Australia. Only one of them will be briefly described here, namely, the Hauraki goldfield, on the Coromandel Peninsula, on the north island of New Zealand, and reaching south as far as Waiorongomai. In this extensive area the Paleozoic rocks, and in part also late Cretaceous and Tertiary strata, according to Park, are intruded or covered by andesites, dacites and trachytes, as well as by their tuffs and breccias[2]. At many places the andesitic rocks are found altered to propylites, which in turn are intruded by younger underlying andesites. Numerous quartz veins traverse both the andesites and the propylites, but are mainly developed as workable gold lodes in the propylite only. Twelve groups of veins are distinguished in the area, among which the Thames group is richest and of the greatest scientific interest.

It lies in a tract about 2.7 km. (1.6 miles) wide, north of Hape creek. The veins striking northeast are intersected by two very large displace-

---

[1] Revised by S. F. Emmons.

[2] James Park: *Papers and Reports* relating to Minerals and Mining, Wellington, 1894, p. 52. K. Schmeisser: 'Australasien,' p. 92, also *Zeit. f. Prak. Geol.*, 1899, p. 366 *et seq.* 'The Hauraki Goldfield, New Zealand,' Waldemar Lindgren, *Engineering and Mining Journal*, Feb. 2, 1905.

ments, the Moanatairi and Collarbone faults, which divide the entire area into three terraces.

The thickness of the veins varies greatly. The largest, the Waiotahi, averages 3.5 meters wide, but is sometimes as much as 13 meters across. The veins are nearly all accompanied by lateral spurs or stringers. The gangue consists mostly of brittle, sugar-like quartz. The gold in it is finely divided, sometimes appearing in the form of wire and plates, and contains 30 to 40% of silver, giving it a whitish-yellow color. With it is found pyrite, and more rarely chalcopyrite, galena, manganese ores, zinc-blende. stibnite and gothite. A quartz vein of the Tararu mine contains much rhodonite, with finely divided gold. The largest lodes are generally poorest, and none have been worked to any great depth.

New Zealand in 1901 had a total production of 412,189 oz. of fine gold.

Instances of the silver-gold ore formation have lately become known also in the Dutch Indies, to wit: veins characterized by a large amount of selenium at the Redjang Lebong mine north-northeast of Benkoelen, in southern Sumatra. The lode, which is over 5 meters thick, and occurs in andesitic rocks, consists mainly of a dense-looking mass rich in silica.

The gold and auriferous silver (electrum), the latter predominating. occur disseminated in minute particles, invisible even under the magnifying glass. The composition of the crude bullion is as follows:

Gold and silver ....91.52 per cent.
Se... 4.35　"
Cu ....　1.82　"
Pb...　1.65　"
Zn ....　0.48　"
Fe... 0.14　"
　　─────
　　99.96　"

The average content of the 12,072 tons of ore produced up to 1900 was 45 gm. of gold and 385 gm. of silver per ton[1].

A similar deposit is worked at Lebong Soelit 7 to 8 k. (4.2 to 4.8 miles) farther west.

This type of vein is also represented in the Philippines[2].

18. *The Fluoritic Gold-Tellurium Veins.*

This group is closely related to the preceding. but is sharply characterized by the appearance of tellurides of gold and silver in a gangue of fluorspar and quartz. The ore minerals consist mainly of native gold, tellurides

---

[1] According to S. J. Truscott, cited in *Zeit. f. Prak. Geol.*, 1902, p. 225.

[2] F. W. Voit: 'Goldvorkommen auf den Philippinen,' *Berg u. Hütten. Zeit.*, Vol. LVII, 1898, p. 251.

and auriferous pyrites; to a slight degree of gray copper, galena, sphalerite, and stibnite.

The best known example is the goldfield of Cripple Creek[1], directly southwest of Pike's Peak, Colorado.

According to Cross a somewhat circular area of 4 to 5 k. (2.4 to 3 miles) in diameter, occurring in the midst of a very extensive tract of porphyritic granite, is covered by volcanic breccias and tuffs, whose material consists mainly of trachy-phonolite fragments, but in part also of fragments of the granite carried up from deep-seated masses. These volcanic accumulations are broken through in places by stock-like masses of trachy-phonolite and phonolite. The last-mentioned rock is exceedingly rare in the United States, being known only at one other place, viz.. the Black Hills, So. Dakota.

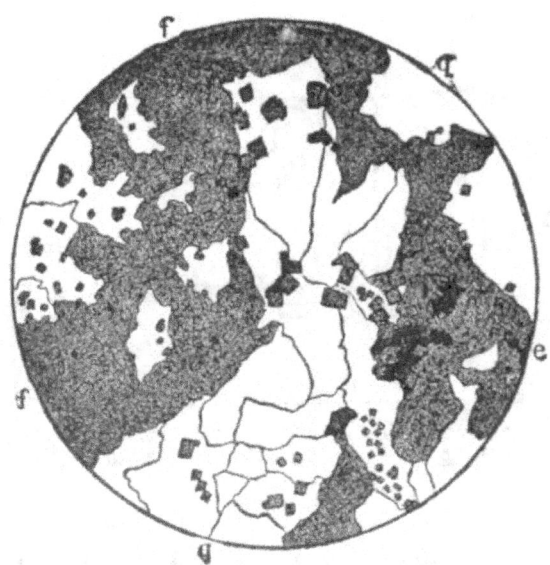

Fig. 179.—Thin section of a 'purple quartz' from Summit mine, Cripple Creek.
f, fluorspar; q, quartz with enclosed fluorspar crystals (outlines drawn by polarized light); e, iron ocher with free gold. (Enlarged fifty times.)

Finally the area is traversed by dikes of basic rocks, nepheline and feldspar basalts, which for the most part strike northeast and northwest.

These recent dikes are often accompanied along one wall by gold veins, while other veins also occur independently of the dikes, cutting not only the andesite, but also the underlying granite. The 'true fissure veins' are mere fissures, often an inch or less thick, and often entirely closed, but not only does this narrow fissure filling carry ore, but the strongly altered country rock on both sides is heavily impregnated with ore. These impreg-

[1] Whitman Cross: 'Geology of Cripple Creek;' and R. A. F. Penrose: 'Mining Industries of Cripple Creek,' 16th *Ann. Rep.* U. S. Geol. Surv., II, 1895. Also W. Lindgren and F. L. Ransome.

nation fissures contain gold tellurides, principally sylvanite and calaverite, rarely petzite, in less amount native gold, as well as auriferous pyrites, while in depth the ores contain gray copper, galena, sphalerite and stibnite. The gangue is quartz and fluorspar intimately coalesced, as shown in Fig. 179. The quartz is of a violet or reddish color, as it is thoroughly permeated with fluorspar; it is locally called 'purple quartz.' The metasomatic alteration of the country rock has been studied by W. Lindgren. The minerals, which are especially common as replacements of the orthoclase of the granite, include quartz, fluorspar, pyrite, sericite, as well as valencianite. The distribution of the tellurides or other ore minerals is very variable and uncertain. Very often ore shoots of very unequal size and shape are found in the same vein.

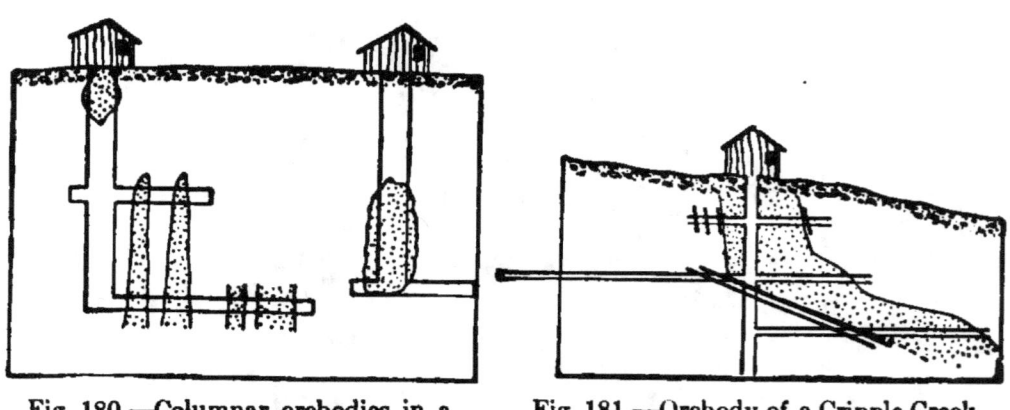

Fig. 180.—Columnar orebodies in a Cripple Creek vein. (R. A. F. Penrose.)

Fig. 181.—Orebody of a Cripple Creek vein. (R. A. F. Penrose.)

Ordinarily the ore shoots are approximately vertical columns varying from a few centimeters to several meters in thickness and up to 100 m. (328 ft.) long (see Fig. 180). The ore shoots sometimes occur directly under the grass roots, while more often their summits are only encountered in depth. In certain cases their location is at once seen to be dependent upon cross fissures, as shown in Fig. 181. These shoots sometimes yield very large amounts of gold.

A unique ore shoot of the Portland mine is called the "Anna Lee Chimney." A volcanic vent in basaltic rock has been worked to a depth of about 300 m. (984 ft.) This chimney, which is 4 to 9 m. across, is filled with rounded basalt fragments cemented by a calcareo-ferruginous material containing auriferous tellurides quite uniformly distributed.

Mention must also be made of the columnar masses of gypsum encountered in the breccias in the Deerhorn shaft. Rickard regards them as deposits of thermal waters[1]. They may represent products of oxidizing sulphides on country rock and calcite.

The Cripple Creek district had long been known, but it was not until 1890 that gold ore was found in large quantities on what was afterward the Gold King mine in Poverty Gulch. Production began in 1891. According to Penrose it amounted to $5,543,967 from 1891 to 1894, inclusive. In 1898 the production was $13,507,349.

Deposits of gold with fluorspar also occur in Montana as described by Weed.[2] The Judith and Mocassin (Warm Springs) districts are situated in the Judith Mountains of Montana, which rise to a height of 6,386 ft. and extend over a distance of 29 km. (17 miles) between the arid grassy plains of the Yellowstone and those of the Missouri river. This mountain mass is an isolated uplift east of and independent of the Rocky Mountains. It consists of Paleozoic and Mesozoic limestones and shales, principally of Carboniferous age, domed by laccolitic intrusive masses. The gold deposits are found at the contact between limestones and syenite porphyries, which are regarded by the author as belonging to the trachyte group. The deposits consist of a contact breccia whose fragments are cemented by fluorspar, quartz, and a little calcite, with finely interspersed particles of tellurides, and the free gold formed by their alteration. The "purple quartz," an intimate mixture of fluorspar and quartz, recurs here.

The characteristic feature of the group of deposits last discussed, namely, the intimate blending of quartz and fluorspar, has been used by Penrose as the basis for a theory of the origin of these deposits. He imagines a reaction of soluble or gaseous fluorides with the calcium carbonate that resulted from the decomposition of calcareous silicates; or which is, as in the last example, supplied directly by the limestone. If the fluorine was introduced as hydro-fluoride of silicon, there might have been a simultaneous precipitation of fluorspar and quartz, such as we commonly have in the purple quartz. The original seat of the fluorine, as well as of the gold, Penrose imagines to have been heated rock masses found at great depths[3].

(η) Antimony Veins.

19. ANTIMONY QUARTZ VEINS.

This class of vein is closely allied to the antimonial gold quartz veins, and represents the extreme or gold-poor facies of the group. It is, how-

[1] T. A. Rickard: 'The Cripple Creek Volcano.' *Trans.* Am. Inst. Min. Eng., February, 1900.

[2] W. H. Weed and L. V. Pirsson: 'Geology and Mineral Resources of the Judith Mountains of Montana.' 18th *Ann. Rep.* U. S. Geolog. Survey, III, 1896–1897, pp. 445–614.

[3] See 'Telluride Veins of America,' Kemp, 'The Mineral Industry,' 1900.

ever, desirable, for practical reasons, to make a separate group of the deposits because they are worked for antimony, not for old. Most of them carry but a very slight amount of gold, whose extraction is unprofitable by reason of existing metallurgic difficulties.

The prevailing gangue consists of quartz, with some calcspar. The ores contain stibnite and its products of decomposition, namely, stibite, antimony-ochre, valentinite and senarmontite, more rarely also antimony blende (pyrostibnite) and sometimes pyrite, bournonite, berthierite, galena, zinc-blende, steinmannite, zinckenite, and cinnabar ochre, rarely native gold. Ordinarily the stibnite is interspersed in the gangue, in rare cases it occurs in compact masses, forming almost all the vein filling, or it is concentrated into ore shoots, while the rest of the vein is lean.

A description of examples will better explain the characteristics of this type.

At Böhmsdorf and Wolfsgalgen, near Schleiz, the Paleozoic schists are traversed by veins of quartz carrying stibnite, with minor amounts of zinc-blende, plumose stibnite, pyrophyllite and ironspar. The veins have sometimes been extensively mined, as at the Halber Mond mine at Bohmsdorf in the fifties of the 19th century.

Among the antimony deposits of Bohemia special mention must be made of that of Pricov, near Selcan, 15 kilometers west of Wotitz, in central Bohemia. At that locality, according to A. Hofmann, a great number of kersantite dikes occurring in granite, are accompanied by veins of hornstone rich in stibnite. The veins were worked for antimony alone for many years, but were abandoned in 1897. The ores held no free gold, and the stibnite contained only traces of that metal. For a depth of 18 m. (59 ft.) below the croppings, the stibnite was changed to antimony ochre (stibiconite). The largest vein, which has a thickness of as much as 20 meters, and forms the rocky crest of the Deschna mountain, does not contain stibnite throughout its entire mass. It is only where the hornstone assumes a light gray color that it is found, when examined under the microscope, to be completely filled with innumerable handsome crystals of ore, and also to carry larger segregations of it. Thus far only the Emil lode, 10 to 50 centimeters thick, has been worked.

Entirely similar, but less extensive, lodes are found at Punnau, near Marienbad. They occur in mica schist and amphibolite near the great granite massive of that locality[1].

In Hungary antimony ores are still extracted from the veins of the Rechnitz Mountains, of the Eisenburger Comitat. Some of the veins have been

[1] J. Schwarz: 'Das Punnauer Antimonbergwerk bei Michaelsberg in Böhmen.' *Oesterr. Z. f. B. u. H.*, 1881, pp. 595–608.

followed for a distance of 3 km. (2 miles), cutting crystalline schists. According to A. Schmidt[1], the veins are especially rich when the country rock is a chloritic or graphitic schist. The vein filling consists of quartz, calcspar and stibnite, with stibiconite and pyrite. The masses of antimony ore proper attain a thickness of 2 to 50 centimeters. The graphitic schists alongside of the lodes, for a distance of 3 to 4 meters from the vein walls, are so richly impregnated with stibnite, together with pyrite and cinnabar, that they also pay working. The pyrite of this ore contains about 0.0021% of gold.

Other veins of stibnite, with a gangue of quartz and carbonates, with small amounts of jamesonite, berthierite, blende and auriferous pyrites, were worked between Aranyidka and Rosenau, in upper Hungary[2].

A noteworthy deposit of antimony occurs at Pereta, south Tuscany. According to Coquand[3] and Toso[4] it consists of a mass of crushed white quartz, whose north end is intercalated in Eocene limy shales, while to the south it cuts Miocene limestones. To the north of the outcrop exhalations of hydrogen sulphide (putizze) occur. The stibnite occurs in the quartz in the form of stringers and pockets, occasionally of large size. Sulphur occurs with it and in the northern part of the deposit considerable amounts have been extracted. In these workings the sulphur-bearing quartz is occasionally seen to be coated with a crust of stibnite, which in turn is studded with small crystals and aggregates of sulphur. Eocene limestones, at the contact with the deposit, are traversed by quartz stringers and largely altered to gypsum or to alum rock.

Stibnite associated with cinnabar is found at other places in Tuscany, notably San Martino, in the Monte Amiata district[5].

In recent years the veins of the northern part of the island of Corsica have become important producers. The veins occur in sericite schist, the gangue materials being quartz, calcite, blende with antimonite and more rarely pyrite, cinnabar and bournonite[6].

In Portugal rich veins of stibnite worked until recently occur 10 km. (24 miles) east of Porto, in the Moinho da Igreja, and San Pedro da Cova, on the Rio Ferreira. They traverse Paleozoic schists.

[1] A. Schmidt: 'Ueber einige Minerale der Umgebung von Schlaining.' Groth's *Z. f. Kryst. u. Min.*, 1898. Vol. XXIX, part 3, p. 194.

[2] G. Faller: 'Reisenotizen.' *Jahrb. d. k. k. Montanlehranst.*, 1867, p. 132.

[3] H. Coquand: 'Solfatares, Alumières, etc., de la Toscane.' *Bull. Soc. Géol. de France*, VI, 1829.

[4] P. Toso: *Rivista del Serv. Min.*, 1899, p. 144.

[5] B. Lotti: 'Zinnober- und Antimonlagerstätten Toscanas.' *Zeit. f. Prak. Geol.*, 1901, pp. 33-46.

[6] M. Nentien: 'Etude sur les gîtes minéraux de la Corse.' *Ann. des Mines*, Vol. XII, 1897, pp. 231-296.

The central plateau of France is very rich in antimony. According to Fuchs and De Launay[1] the most important lodes are found in various rocks, viz.:

Mica schist in the worked-out mines of Nades in the Bourbonnais; in granite at Bresnay; on the contact between granite and gneiss at Montignat (Allier) at Villerange, in the Culm graywackes, south of Saint-Yrieux (Haute-Vienne). Great numbers occur in mica schist and amphibole schist, at Freycenet, and elsewhere in the cantons of Puy-de-Dôme, Cantal, and Haute Loire, where they are still extensively mined for antimony ore. Quartz is the main gangue, accompanied at Malbose (Ardèche) by a little calcite and barite.

The stibnite is accompanied by arsenopyrite, occurring in small quartz veinlets in Carboniferous aplite dikes, forming veritable stockworks at Montignat (Allier)[2].

Similar deposits are known in Australia, viz., on Macleay river, South Australia, on Donovan's creek and on the Upper Yarra, near Sunburg, Victoria, and in the Hillgrove district, New South Wales.

Mexico contains important antimonial veins, traversing quartzites and limestones at La Sonora. Deposits in Asia Minor are described by K. E. Weiss[3].

Much discussion, especially upon the mineralogic associations, has been aroused by the stibnite veins of Japan. According to K. Yamada[4] they are especially abundant on the island of Shikoku; they are not thick, and occur mostly in schists and other Paleozoic rocks. Quartz is usually the only gangue, the stibnite being sometimes accompanied by pyrite.

The largest and most famous antimony mine of Japan is that of Itshinokawa on Shikoku, which furnishes large stibnite crystals.

The country rock is a sericite schist, which is impregnated with pyrite near the veins. Four east-west lodes are known, three of which dip about 80° south, the other but 25° in the same direction, this being the one containing the large crystals up to 0.5 meter long. The gangue is quartz with a little calcspar; the vein filling is banded with druses containing the large crystals. With one exception, the Nakase mine, the antimony ores do not carry gold. The Itshinokawa district produced 799 tons of antimony sulphide and 83.8 tons of refined antimony in 1897[5].

Great quantities of antimony ore have also been from time to time ex-

[1] Fuchs and De Launay: 'Traité des Gîtes Minéraux.' Vol. II, 1893, pp. 193–199; with full bibliography.

[2] L. De Launay: *Compte-Rendu* du VIII Géol. Congrès Inter., 1900.

[3] K. E. Weiss: 'Lagerstätten im West. Anatolien,' *Zeit f. Prak. Geol.*, 1901, part 7.

[4] Letter to the author.

[5] 'Les Mines des Japan,' Paris Exposn., 1900, p. 10.

ported from Borneo. The ore comes from veins at Tambusan and Tagui, in the Sarawak district, in the northern part of the island. They occur in limestone and slate and carry stibnite, oxidized antimony ores, and as a rarity native antimony, in a quartzose gangue.

While in most cases the antimony veins may be regarded as extreme members of the antimonial gold quartz veins, in which the gold content is nil, and vice versa, there are, on the other hand, examples of antimony veins which may be regarded as extreme developments of the class of rich silver veins. An example of this kind was mined until the sixties at Mobendorf, Saxony, seven miles northwest of Freiberg[1], which Freiesleben has designated as the Mobendorf type. According to H. Müller[2], these veins occur in gneiss, have a steep dip and but rarely exceed 5 cm. in thickness. Their filling consists of stibnite, with some berthierite, bournonite, striated kaolin, steinmannite (antimonial galena), zincite, kermesite, stibiconite, pyrite, quartz and some brownspar. A few kilometers farther east-southeast at Bräunsdorf the same ores are found intimately associated with rich silver ores in the rich silver quartz veins.

(δ) **Cobalt-Nickel and Bismuth Veins.**

Two classes of veins, quite different from each other, are included under this heading, viz.:

A. Veins composed essentially of cobalt-nickel and bismuth sulphides.

B. Veins consisting of nickel-magnesia hydrosilicates. The first group is itself subdivided into a class characterized by a gangue composed mainly of a carbonspar gangue. with pure cobalt-nickel ores, and a second class characterized by a gangue of quartz, and other siliceous minerals in which carbonates are subordinate, containing bismuth ores, besides cobalt-nickel ores.

Accordingly the various groups may be designated as follows:

1. Cobalt-nickel deposits with carbonate gangue.
2. Quartzose cobalt-nickel-bismuth deposits.
3. Hydrosilicate nickel deposits.

Further differences between these three groups will be given in the description of various examples.

---

[1] J. C. Freiesleben: 'Magazin für Oryktographie von Sachsen.' 1. Appendix, p. 77 2. Appendix, p. 179.

[2] H. Müller: 'Erzlagerstätten bei Freiberg.' Cotta's Gangstudien, I, 1850, p. 104.

20. COBALT-NICKEL VEINS WITH CARBONATE GANGUE.

This type is most pronounced in the cobalt veins of Dobschau, Hungary, which represent an extreme facies of the spathic iron-ore veins.

According to Voit,[1] chloritic-talcose and quartzose Palcozoic slates are intruded by a sheet of diorite, which follows the contact with a stock of garnetiferous serpentine. The diorite is badly altered and shows transitions from a hornblende diorite to a hornblende granitite. The basic contact facies next to the clay slate has been altered to a chloritic schist rich in epidote. The cobalt veins on the south flank of the Langenberg, four miles south of Zemberg, cut the diorite close to its contact with the underlying chloritic schists. Three lodes composed of veins with steep southerly dip are recognized, each one dividing into fanlike stringers above, while downward they disappear, or at any rate grow barren, at a depth of 180 to 200 m. (590 to 655 ft.) The fissures are not over 3 m. (10 ft.) thick, and consist mainly of siderite, calcite, ankerite and some quartz. In the southern veins tourmaline needles also occur, usually in the quartz, more rarely in the siderite. Many fragments of decomposed country rock also occur in the vein. The ore minerals occur as in pockets, parallel layers, and large compact bodies, irregularly scattered through the vein. The common ore is a compact, very finely crystalline mixture of smaltite (cobalt pyrite) and Rammelsbergite ($NiAs_2$). The richer ores yield 8 to 10% of cobalt and about 17% of nickel, others only 4% of cobalt, but hold 22% of nickel. This ore is usually fissured, the fractures showing slickensides, whose glassy surface is coated with carbonaceous material. Rich orebodies, consisting of tetrahedrite, sometimes occur in the upper parts of the vein. Chalcopyrite, bornite, arsenopyrite, lollingite ($FeAs_2$) and niccolite are rarely observed. Secondary ores, erythrine, malachite, etc., also occur.

The cobalt veins broaden upwards into trumpet-shaped expansions of coarsely crystalline siderite as much as 100 feet thick, which are exposed by great open-cuts, at Langenberg, Bingarten and Massörtern. These deposits completely fill irregular depressions on the surface of the diorite, and are in turn overlain by Carboniferous sandstones, slates and limestones carrying segments of crinoid stems and other fossils. These deposits seem to be connected with the lodes, for the siderite contains scattered nests, or even distinct layers in its lower layers, of nickel ores, together with some tetrahedrite, chalcopyrite, arsenopyrite and brown hematite. The solutions rising in the ore-bearing fissures seem to have metasomatically replaced part of the Carboniferous limestone overlying the diorite.

[1] F. W. Voit: 'Geognost. Schild. d. Lagerst.-Verh. von Dobschau in Ungarn.' *Jahrb. d. k. k. geol. Reichsanst.*, 1900, Vol. L, part 1.